FRITZ SCHIDER

AN ATLAS OF ANATOMY FOR ARTISTS

REVISED BY PROFESSOR DR. M. AUERBACH

AND TRANSLATED BY BERNARD WOLF, M.D.

NEW BIBLIOGRAPHY BY ADOLF K. PLACZEK,
COLUMBIA UNIVERSITY

ADDITIONAL ILLUSTRATIONS FROM THE
OLD MASTERS AND HISTORICAL SOURCES

WITH A NEW SECTION ON HANDS SELECTED BY
HEIDI LENSSEN

THIRD AMERICAN EDITION

DOVER PUBLICATIONS, INC.
NEW YORK

Published in Canada by General Publishing Company, Ltd., 30 Lesmill Road, Don Mills, Toronto, Ontario.
Published in the United Kingdom by Constable and Company, Ltd., 10 Orange Street, London WC 2.

An Atlas of Anatomy for Artists is a new English translation of the sixth (1929) edition of *Plastisch-Anatomischer Handatlas für Akademien, Kunstschulen und zum Selbstunterricht 5. Aufl.* published by E. A. Seeman.

Standard Book Number: 486-20241-0
Library of Congress Catalog Card Number: 58-3622

Manufactured in the United States of America
Dover Publications, Inc.
180 Varick Street
New York, N. Y. 10014

PREFACE

TO THE THIRD AMERICAN EDITION

This third revised American edition is augmented by 10 illustrations from Jules Cloquet's *Anatomie de l'Homme* (plates 157-166), 16 illustrations from Jeno Barcsay's *Anatomy for the Artist* (plates 171-176), and a new section on hands selected by Heidi Lenssen (plates 97-106).

1957 Dover Publications, Inc.

PREFACE

TO THE SECOND AMERICAN EDITION

In this second revised American edition, the publishers have aimed to increase the usefulness of a book that has been standard for many years. The book has been expanded by the addition of the following material:

(1) A new bibliography.

(2) A wide selection of illustrations from historical sources: Vesalius, Leonardo, Goya, Degas, and others.

(3) Photographic illustrations of interest to the artist which are reproduced for the first time in this book: the Nancy Bayley photographs of growing children and the Muybridge action studies.

Although Schider has always been a valuable book for the study of anatomy, it is hoped that the added sections will encourage the student to study life drawing from the rich repository of material that is readily available in the great libraries and museums of the world. Rimmer and Muybridge, for example, were great teachers and students of the human figure during the nineteenth century; yet, their books are out of print at the present time. If this book introduces to the student such works as these and encourages him to investigate the artistic and photographic resources that are available, much of the purpose of the book will have been achieved.

Schider has been particularly useful in that he has never encouraged the student to follow any style other than his own. He has concentrated primarily on presenting the essential facts of anatomy in a straightforward manner leaving the student in less danger of imitating particular styles or mannerisms. This aspect of the book has not been altered; rather, the introduction of the historical material should make the student continuously aware of the variety of style and approach that is possible.

1954 Dover Publications, Inc.

INTRODUCTION

PLATES 1 and 2. The Skeleton.

PLATES 1 and 2 show the skeleton of a young man from the front, side, and back.

NOTE: The female skeleton is clearly differentiated from the male by the small face and skull, the narrow, short thorax, and particularly the more rounded pelvis (compare the drawings).

PLATE 3. The Various Shapes of Bones.

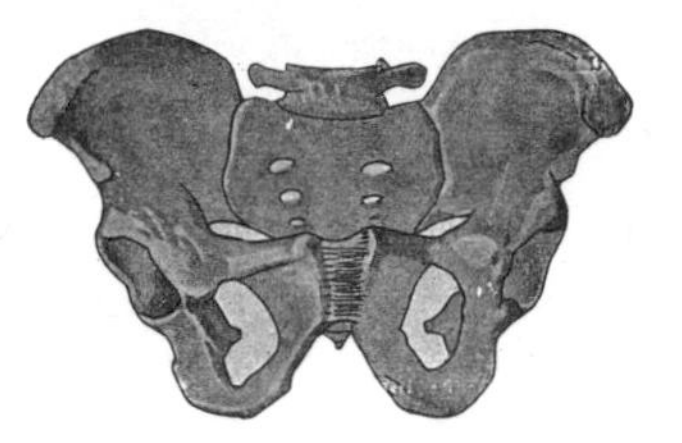

Male Pelvis

Fig. 1 demonstrates the groove between the two tuberosities at the upper end of the humerus, a typical bone groove, and the oval rough area of the humerus (insertion of the deltoid muscle).

Fig. 2 demonstrates the linea aspera, the rough line on the posterior aspect of the femur (origin and insertion of thigh muscles), a typical bone ridge; the head of the femur, the upper cartilage-covered end of the femur, with the femoral neck and the two femoral trochanters.

Fig. 3 demonstrates the crest of the tibia, the upper portion of the S-shaped edge of the tibia, a typical bone edge.

Fig. 4 demonstrates the ischial spine, the pointed process of the ischium, and the acetabulum which serves to receive the head of the femur.

Fig. 5 shows a tubular bone sawn across with its marrow cavity.

PLATE 4. The Types of Joints.

The various joints are classified according to the shape of the articular surfaces.

A. Ball and Socket Joints.

Fig. 1. The ligaments between the humerus and scapula form the joint capsule.

Fig. 2. The ligaments between the femur and pelvis.

The ball and socket joint consists of a spherical head which fits into a cavity, the acetabulum, and which allows motion in all directions. Flexion, extension, adduction, and circumduction are possible in this type of joint.

B. Hinge Joints.

Fig. 3. The joints of the fingers, the interphalangeal joints, are shown as examples of this type.

In a hinge joint, one bone has a transverse convex cylindrical surface and the other bone shows the reciprocal contour. Only flexion and extension are possible in such a joint.

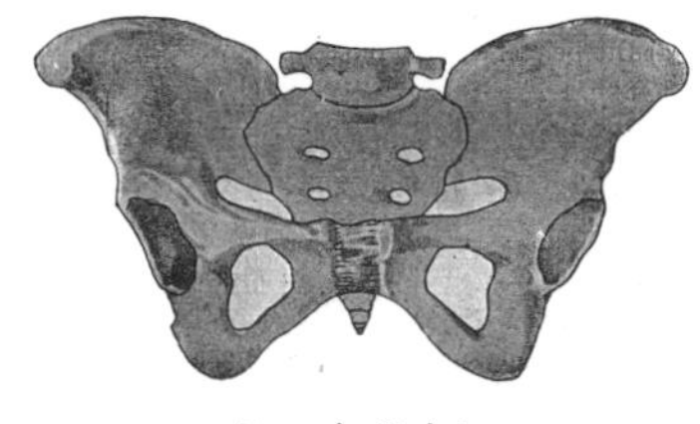

Female Pelvis

C. Combination Type of Joint.

Fig. 4. The elbow joint is shown as an example of this type of joint. Three or more articular surfaces with various shapes are involved: the joint between the ulna and the humerus forms a hinge joint while the joint between the radius and the humerus is of the ball and socket type. In addition, there is a special joint between the ulna and radius. In this combined joint, pronation and supination, flexion and extension are possible. (Pronation refers to the motion of rotating the palm of the hand inwards towards the body; the pronated posi-

tion of the forearm and hand is the position assumed after maximum inward rotation—the palm then faces outwards. Supination refers to the opposite motion, i.e. rotating the palm outwards away from the body; the supinated position is the position assumed after maximum outward rotation—the palm faces forward and slightly outwards.)

D. Immobile Type of Joint.

Fig. 5. The joints between the individual wrist (carpal) and ankle (tarsal) bones and between the carpal and metacarpal, tarsal and metatarsal bones are examples of this type.

PLATE 5. Schematic Cross-section Through a Joint.

The important features are clearly labeled on the plate.

THE BONES OF THE HUMAN BODY

I. The Bones of the Skull.

PLATES 6 and 7.

Plate 6, Fig. 1 is a view of the skull from below; Fig. 2, from the front.

In Fig. 1, note:

A. The two occipital condyles with joint surfaces which articulate with concave facets on the first cervical vertebra.

B. The two mandibular fossae in which the articular processes of the mandible move.

C. The occipital protuberance to which the ligamentum nuchae ("ligament of the neck") is attached.

D. The mastoid processes, the styloid processes, and the external occipital crest which serve for the origin or insertion of muscles.

E. The foramen magnum is the connection between the cranial cavity and the vertebral (spinal) canal.

Plate 6, Fig. 2. In this drawing, significant features as far as external appearance is concerned are:

A. The two frontal prominences — rounded protuberances more definitely marked in children and women than in men;

B. The two superciliary arches — slender ridges above the orbits more distinctly marked in men than in women or children;

C. The glabella — a small, flat surface between the superciliary arches;

D. The temporal lines — characteristically individual lines which form the lateral margins of the forehead;

E. The nasal bones;

F. The zygomatic bones with their very prominent zygomatic processes forming the anterior portions of the zygomatic arches;

G. The chin formed by the central part of the mandible.

Plate 7, Figs. 1 and 2 show the skull of the newborn, viewed from above and from the left side. Sutural lines have not formed as yet. Instead, membrane-covered spaces are present between bones concerned. The frontal bone consists of two portions, unfused as yet.

Fig. 3 demonstrates the senile skull. As a result of the teeth falling out, the mandible is thinned, the angle of the mandible obtuse, the mandible extends beyond the maxilla, and the chin protrudes.

Figs. 4 to 6 demonstrate the contours of three different skulls with their sutures.

II. The Bones of the Trunk.

PLATES 8, 9, and 10.

These plates include the bones of the trunk consisting of the spinal column and the thoracic cage.

A. The spinal column of the adult consists of 24 distinct (true) vertebrae, the sacrum, and the coccyx. The 24 true vertebrae are made up of 7 cervical vertebrae, 12 thoracic vertebrae, and 5 lumbar vertebrae. The sacrum consists of 5 fused (false) vertebrae; the coccyx, 4 fused vertebrae.

B. The thoracic cage includes the sternum and 12 pairs of ribs. The upper 7 ribs (true ribs) are directly connected to the sternum by their costal cartilages; of the lower 5 ribs (false ribs), the eighth, ninth, and tenth are attached by their costal cartilages to the costal cartilage of the seventh rib, forming thereby the inferior thoracic margin, clearly indicated in the living. The eleventh and twelfth ribs lie within the posterior abdominal wall with their anterior ends unattached ("floating" ribs).

PLATE 11. Types of Vertebrae.

Fig. 1. The first cervical vertebra (atlas); note the concave articular facets into which the occipital condyles fit.

Fig. 2. The second cervical vertebra (axis); note the tooth-shaped process (the dens).

Fig. 3. The first and second cervical vertebrae, articulated.

Fig. 4. The seventh cervical vertebra; note a bifurcated spinous process and perforated transverse process.

(N.B. the spinous process of the seventh cervical vertebra is rarely bifurcated but usually presents a single tubercle easily palpable beneath the skin, as indicated by the name "vertebra prominens" sometimes used for this vertebra.)

Figs. 5 and 6. The first thoracic vertebra; note the articular facets for the ribs.

Fig. 7. The fifth lumbar vertebra; note the massive body and strong spinous process.

PLATES 12 and 13. Movements of the Spinal Column.

The nodding motion between the head and first cervical vertebra and the rotatory motion between first and second cervical vertebrae are not shown. Only the movements of the spinal column from the third cervical vertebra to the sacrum are illustrated: forward and backward flexion, lateral (right and left) flexion, and rotation about the longitudinal axis.

Forward and backward flexion are performed predominantly in the cervical and lumbar portions. For this purpose, the thoracic portion of the spine with the thorax may be considered as fixed. Also, lateral flexion occurs in the main in the cervical and lumbar portions. Rotation about the longitudinal axis occurs, on the other hand, predominantly in the thoracic portion of the spine and particularly in its lower part. Rotation from the eighth to the twelfth thoracic vertebrae may be as much as 28 degrees. Total amount of rotation from the third cervical vertebrae to the sacrum is about 47 degrees.

III. The Bones of the Upper Extremity. PLATES 14 and 15.

The bones of the upper extremity may be said to include:

A. The clavicle and the scapula which together form the shoulder girdle;

B. The humerus;

C. The two bones of the forearm (ulna and radius);

D. The bones of the wrist;

E. The bones of the palm and fingers.

Subdivisions B, C, D, E together make up the upper extremity proper or the "free" portion of the upper extremity.

PLATE 14, Fig. 1 illustrates the bones of the upper extremity as seen from the front with the forearm hanging down at the side naturally.

NOTE: This position of the forearm is midway between supination and pronation. For purposes of strict anatomical description, the "anterior view" of the forearm is the anterior aspect of the supinated forearm with palm facing directly forward. The outer aspect of the forearm and hand in this position is also called the lateral or radial side. The inner aspect is also called the medial or ulnar side. "Radial" and "ulnar" refer to the two bones of the forearm.

Note the S-shaped clavicle, the apex of the shoulder formed by the acromion process of the scapula, the coracoid process of the scapula, the humerus with its characteristic joint surfaces, the bones of the forearm articulating with the humerus, and finally, below the forearm, the bones of the wrist, palm, and fingers (carpal bones, metacarpal bones and phalanges).

PLATE 14, Fig. 2 illustrates the bones of the upper extremity, with pronated forearm, as seen from the medial (or inner) side. Note the foreshortened clavicle and acromion process, the medial epicondyle of the humerus, the crossed bones of the forearm, and the lateral aspect of the wrist and hand.

PLATE 15, Fig. 1 illustrates the bones of the upper extremity with forearm pronated, as seen from the lateral (or outer) side. Note the axillary border of the scapula, the foreshortened clavicle, the clearly demonstrated head of the humerus and lateral epicondyle of the humerus, the adjacent S-shaped bones of the forearm, and the side view of the wrist and hand.

PLATE 15, Fig. 2 shows the bones of the upper extremity, forearm pronated, as seen from behind. Note that the scapula is seen in its entire extent and that both epicondyles of the humerus are well demonstrated. (Extensor muscles are attached to the lateral epicondyle; flexor muscles to the medial epicondyle.) The ulna is well seen, especially its upper end or olecranon, and its lower end, the styloid process and the head which form a prominence just above the wrist.

IV. The Bones of the Lower Extremity. PLATES 16 and 17.

The bones of the lower extremity consist of:

A. The two innominate bones (Each innominate

bone is made up of three bones distinct in development but fused in the adult — the pubis, ischium, and ilium. The innominate bones, the sacrum, and coccyx, together, form the pelvis, sometimes called the pelvic girdle.);

B. The femur;

C. The leg bones (tibia and fibula);

D. The bony structure of the foot.

Subdivisions B, C, and D make up the lower extremity proper or "free" lower extremity.

PLATE 16, Fig. 1 shows the bones of the lower extremity as seen from the front. Note the half-pelvis, the innominate bone with well-marked anterior superior and inferior spines, the femur with its well-developed ends, the patella, the two leg bones, and the bones of the foot viewed from above and in front. The bones of the foot consist of the tarsal bones, the metatarsal bones, and the bones of the toes (phalanges).

PLATE 16, Fig. 2 shows the bones of the lower extremity as seen from behind. Note the half-pelvis, the innominate bones with well-marked posterior superior and inferior spines, the ischium with its tuberosity and spine, the femur with the two trochanters at its upper end and the two condyles at its lower end, the tibia articulating with the femur at the knee joint, the fibula, and the bones of the foot.

PLATE 17, Fig. 1 shows the bones of the lower extremity from the medial aspect. Note the foreshortened pelvis, the medial condyle of the femur, the prominent tibial tuberosity at the upper end of the tibial crest, and the medial aspect of the bones of the foot.

PLATE 17, Fig. 2 shows the lateral view of the bones of the lower extremity. Note the half-pelvis with prominent iliac crest, the femur and patella, the tibia with its tuberosity, the fibula with the fibular head at its upper end, and the lateral aspect of the bones of the foot.

V. The Articulations of the Human Body. PLATES 18 and 19.

PLATE 18 shows the ligamentous capsule of the hip joint and the ligaments of the elbow joint.

PLATE 19 shows the knee joint with and without its capsule. The capsule is re-enforced by accessory ligaments, not only on the outside of the joint but also within the joint as cruciate ligaments. Note the position of the two fat pads below the patella. These fat pads determine to a considerable extent the external appearance of the knee.

THE MUSCLES OF THE HUMAN BODY

I. General Considerations on the Types of Muscles and Tendons. PLATE 20.

A typical muscle may be said to consist of a central, red, fleshy or "muscular" portion (the belly) which changes its length and a tendon which does not alter its length but is stretched when the muscular portion is shortened.

PLATE 20, Fig. 1 is a schematic representation of a muscle to clarify the terminology used. In general, the "origin" of a muscle is the uppermost muscle attachment or the muscle attachment nearest the midline of the body, while the opposite muscle attachment is called the "insertion." Some muscles are subdivided into a "head," any expanded portion at the origin, a central portion (the muscle belly), and a terminal portion or tail. Tendons appear in several forms:

A. As terminal tendons, attached at the end of the muscle, e.g. the gastrocnemius muscle and the Achilles' tendon (Fig. 3).

B. As interstitial tendons, inserted in the substance of the muscle belly, e.g. the tendinous "inscriptions" of the rectus abdominis muscle Fig. 4).

C. As sheets, bands, or strands which frequently extend from the origin or insertion deep into the muscle substance, e.g. the tendinous strands in the extensor carpi ulnaris muscle (Fig. 2).

D. As aponeuroses—the term used for broad extensive tendon sheets, e.g. the aponeurosis of the external oblique muscles (overlying the right rectus abdominis muscle in Fig. 4).

E. As tendinous sheets or bands which cover a portion of the muscle belly, e.g. the tendons of the gastrocnemius muscle (Fig. 3).

It is also possible to distinguish several types of muscle bellies:

A. Muscles with two, three, or more heads which arise at different sites and fuse into one belly, e.g. the biceps muscle and the triceps muscle of the arm and the quadriceps muscle of the thigh.

B. Muscles with a single belly which divides into several slips which insert independently, e.g. the flexors and extensors of the fingers and toe.

C. Broad muscles which, besides contracting, serve also to cover or protect body cavities, e.g. the pectoralis major muscle, and the external oblique abdominal muscle.

D. Ring-shaped muscles, e.g. the circular muscles surrounding the eyes and mouth.

E. "Skin" muscles which arise from some deeper site but insert into the skin, e.g. the platysma muscle in the neck.

NOTE: The term fascia is applied to a membranous connective tissue sheet which surrounds a muscle or muscle group.

II. The Muscles of the Head. PLATES 21 through 24.

The muscles of the head may be divided into:

A. The muscles associated with the lids;
B. The muscles associated with the mouth;
C. The muscles for the nose;
D. The muscles over the top of the skull;
E. The muscles associated with the lower jaw.

A. Muscles associated with the lids. Orbicularis oculi muscle (PLATE 21).

Origin: Medial angle of the eye, lachrymal bone and medial ligament of the lid.
Insertion: Interdigitates with fibers at origin.
Action: Closes the eyelids.

B. Muscles associated with the mouth.

1. Orbicularis oris muscle (PLATES 21, 22, Fig. 2, and PLATE 24).
Origin and insertion: Consists of prolongations from all of the adjacent muscles on each side of the face.
Action: Closes the mouth.

Insertion: Corner of the mouth.
Action: Depresses the corner of the mouth.

4. Caninus muscle (levator anguli oris muscle).
Origin: Canine fossa of the maxilla.
Insertion: Into the orbicularis oris muscle.
Action: Elevates the corner of the mouth.

5. Risorius muscle.
Origin: Subcutaneous tissue and as a prolongation of the platysma muscle, overlying the masseter muscles.
Insertion: Skin and mucous membrane at the corner of the mouth.
Action: Pulls the corner of the mouth laterally, producing a dimple in the cheek.

6. Quadratus labii superioris muscle, infraorbital head.
Origin: Lower margin of the orbit.
Insertion: Skin of the upper lip.
Action: Raises the upper lip.

7. Quadratus labii superioris, angular head.
Origin: Medial angle of the eye and nasal process of the maxilla.
Insertion: Anterior limb inserts into skin and alar cartilage of nose; posterior limb inserts into the skin of the upper lip.
Action: Raises the nasal alar cartilage and upper lip.

8. Zygomaticus muscle (zygomaticus major mus-

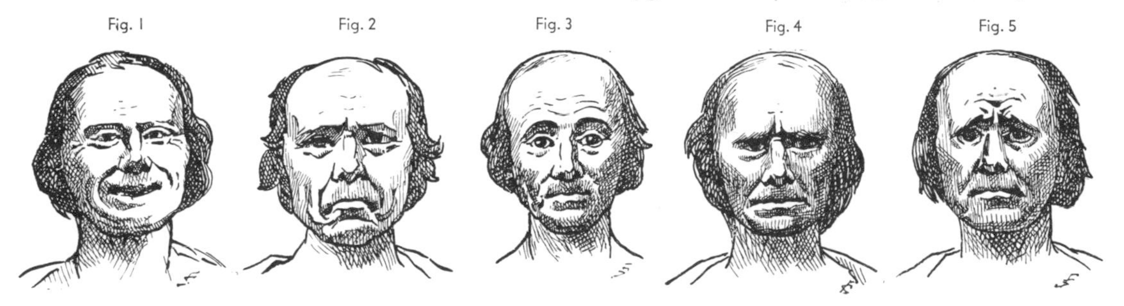

Physiognomy (after Duchenne)

2. Buccinator muscle (PLATES 21, 22, Fig. 2, and PLATE 24).
Origin: External surface of the maxilla and external oblique line of the mandible.
Insertion: The muscle bundles cross at the corners of the mouth and extend into the orbicularis oris muscle.
Action: Draws the corner of the mouth laterally.

3. Triangularis muscle (depressor anguli oris muscle) (PLATES 21, 22, Fig. 3).
Origin: External surface of mandible.

cle) (PLATE 21).
Origin: External surface of zygomatic bone.
Insertion: Corner of the mouth.
Action: Pulls the corner of the mouth upward and laterally.

9. Quadratus labii superioris muscle, zygomatic head (zygomaticus minor muscle) (PLATE 21).
Origin: External surface of zygomatic bone.
Insertion: Lateral portion of the upper lip.
Action: Pulls the corner of the mouth upward and laterally.

10. Quadratus labii inferioris muscle (depressor

labii inferioris muscle) (PLATE 21).

Origin: Border of mandible between mental foramen and mental tubercle.

Insertion: Skin of lower lip.

Action: Pulls the lower lip downward and laterally.

11. Mentalis muscle.

Origin: From mandible between the canine teeth.

Insertion: Skin of the chin.

Action: Draws up the skin of the chin and causes lower lip to protrude.

C. Muscles for the nose.

Procerus muscle (pyramidalis nasi muscle) (PLATES 21 and 24).

Origin: Root of the nose.

Insertion: Skin over the bridge of the nose.

Action: Draws the skin of the nose upward and assists in widening the nostril.

The other nasal muscles are: transverse portion of the nasalis muscle (compressor naris muscle); alar portion of the nasalis muscle (depressor alae nasi muscle); the depressor septi nasi and the dilator naris muscles.

D. The muscles over the top of the skull.

1. Occipitalis muscle.

Origin: Occipital bone, above the superior nuchal line.

Insertion: Into the galea aponeurotica, i.e. the aponeurosis covering the top of the skull.

Action: Draws backward the skin over the head.

2. Frontalis muscle.

Origin: Root of the nose and superciliary arches.

Insertion: Anterior margin of the galea aponeurotica (cranial aponeurosis).

Action: Draws forward the skin over the head, elevates the eyebrows and wrinkles the forehead.

E. Muscles associated with the mandible.

1. Masseter muscle (PLATE 22, Figs. 1 and 2).

Origin: Lower border of the zygomatic arch.

Insertion: Angle of the mandible.

Action: Elevates the mandible and presses the lower and upper teeth together.

2. Temporalis muscle (PLATE 22, Figs. 1 and 2).

Origin: Superior temporal line, external surface of the temporal bone and anterior border of the temporal fossa.

Insertion: Through a strong tendon, which passes deep to the zygomatic arch, into the coronoid process of the mandible.

Action: Pulls the mandible upward.

NOTE 1. Correlation of muscular action and facial expressions: The muscles noted above are those which alter the facial expression in accordance with the emotions, e.g. (see illustrations):

Happiness, laughter—zygomaticus muscle.
Contempt, discontent—triangularis muscle.
Attention, astonishment—frontalis muscle.
Meditation—upper portion of orbicularis oculi muscle.
Pain—corrugator (supercilii) muscle.

PLATE 23 shows the eye, nose, and (external) ear from the front and side with detailed features labeled.

III. The Muscles of the Trunk. PLATES 22 through 27.

The muscles of the trunk may be divided into:

A. The neck muscles;
B. The thoracic muscles;
C. The muscles of the abdominal wall;
D. The muscles of the back.

A. The neck muscles.

1. Platysma muscle (PLATE 22, Fig. 3).

Origin: Fascia covering the pectoral and deltoid muscles.

Insertion: Lower border of mandible and the skin of the face.

Action: Stretches the skin of the neck.

2. Sternocleidomastoid muscle (PLATES 21 and 22, Fig. 3).

Origin: The medial head arises from the anterior surface of the manubrium sterni; the lateral head arises from the medial third of the clavicle.

Insertion: Wide portion of the mastoid process and the superior nuchal line.

Action: If the muscle of one side contracts, the head is rotated to the opposite side. If the muscles of each side contract simultaneously, the chin is raised.

3. Sternothyroid muscle (PLATE 21).

Origin: Posterior surface of the manubrium sterni.

Insertion: Body of the hyoid bone.

Action: Pulls the hyoid bone down.

4. Omohyoid muscle (PLATE 21).

Origin: Superior border of the scapula.

Insertion: Body of the hyoid bone.

Action: Pulls the hyoid bone down and back.

In addition, this group of muscles includes the sternothyroid muscle and the thyrohyoid muscle (Deep layer of the neck muscles).

5. Scalene muscles; three in number, the anterior, middle, and posterior.
 Origin of scalenus anterior muscle: From the anterior tubercles of the transverse processes of the third to sixth cervical vertebrae. *Origin of scalenus medius muscle:* From the posterior tubercles of the transverse processes of all of the cervical vertebrae. *Origin of scalenus posterior muscle:* From the posterior tubercles of the transverse processes of the three lower cervical vertebrae.
 Insertion of scalenus anterior and medius muscles: On the superior surface of the first rib. *Insertion of scalenus posterior muscle:* On the external surface of the second rib.
 Action: Elevate the first two ribs.
6. Digastric muscle (PLATE 24).
 Origin: From the inner surface of the mastoid process.
 Insertion: On the superior border of the hyoid bone and the posterior aspect of the chin.
 Action: With fixed hyoid bone, pulls the mandible down.

In the study from life, PLATE 25, the following structures are well seen: the contour of the anterior portion of the digastric muscle, and, below this, the anterior surface of the hyoid bone, the laryngeal prominence (Adam's apple), and the inferior cervical fossa or jugular notch just above the manubrium sterni. The lateral view shows the sternocleidomastoid muscle with its two sites of origin, the sternum and the clavicle.

PLATE 24 shows the muscles of the trunk as seen from the front.

B. The thoracic muscles.
1. Pectoralis major muscle.
 Origin: The clavicle and the entire anterior surface of the sternum down to the level of the sixth costal cartilage.
 Insertion: The bony ridge below the greater tuberosity of the humerus.
 Action: Pulls the humerus toward the anterior aspect of the thorax and rotates it inward.
2. Pectoralis minor muscle (PLATES 21 and 28, Fig. 2).
 This muscle is almost entirely covered by the pectoralis major muscle.
 Origin: From the third, fourth, and fifth ribs.
 Insertion: Coracoid process of the scapula.
 Action: Pulls the scapula down.
3. Serratus anterior muscle (PLATE 28, Fig. 1).
 Origin: Through nine slips (digitations) from the external surface of the eight upper ribs.
 Insertion: Vertebral border of the scapula.
 Action: Pulls the scapula forward and pulls it tightly against the trunk.

The deep layer of thoracic muscles, covered for the most part by the muscles noted above, includes the intercostal muscles (PLATE 28, Fig. 1), the subclavius muscle, and the intertransverse muscles.

C. Muscles of the abdominal wall.
1. External oblique abdominal muscle.
 Origin: Through eight fleshy slips or digitations from the eight lower ribs.
 Insertion: The three last digitations extend to the iliac crest; the other slips end in a broad but thin aponeurosis which inserts partly into the inguinal ligament and, in the midline, fuses with the corresponding aponeurosis of the opposite side to form the "linea alba" or white line.
 Action: Decreases the size of the abdominal cavity, increasing the intra-abdominal pressure.
2. Internal oblique abdominal muscle-covered by the external oblique muscle. PLATE 24 shows its appearance after removal of the external oblique muscle.
 Origin: Anterior superior iliac spine and iliac crest.
 Insertion: On the last three ribs and in the region of the linea alba.
 Action: Same as the external oblique muscle.
3. Transversus abdominis muscle-covered by both the external and internal oblique muscles.
 Origin: From the lower costal cartilages and the crest of the ilium.
 Insertion: Into the linea alba.
 Action: Same as external and internal oblique muscles.
4. Rectus abdominis muscle.
 Origin: Pubis, from pubic spine to symphysis.
 Insertion: External surface and lower border of the fifth to seventh costal cartilages.
 Action: Same as the other abdominal muscles.

PLATE 25 is a study from life of the surface anatomy of the human body, demonstrating the entire course of the left clavicle below the skin, the depression between the two major pectoral muscles corresponding to the sternum, the inferior thoracic margin, the jugular notch above the manubrium sterni between the two sternocleidomastoid muscles, and the delto-pectoral or inferior clavicular fossa between the deltoid muscle and the pectoralis major muscle. The pectoralis major muscles form prominent masses. The tendinous inscriptions of the rectus abdominis muscle above the navel de-

lineate six, approximately quadrilateral areas; below the navel, the external oblique muscle is recognizable. The inguinal (Poupart's) ligament, with its graceful curved course, is well delineated and forms the boundary between trunk and thigh.

D. Muscles of the back (PLATES 29 and 47).

1. Trapezius muscle.
 Origin: Superior nuchal line of the occipital bone, spinous processes of all cervical and thoracic vertebrae.
 Insertion: Superior surfaces of the outer third of the clavicle and the spine of the scapula.
 Action: Pulls the scapula backward and assists in raising the arm by raising the scapula.
2. Latissimus dorsi muscle.
 Origin: From the spinous processes of the lower six or seven thoracic vertebrae and the spinous processes of all lumbar and sacral vertebrae; also three or four slips arise from the lower three or four ribs as digitations between those of the external oblique muscle.
 Insertion: Crest of lesser tubercle of humerus and intertubercular (bicipital) groove.
 Action: Pulls the arm back and down and rotates it medially.
3. The rhomboid muscles (PLATE 47, Fig. 1).
 Origin: The spinous processes of the two lower cervical and four upper thoracic vertebrae.
 Insertion: Vertebral border of the scapula.
 Action: Pulls the scapula upward and toward the spine.
4. Levator scapulae muscle (PLATE 47, Fig. 1).
 Origin: Through four slips from the posterior tubercles of the transverse processes of the four upper cervical vertebrae.
 Insertion: Superior angle of the scapula.
 Action: Raises the scapula.
5. Sacrospinalis muscle (erector spinae muscle) (PLATE 47, Fig. 2).
 Origin: The lateral portion of this combined muscle, the iliocostalis lumborum muscle, arises from the posterior portion of the iliac crest; the medial portion, lying next to the midline, the longissimus dorsi muscle, arises from the sacrum.
 Insertion: The iliocostalis lumborum muscle inserts through 12 slips on the lower border of the 12 ribs at their angles and another slip goes to the transverse process of the seventh cervical vertebra. The lateral portions of the longissimus dorsi muscle insert on the transverse processes of the lumbar vertebrae and the lower borders of the lower ten ribs; the medial portion inserts on the lumbar and thoracic vertebrae.
 Action: Straightens and extends the spinal column.

This group of muscles also includes several other muscles adjacent to the spine as well as muscles extending between the occipital bone and the first cervical vertebra.

PLATE 30, the study from life of the surface anatomy, shows the prominent spinous process of the seventh cervical vertebra, the spine of the scapula, and the iliac crest. The tendinous area of the trapezius muscle about the seventh cervical spine appears as a flat, moderately depressed area. The triangular tendon of origin of the trapezius muscle from the spine of the scapula produces a small fossa. The course of the trapezius muscle below the skin is well delineated as well as that of the latissimus dorsi muscle. Below the lower border of the latter muscle is seen the inferior angle of the scapula. Below, the sacrospinalis muscle of each side forms a prominent mass next to the midline.

PLATE 31 shows a cross-section through the neck of a 20-year-old male. The positions and relationships of the bones and muscles as well as the large blood vessels and nerves, the larynx and esophagus, are clearly seen.

PLATES 32 and 33 show a lateral view of the bone and muscle relationships in "The Fighter" by Borghese. The work of Salvage, "Le Gladiateur Combattant," Folie, 1812, Paris, was used to obtain the general outlines.

PLATES 35, 37, and 39-44 are photographs of a splendidly developed male body. Surface anatomical features are clarified by the accompanying sketches.

PLATES 36 and 38 serve as examples of the "Hercules" type athlete. Accompanying diagrams sketch the muscles demonstrated. In the anterior view may be seen the clavicle, iliac crest, sternum (as a deep longitudinal depression in the midline of the chest) with adjacent prominent pectoral muscles, and the triangular delto-pectoral fossae. Below the elevated right arm, five digitations of the serratus anterior muscle may be identified. The portion of the oblique abdominal muscles which inserts into the iliac crest is clearly delineated. The inferior thoracic border is well marked as well as the grooves due to the three tendinous inscriptions of the rectus abdominis muscle above the navel. In the posterior view, one can identify the two borders of the scapula, and in the midline, about the seventh cervical spinous process, the somewhat tri-

angular tendinous area of the trapezius muscle. Below is seen the curved line due to the iliac crest. The trapezius tendon attached to the scapular spine produces a deep triangular fossa. The inferior angle of the scapula is visible below the prominent border of the latissimus dorsi muscle. The well-developed sacrospinalis muscle can be followed some distance upward from the sacrum and ilium.

PLATES 45 and 46 show anatomical demonstrations of the human body in various positions as seen from the front, side, and back.

IV. The Muscles of the Upper Extremity. PLATES 48-69.

These muscles may be divided into:

A. The shoulder muscles (PLATES 48-53);
B. The muscles of the (upper) arm (PLATES 48-53);
C. The muscles of the forearm (PLATE 54);
D. The muscles of the hand (PLATES 56-59).

A. The shoulder muscles.

1. Deltoid muscles.
Origin: Lower border of the scapular spine, outer border of the acromion and lower border of the clavicle.
Insertion: The triangular rough area on the lateral side at the middle of the shaft of the humerous.
Action: Elevates the arm.

2. Infraspinatus muscle.
Origin: From the entire extent of the infraspinous fossa with exception of the axillary border and inferior angle of the scapula.
Insertion: Greater tuberosity of the humerus.
Action: Elevates the arm.

3. Teres minor muscle.
Origin: Middle portion of the axillary border of the scapula.
Insertion: Greater tuberosity of the humerus.
Action: Rotates the arm outward.

4. Teres major muscle.
Origin: Inferior angle of the scapula.
Insertion: Medial lip of intertubercular groove.
Action: Depresses the arm and rotates it inward.

B. The muscles of the arm.

1. Anterior arm muscles.

a. Biceps muscle.
Origin: Short head arises from the coracoid process of the scapula; long head arises from the superior border of the glenoid cavity.
Insertion: Radial tuberosity.
Action: Flexes the forearm at the elbow joint and is the strongest supinator of the forearm.

b. Coracobrachialis muscle.
Origin: Coracoid process of the scapula.
Insertion: Middle of the shaft of the humerus.
Action: Elevates the arm.

c. Brachialis muscle.
Origin: Anterior surface of humerus, surrounding the site of insertion of the deltoid muscle.
Insertion: Ulnar tuberosity.
Action: Flexes the forearm and puts tension on the medial portion of the capsule of the elbow joint.

2. Posterior arm muscles.

a. Triceps muscle.
Origin of long head: Axillary border of scapula.
Origin of lateral head: Along a line which extends from the site of insertion of the infraspinatus muscle to the lower third of the humerus.
Origin of medial head: Below the site of insertion of the teres major muscle.
Insertion: Upper part of olecranon.
Action: Extends the arm.

b. Anconeus muscle.
Origin: Lateral epicondyle of the humerus.
Insertion: Lateral aspect of olecranon.
Action: Extends the forearm.

C. Muscles of the forearm.

1. Muscles on the flexor aspect.

a. Superficial layer.

1.) Pronator teres muscle.
Origin: Medial epicondyle of the humerus.
Insertion: Rough area in middle of shaft of radius.
Action: Pronates and flexes the forearm.

2.) Flexor carpi radialis muscle.
Origin: Medial epicondyle of the humerus.
Insertion: Anterior surface of base of second metacarpal.
Action: Flexes the hand at the wrist.

3.) Palmaris longus muscle.
Origin: Medial epicondyle of humerus.
Insertion: Into palmar fascia.
Action: Flexes forearm and hand.

4.) Flexor carpi ulnaris muscle.
Origin: Medial epicondyle of humerus.

Insertion: Pisiform bone.
Action: Flexes the forearm and hand.

5.) Flexor digitorum sublimis muscle.
Origin: Medial epicondyle of humerus.
Insertion: Through four strong tendons on the middle phalanges of the second to fifth fingers.
Action: Flexes the middle phalanges.

b. Deep layer (PLATE 54, Fig. 2).

1.) Flexor digitorum profundus muscle.
Origin: Medial and anterior surfaces of the ulna and the interosseous membrane.
Insertion: Through four tendons on the terminal phalanges of the second to fifth fingers.
Action: Flexes the fingers, particularly the terminal phalanges.

2.) Flexor pollicis longus muscle.
Origin: Anterior surface of radius.
Insertion: Terminal phalanx of thumb.
Action: Flexes the terminal phalanx of the thumb.

3.) Pronator quadratus muscle.
Origin: Lower fourth of the ulna.
Insertion: On the radius at the same level as origin from ulna.
Action: Pronates the forearm.

2. Muscles on the extensor aspect of the forearm.

a. Superficial layer.

1.) Brachioradialis (supinator radii longus).
Origin: Bony ridge on the lateral epicondyle of the humerus.
Insertion: Lateral side of base of styloid process of radius.
Action: Flexes the forearm. Acts as supinator only when forearm is extended and pronated.

2.) Extensor carpi radialis longus muscle.
Origin: Below the brachioradialis muscle.
Insertion: Base of second metacarpal bone.
Action: Extends and abducts the hand.

3.) Extensor carpi radialis brevis muscle.
Origin: Lateral epicondyle of the humerus.
Insertion: Metacarpal of middle finger.
Action: Extends the hand radialward.

4.) Extensor digitorum communis muscle.
Origin: Lateral epicondyle of the humerus.
Insertion: By four tendons on the bases of the phalanges of the fingers.
Action: Extends the fingers.

5.) Extensor carpi ulnaris muscle.
Origin: Lateral epicondyle of the humerus.
Insertion: Base of fifth metacarpal bone.
Action: Extends and abducts the hand ulnarward.

6.) Supinator (brevis) muscle (lies deeply concealed under the brachioradialis muscle).
Origin: Lateral epicondyle of the humerus.
Insertion: Medial surface of the radius.
Action: Supinates the forearm.

7.) Abductor pollicis longus muscle.
Origin: Lateral surface of the ulna, the interosseous membrane, and the radius.
Insertion: Base of first metacarpal bone.
Action: Abducts the thumb.

8.) Extensor pollicis brevis muscle.
Origin: Below the abductor pollicis longus muscle.
Insertion: On first phalanx of thumb.
Action: Extends the first phalanx of the thumb.

9.) Extensor pollicis longus muscle.
Origin: Interosseous membrane and ulna.
Insertion: The tendon of this muscle fuses with the tendon of the extensor pollicis brevis muscle.
Action: Extends both phalanges of the thumb.

10.) Extensor indicis proprius muscle.
Origin: Ulna and interosseous membrane.
Insertion: The tendon of this muscle fuses on the dorsal surface of the hand with the tendon to the index finger from the extensor digitorum communis muscle.
Action: Extends the finger.

V. Life Study of the Upper Extremity. PLATES 48, 50-55.

GENERAL NOTE: In these plates, a markedly well-developed upper extremity of a middle-aged man is drawn from life. The muscular prominences are natural, but the transitions from muscles to tendons are in many places accentuated.

PLATE 48 shows the anterior view of the upper extremity with forearm pronated. In this position, the bones of the forearm are crossed due to the rotation of the radius about the fixed ulna.

Above the deltoid muscle, the clavicle and acromion are clearly seen. Comparison with the accompanying figures, showing the superficial muscles of the upper extremity, will clarify the drawings. The site of crossing of the extensor carpi radialis longus and the brachioradialis muscles is of importance in determining the lateral contour of the forearm. The same is true of the crossing of the pronator teres and the brachialis muscles in relation to the medial contour. The biceps muscle extends with its tendon into the depths of the bend of the elbow.

The veins of the upper extremity are very prominent, particularly the cephalic vein over the biceps.

PLATE 50 shows the medial view of the upper extremity.

Of the bony parts, the medial epicondyle of the humerus is very prominent at about the center of the figure near the bend of the elbow and below the head of the ulna at the wrist.

On the lateral contour, the deltoid muscle is seen crossing the biceps. The biceps is crossed also lower down above the bend of the elbow by the brachioradialis muscle.

The venous network of the flexor side of the upper extremity is clearest in this view. A large vein is seen crossing the bend of the elbow obliquely, the so-called median vein. The continuation of this vein on the medial aspect of the arm is called the basilic vein which empties above the elbow joint into the deep veins of the arm. These, along with the brachial artery and nerves, extend in a bundle in a groove between the biceps and triceps muscles to the axilla. In the middle of the arm, these vessels lie almost directly beneath the skin.

PLATE 51 shows the lateral view of the upper extremity. Of the bony structures, the olecranon and the head of the ulna are most prominent. The bones of the arm and forearm are covered with muscles. The lateral epicondyle of the humerus, emphasized in the figures demonstrating the muscles, is covered by the brachioradialis and extensor carpi radialis longus muscles. The acromion and clavicle are clearly seen.

Note how the deltoid muscle is inserted between the biceps and brachialis muscles. Characteristic features of the medial contour are the transition of the triceps muscle into its tendon, the attachment of the lateral head of this muscle to the olecranon, and the crossing of the brachioradialis and extensor carpi radialis longus muscles over the brachialis and biceps muscles. Between the extensor carpi ulnaris and flexor carpi ulnaris muscles, which form the gently curved external contour from olecranon to ulnar head, can be noted the ulna throughout its entire length forming a well-marked boundary between flexors and extensors.

The anconeus muscle presents a well-demarcated triangular elevation and crosses the extensor carpi ulnaris muscle, below the bend of the elbow. Of importance in determining the external contour above the wrist is the crossing of the long muscles of the thumb over the extensor carpi radialis brevis muscle.

PLATE 52 shows the posterior view of the upper extremity. Of the bony structures, the olecranon, the medial epicondyle of the humerus, and the head of the ulna are most prominent.

The spine of the scapula forms a depression between the trapezius and deltoid muscles. Clearly seen is the transition of the triceps muscle into its tendon.

PLATE 53 shows the muscles which form the axilla and the deep layer of muscles of the arm.

PLATE 54 shows the deep layer of the muscles of the forearm.

PLATE 55 shows the ulna on the medial side of the arm.

D. The muscles of the hand (PLATES 56-59).

1. The muscles of the ball of the thumb.

a. Abductor pollicis brevis muscle.
Origin: Transverse carpal (anterior annular) ligament and the greater multangular bone (trapezium).
Insertion: Basal phalanx of the thumb.
Action: Abducts the thumb.

b. Flexor pollicis brevis muscle.
Origin: Superficial portion from the transverse carpal ligament and greater multangular bone; deep portion from the os multangulum minus (trapezoid) and the os capitatum (magnum).
Insertion: Base of the first phalanx of the thumb.
Action: Flexes the proximal phalanx of the thumb.

c. Opponens pollicis muscle.
Origin: Transverse carpal ligament and greater multangular bone (trapezium).
Insertion: Along the entire length of the lateral border of the first metacarpal bone.
Action: Flexes, adducts, and rotates medialward the first metacarpal bone. The volar surface of the thumb is thus brought in apposition with the volar surfaces of the other fingers.

d. Adductor pollicis muscle.

Origin: Second and third metacarpal bones and deep carpal ligaments.
Insertion: Medial side of base of first phalanx of the thumb.
Action: Adducts the thumb.

2. The muscles of the ball of the little finger.
 a. Abductor digiti quinti muscle.
 Origin: Pisiform bone.
 Insertion: Ulnar surface of the first phalanx of the little finger.
 Action: Abducts the little finger.
 b. Flexor digiti quiti brevis muscle.
 Origin: Transverse carpal ligament.
 Insertion: Ulnar surface of the first phalanx of the little finger.
 Action: Flexes the little finger.
 c. Opponens digiti quiti muscle.
 Origin: Transverse carpal ligament.
 Insertion: Ulnar border of fifth metacarpal bone.
 Action: Brings the fifth finger into apposition with the thumb.
3. The muscles of the palm.
 a. Interosseous muscles (dorsal).
 Origin: Borders of the metacarpal bones.
 Insertion: The first of these muscles goes to the radial side of the basal phalanx of the index finger; the second, similarly to the middle finger; the third, to the ulnar side of the middle finger; and the fourth, to the ulnar side of the ring finger.
 b. Interosseous muscles (volar).
 Origin: The ulnar side of the index finger and the radial sides of the fourth and fifth fingers.
 Insertion: On the borders of the basal phalanges.
 Action: Adduct the fingers toward the middle finger.
4. Lumbrical muscles.
 Origin: In the palm, from the tendons of the deep flexors.
 Insertion: Cross on the radial side of the four fingers from the palm to the back of the fingers.
 Action: Flex the fingers at the basal phalanges.

VI. Life Study of the Hand. PLATES 56-60.

For this purpose, the hand of an old individual is chosen, since the various structures are more distinctly seen.

PLATE 56 shows the dorsal surface of the hand. Note the prominent head of the ulna on the external contour. The heads of the metacarpal bones protrude as the "knuckles." Characteristic dimples are formed about the knuckles by collections of fat in the female and in children. Of the hand muscles, note the abductor digiti quinti muscle forming the graceful curve on the medial border of the hand and the prominence on the lateral border of the hand formed by the interosseous muscle of the index finger.

The tendons of the extensor digitorum communis muscle and the veins are clearly seen in this view.

The skin on the back of the fingers is stretched and shows over the joints between the first and second phalanges three characteristic folds: the central fold is straight, the proximal fold (toward the wrist) is convex upward, and the distal fold is convex downward. These folds are particularly well-marked in the thumb.

PLATE 57 shows the hand as seen from the radial side. Again, the heads of the metacarpals and phalanges are clearly seen.

The tendons of the long and short extensors of the thumb form a characteristic triangle before joining one another distally.

In hyperextension, a well-marked depression is formed between the stretched tendons—the "tabatière" or snuff-box.

Of the hand muscles, the interosseous muscle of the index finger and the abductor pollicis brevis muscle are important features of the external contour. Another muscle, the adductor pollicis, clearly demonstrates its triangular shape in this view. Small veins which join proximally to form larger vessels, are present over the radial side of the hand.

Between the abducted thumb and the index finger, the skin forms a prominent fold called the "web."

PLATE 58 shows the palm. Of the bony parts, note the styloid process of the radius on the lateral contour just above the wrist, the pisiform bone as a small prominent elevation, and the very prominent head of the first metacarpal bone.

The muscles of the ball of the thumb form a well-marked egg-shaped elevation, considerably more prominent than the muscles of the ball of the little finger.

Three tendons of muscles of the forearm are clearly seen above the wrist: medially, the tendon of the flexor carpi ulnaris muscle; in the center, more prominent than the others, the tendon of the palmaris longus; finally, immediately deep to the palmaris longus tendon, the tendon of the flexor

carpi radialis muscle. The palmaris longus tendon passes over the transverse carpal ligament and joins the palmar fascia. The tendons and muscles of the palm are poorly seen because of surrounding collections of fat and are covered by the palmar fascia.

In the figure demonstrating the muscles, the palmar fascia is removed. Above the transverse carpal ligament, there is a vein which can always be seen if the skin is thin enough. The skin of the palm covers a layer of fibrous fat. Small fat pads are present immediately over the basal phalanges of the fingers.

The skin of the palm shows three prominent creases, and the fingers and thumb are crossed by three transverse folds.

PLATE 59 shows the hand as seen from the ulnar side. Most prominent of the bony structures is the head of the ulna. In thin individuals, the ligaments about the phalangeal joints and the pisiform bone may be seen, the latter, as a small rounded elevation.

On the external contour, the abductor and adductor of the thumb and abductor of the little finger can be identified. The thin palmaris brevis muscle, arising from the palmar fascia, passes transversely from its origin to the abductor digiti quinti muscle which it partially covers.

PLATE 60 shows the veins of the hand which are prominent in the surface anatomy (PLATE 56).

VII. Demonstrations. PLATES 61-69.

PLATES 61-63 are drawings of various positions of the flexed upper extremity with corresponding studies from life of the surface anatomy. The purpose of these plates is to demonstrate the anatomy of the upper extremity held in various positions. The drawings are, therefore, not meant to be of artistic value but rather to indicate the anatomical relationships.

PLATE 64 shows movements of the elbow joint. Since the movements at the elbow joint have great influence on the appearance of the upper extremity, Plate 64 illustrates various positions of the arm in relation to the forearm.

This combined joint is made up of three joints: articulation between ulna and humerus, articulation between radius and humerus, articulation between ulna and radius—the radio-ulnar joint.

PLATES 65-66 demonstrate various movements at the shoulder and elbow joints. In raising the arms, the inferior angle of the scapula moves outward, the clavicle moves up and out. In crossing the arms over the chest, the scapulae move toward the sides of the thorax. In crossing the arms over the back, the scapulae approach each other.

Of the elbow joint movements, the most important is rotation of the radius about the ulna, i.e. pronation. This produces marked alterations in the appearance of the forearm. When the palm faces forwards, the ulna and radius lie in the same vertical plane adjacent to each other, i.e. the forearm is in the position of supination and the external shape of the forearm appears flattened from front to back. As the radius rotates about the ulna, the shafts of the bones cross each other, i.e. the forearm assumes the position of pronation in which its external shape is rounded. (See also Plates 45 and 46.)

PLATES 67-69 are demonstrations of hand photographs. In the male, joints and extensor tendons are prominent. In the female and in children, dimples produced by collections of fat about the knuckles are characteristic.

VIII. The Muscles of the Lower Extremity. PLATES 70-81.

These muscles are divided into:

A. The muscles of the hip and buttock;
B. The thigh muscles;
C. The leg muscles;
D. The foot muscles.

A. The-muscles of the hip and buttock.

1. Superficial layer.
 a. Gluteus maximus muscle.
 Origin: From the outer surface of the ilium in the region of the posterior superior iliac spine, from the sacrum and coccyx, and from the ischial tuberosity.
 Insertion: On the gluteal tuberosity of the femur and the iliotibial band.
 Action: Most powerful extensor of the thigh.
 b. Gluteus medius muscle.
 Origin: Outer surface of ilium between the anterior and posterior gluteal lines and the anterior three-fourths of the iliac crest.
 Insertion: Anterior border of the great trochanter.
 Action: Abducts the thigh.
 c. Tensor fasciae latae muscle.
 Origin: Anterior superior iliac spine.
 Insertion: Inserts into iliotibial band and thereby indirectly into the lateral femoral condyle.
 Action: Rotates medially, flexes, and abducts the thigh.

2. Deep layer.
 a. Gluteus minimus muscle.
 Origin: External surface of ilium.
 Insertion: Anterior border of great trochanter.
 Action: Abducts the thigh and rotates it medialward.
 b. Piriformis muscle.
 Origin: Anterior surface of sacrum.
 Insertion: Apex of great trochanter.
 Action: Extends, abducts, and rotates the thigh laterally.
 c. Obturator internus muscle.
 Origin: Inner surface of ischium and pubis.
 Insertion: Great trochanter.
 Action: Rotates the thigh laterally.
 d. Quadratus femoris muscle.
 Origin: Tuberosity of ischium.
 Insertion: On vertical ridge below the great trochanter.
 Action: Rotates the thigh laterally.
 e. Psoas major muscle.
 Origin: Twelfth thoracic and lumbar vertebrae.
 Insertion: Lesser trochanter.
 Action: Flexes the thigh.
 f. Iliacus muscle.
 Origin: Inner surface of the ilium.
 Insertion: Lesser trochanter.
 Action: Flexor of thigh.

B. The muscles of the thigh.
1. Anterior group.
 a. Sartorius muscle.
 Origin: Ilium below the anterior superior iliac spine.
 Insertion: Medial surface of tibia near tuberosity.
 Action: Rotates leg medially when knee is flexed.
 b. Quadriceps femoris muscle. This muscle is composed of four large separate muscles (heads), three of which are partially united to each other.
 1.) Rectus femoris muscle.
 Origin: Anterior inferior iliac spine.
 2.) Vastus lateralis muscle.
 Origin: Great trochanter and lateral lip of linea aspera.
 3.) Vastus medialis muscle.
 Origin: Medial lip of linea aspera.
 4.) Vastus intermedius (crureus) muscle.
 Origin: Anterior surface of femoral shaft.
 Insertion: Through a tendon arising from the four muscles into the tibial tuberosity. The patella is a sesamoid bone in the substance of this tendon.
 Action: Extends the leg.
2. Muscles on the medial aspect of the thigh.
 a. Gracilis muscle.
 Origin: Symphysis pubis.
 Insertion: Into tibia below the medial condyle.
 Action: Adducts and flexes the thigh; flexes the leg.
 b. Adductor longus muscle.
 Origin: Superior border of symphysis pubis.
 Insertion: Middle third of femoral shaft.
 Action: Adducts the femur and crosses one thigh over the other.
 c. Adductor brevis muscle.
 Origin: Inferior ramus of pubis.
 Insertion: Upper third of linea aspera.
 Action: Adducts the femur.
 d. Pectineus muscle.
 Origin: Crest of pubic bone.
 Insertion: Behind lesser trochanter.
 Action: Flexes and adducts the thigh.
 e. Adductor magnus muscle.
 Origin: Inferior ramus of pubis and superior ramus of ischium.
 Insertion: Femoral shaft from lesser trochanter to medial condyle.
 Action: Strongest of the adductor muscles. When the thigh is fixed, the adductor muscles and the pectineus muscle assist in maintaining the erect position of the trunk and may incline the trunk forward.
3. Muscles on the posterior aspect of the thigh.
 a. Semitendinosus muscle.
 Origin: Ischial tuberosity.
 Insertion: Below the medial condyle of the tibia.
 Action: Flexes the knee.
 b. Semimembranosus muscle.
 Origin: Ischial tuberosity.
 Insertion: Posterior surface of medial condyle of tibia.
 Action: Flexes the leg.
 c. Biceps femoris muscle.
 1.) Long head.
 Origin: Ischial tuberosity.
 2.) Short head.
 Origin: Linea aspera.
 Insertion: Head of the fibula.
 Action: Flexes the knee.

C. The muscles of the leg.
1. Anterior group.
a. Tibialis anterior muscle.
Origin: Lateral condyle and lateral surface of tibia and the interosseous membrane.
Insertion: First cuneiform bone and metatarsal of the great toe.
Action: Flexes the foot and elevates the inner border of the foot.
b. Extensor digitorum longus muscle.
Origin: Lateral condyle of the tibia, interosseous membrane, head and anterior border of fibula.
Insertion: Second to fifth toes.
Action: Extends the lateral four toes.
c. Peroneus tertius muscle.
Origin: Lower third of fibular shaft.
Insertion: Base of fifth metatarsal bone.
Action: Flexes the foot and assists in elevating the lateral border of the foot.
d. Extensor hallucis longus muscle.
Origin: Anterior portion of medial surface of fibula and interosseous membrane.
Insertion: Base of second phalanx of big toe.
Action: Extends big toe.
2. Muscles on lateral aspect of leg.
a. Peroneus longus muscle.
Origin: Head of fibula.
Insertion: First cuneiform; first and second metatarsal bones.
Action: Extends the foot and raises its lateral border.
b. Peroneus brevis muscle.
Origin: Lower half of fibula.
Insertion: Tuberosity of fifth metatarsal bone.
Action: Extends the foot and raises its lateral border.
3. Muscles on the posterior aspect of the leg.
a. Gastrocnemius muscle.
Origin: Lower end of femur.
Insertion: Through the Achilles' tendon into the calcaneus.
Action: Extends the foot; also flexes the leg at the knee.
b. Soleus muscle.
Origin: Back of the fibular head and upper third of posterior surface of fibular shaft and from the tibia.
Insertion: By fusion with the Achilles' tendon into the calcaneus.
Action: Extends the foot.
c. Plantaris muscle.
Origin: Lowermost portion of lateral lip of linea aspera.
Insertion: Calcaneus.
Action: No known function in man.
4. Deep layer.
This group includes:
a. Tibialis posterior muscle;
b. Popliteus muscle;
c. Flexor hallucis longus muscle;
d. Flexor digitorum longus muscle.
D. The muscles of the foot.
1. Dorsal group.
a. Extensor digitorum brevis muscle.
Origin: Superior surface of the calcaneus (near lateral malleolus).
Insertion: Through four tendons on the bases of the second phalanges of the lateral four toes.
Action: Extends the toes.
b. Dorsal interosseous muscles.
Origin: Medial surfaces of the four lateral metatarsal bones.
Insertion: Bases of first phalanges; the second toe has two interosseous muscles, the third and fourth toes only one.
Action: Spread the toes.
c. Plantar interosseous muscles.
Origin: Medial surfaces of the three lateral metatarsal bones.
Insertion: Medial aspects of bases of first phalanges of second to fourth toes.
Action: Pull the toes closer to one another.
2. Muscles on the sole of the foot.
a. Abductor hallucis muscle.
Origin: Medial surface of calcaneus, the navicular, the first cuneiform and first metatarsal bones.
Insertion: Medial side of the base of the proximal phalanx of the big toe and sesamoid bones.
Action: Abducts the big toe.
b. Abductor digiti quinti muscle.
Origin: Posterior portion arises from the calcaneus, the anterior portion from the tubercle of the fifth metatarsal bone.
Insertion: Lateral side of the bases of the proximal phalanx of the little toe.
Action: Abducts the little toe.
3. Deep layer.
This group includes:
a. Flexor hallucis brevis muscle;
b. Adductor hallucis muscle;
c. Flexor hallucis longus muscle;
d. Flexor digiti quinti muscle;

e. Flexor digitorum brevis muscle;

f. Flexor digitorum longus muscle;

g. Lumbrical muscles.

These muscles, however, have no influence on the external appearance of the foot.

PLATES 70-73.

PLATE 70 shows the anterior view of the lower extremity.

Bony structures visible: Heart-shaped patella, the head of the fibula; a portion of the medial surface of the tibia covered only by skin and fascia is apparent from the knee to the medial malleolus.

Muscular details: On the lateral contour may be seen the gluteus maximus and medius muscles, the crossing of the gluteus maximus and medius muscles as well as the crossing of the tensor fasciae latae muscle over the vastus lateralis muscle. On the medial contour, the course of the sartorius muscle clearly demarcates the group of extensor muscles on the anterior aspect of the thigh from the adductor muscles on the medial aspect. Above the patella is seen the common tendon of the four heads of the quadriceps femoris muscle which by insertion into the patella and prolongation into the patellar tendon, is inserted into the tibial tuberosity. On each side of the patellar tendon, fat pads are present. The development of these fat pads is related to the amount of pressure transmitted through the knee joint.

From the anterior superior iliac spine, the sartorius muscle may be followed in its medial and downward course to the tibia and the tensor fasciae latae muscle in its lateral and downward course to the lateral condyle of the femur. Both of these muscles at their origins cover the rectus femoris muscle. As they deviate from one another, a small groove is formed.

PLATE 71 shows the lateral view of the lower extremity.

Bony structures visible: The great trochanter, the patella, and the head of the fibula.

The gluteus maximus and medius muscles are clearly evident throughout their entire extent. The position of the tensor fasciae latae muscle on the outer contour as it crosses the vastus lateralis muscle and the continuation of the tensor muscle through the iliotibial band to the lateral condyle of the tibia are evident. Note the demarcation between the biceps femoris muscle on one side and the semitendinosus and semimembraneous muscles on the other as they descend to form the superior borders of the popliteal fossa. The gastrocnemius muscle forms a marked prominence at the base of the popliteal fossa.

On the lateral aspect of the leg, the individual muscles as well as the triangular uncovered lower end of the fibula with its lateral malleolus are evident.

PLATE 72 shows the posterior view of the lower extremity. Bony structures visible: Great trochanter, the head of the fibula and the two malleoli.

Of the buttock muscles, note particularly the gluteus maximus and medius muscles.

Half of the gluteus maximus muscle goes to the femur and half to the fascia of the thigh. In the living, the lower border of the gluteus maximus is completely covered by fat, so that the buttock is considerably more rounded than the gluteus maximus muscle itself. In addition, while the lower border of the muscle goes obliquely downward and outward, the lower border of the buttock is transverse.

The three flexor muscles—the biceps femoris, the semimembranosus and semitendinosus muscles —form a prominent mass below the gluteal muscles and, as they gradually deviate to either side, form the popliteal fossa behind the knee. In anatomical dissections, the popliteal fossa appears as a deep rhomboid depression, but, in the living, when the leg is extended, the popliteal fossa is made prominent by a collection of fat.

Most prominent of the muscles of the leg is the gastrocnemius muscle. The size of this muscle varies from one individual to another and determines the contour of the leg in this view.

The lateral head is smaller than the medial and does not extend as far distally. The gastrocnemius muscle, along with the soleus muscle, ends in the common Achilles' tendon—the strongest tendon in the body.

PLATE 73 shows the medial view of the lower extremity. Bony structures visible: Profile view of patella, medial surface of tibia from knee to medial malleolus.

The course of the band-like sartorius muscle, from the ilium around the medial margin of the thigh to its insertion on the upper end of the tibia is clearly marked. In the leg, the gastrocnemius and soleus muscles, immediately behind the tibia, form with the flexor digitorum longus muscle an elongated prominence.

The smoothly convex anterior prominence of the leg below the tibial tuberosity is formed by the tibialis anterior muscle. The course of the great saphenous vein is easily followed in the leg.

PLATES 74-77.

These plates are drawings of the musculature from a model of "The Fighter" by Borghese—medial and lateral views of the right lower extremity.

PLATE 76 demonstrates the deep layer of muscles of lower extremity.

PLATE 77 shows cross-sections:

Fig. 1. Through the thigh.

Fig. 2. Through the leg.

IX. Drawings From Life.
PLATES 78-81.

PLATE 78 shows the anterior view of the foot.

Bony structures: Note the prominent lateral and medial malleoli: the lateral malleolus is on a lower horizontal plane than the medial.

The lateral border of the extensor digitorum brevis muscle is easily recognizable. It forms the smooth prominence called the arch of the foot and the convex lateral border of the foot immediately below the ankle.

Immediately beneath the skin, the tendons of the extensor digitorum longus and extensor hallucis longus muscles are easily followed to the toes.

The veins of the dorsum of the foot are clearly evident. The largest vein goes upwards to the medial side of the leg.

PLATE 79 shows the posterior view of the foot.

The positions of the malleoli are well seen in this view.

The abductor digiti quinti muscle, particularly when contracted, is of importance in determining the lateral contour of the foot.

The Achilles' tendon is particularly important in determining the appearance of the lower leg and foot in this view. Above, this tendon is broad but it narrows below as it approaches its insertion on the calcaneus.

PLATE 80 shows the lateral view of the foot.

Bony structures: The lateral malleolus, the joints and phalanges of the outer four toes.

Note particularly the curvature of the four outer toes as compared with the straight course of the big toe. This is evident in all three figures.

The exposed triangular surface of the lateral malleolus is bounded posteriorly by the peroneus longus and brevis muscles and anteriorly by the extensor digitorum longus muscle.

The extensor digitorum brevis muscle forms a prominent smooth mass at its origin from the superior surface of the calcaneus in front of the malleoli. The muscle belly decreases in size gradually as it extends forward under the tendons of the extensor digitorum longus muscle.

The abductor digiti quinti muscle forms the elevation along the calcaneus and the second phalanx of the little toe.

The veins of the dorsum of the foot as well as a vein over the lateral malleolus are clearly seen in this view.

PLATE 81 shows the medial view of the foot.

Bony structures: The medial malleolus and the joint between the big toe and the first metatarsal bone.

The abductor hallucis muscle in its course from calcaneus to the first phalanx of the big toe is clearly seen.

Of the tendons, those of the tibialis anterior and the extensor hallucis longus muscles stand out prominently. The posterior contour is formed by the vertical Achilles' tendon. Near the border of the medial malleolus is the great saphenous vein.

PLATE 82 shows the veins of the dorsum of the foot after removal of the skin.

PLATE 83 shows sketches of the knee in various positions. Extension and flexion of the leg produces marked changes in the shape of the knee and its surroundings.

In marked extension, the patella moves proximally, the patellar ligament is stretched, and the fat pads on each side of the quadriceps tendon appear as rounded prominences.

In marked flexion, a cleft appears between the femoral condyles and the upper end of the tibia. The patella is pulled down into this space. The fat pads, capsule, and skin are pressed against the rounded surfaces of the femoral condyles producing the rounded external appearance of the flexed knee.

PLATES 84 and 85 show various motions at the knee joint.

PLATES 86-92.

PLATE 86 shows the relative proportions of the parts of the body of the adult, according to Michelangelo. This divides the total length of the body into eight "head lengths" plus the distance from the malleoli of the ankle to the sole of the foot. In contrast with the older conception, which placed the center of the body at the lower border of the symphysis pubis, the lower extremities are relatively lengthened and the trunk elevated.

PLATE 87. Proportions of the human figure as given by Paul Richer, 1849-1933. *Anatomie Artistique, Description des formes extérieurs du corps*

humain au repos et dans les principaux mouvements. Paris, 1890.

PLATES 88 and 89. Phases in the development of a growing boy. The pictures were taken at approximately two-year intervals under precisely the same conditions — the same distance, standard frame, etc. They are therefore directly comparable for proportions. These photographs, and those of the girl following, were taken at the Institute of Child Welfare at the University of California, and are used through the courtesy of Dr. Nancy Bayley.

PLATES 90 and 91.— Phases in the development of a growing girl. The scale of this series is slightly larger than that of the boy. Within the series, the pictures are again directly comparable.

PLATE 92 shows the proportions of the child according to Kollmann. No general rules can be given since the relative proportions of the body vary markedly with age. On the average, the total body length of a new-born child is equal to four "head lengths"; at the sixth year, about six "head lengths"; at the twelfth year, about seven "head lengths." If one divides the total body length into 100 parts, the relative proportions in children from two to four years according to Kollmann are indicated by the following figures:

From skull to navel (variable)	54
From navel to sole of foot	46
From skull to perineum (sitting height)	61
Length of the lower extremity	39
Thigh length (only the mobile portion)	16
Leg length	19
Height of the foot	4
Length of the upper extremity (from acromion to the tip of the middle finger)	42
Arm length	19
Forearm (elbow-joint to wrist joint)	13
Hand length	10
Head length (top of skull to lower border of chin)	22
Widest diameter of head	15
Transverse diameter of the pelvis	15

The figures given above vary, of course, between individuals and are only approximate.

PLATE 93. Photographs showing folds and wrinkles around the eyes.

PLATES 94-96.

PLATES 94-96 show various types of breasts. The mature type includes the body of the breast, the areola, and the nipple. In the body of the breast, surrounded by more or less fatty tissue, are found the lobules of the mammary glands which during nursing secrete milk into canaliculi emptying at the nipple. The nipple and the areola are more darkly pigmented than the skin in general.

Normally, the breasts extend from the third to the sixth ribs although they may be above or below this—"high" or "low". In general, "high" breasts are considered much more beautiful. A well-built breast is fixed and tense, not pendant, and its lower margin should join the body without any marked fold. Marked decrease in the fat content of the breast produces a flabby pendant breast.

The breast is not completely developed until some time after adolescence and does not reach full development until the nursing stage.

In girls under seven years, the breast is indistinguishable from the male breast except that the nipple is more prominent. Later the areola becomes prominent and finally the body of the breast itself. In later life, particularly in the white race, the prominence of the areola may become less marked.

A Selection of Hands by Heidi Lenssen.

PLATES 97-98. Hands of the violinist Isaac Stern. Courtesy of S. Hurok.

PLATE 99. Hands of Diego Rivera. Hands of an old laborer (one of the last surviving slaves). Photos by Fritz Henle.

PLATE 100. Fine mechanic at work. Hands of a silversmith. Photos by Fritz Henle.

PLATE 101. Hands of a young man. Greek (*c.* 350 B.C.). Photos by C. W. Huston. Donatello, 1386(?)-1466. "St. John" (detail).

PLATE 102. Andrea del Verrocchio, 1435-1488. Maria and Child" (detail). "St. John and an Angel" (detail).

PLATE 103. Hugo van der Goes, 1452-1482. "Adoration of Jesus" (detail). South German (unknown), *c.* 1470, (detail).

PLATE 104. Giambattista Moroni, 1525(?)-1578. "Gentleman in Adoration" (detail). Jacopa Bassano, 1510-1592, (detail).

PLATE 105. Albrecht Dürer, 1471-1528. "Praying Hands." Courtesy of E. S. Herrmann, Inc.

PLATE 106. El Greco, 1541-1614. "Incognito" (detail). Velázquez, 1599-1660. "Portrait of Innocentius Pope" (detail). Gilbert Stuart, 1755-1828. Portrait (detail).

Illustrations from Historical Sources.

PLATES 107-189.

PLATE 107. Antonio Pollaiuolo, 1432(?)-1498. "Battle of the Naked Men." Pollaiuolo, a pupil of

Donatello, seems to have been the first to dissect bodies in order to make studies of the anatomy for artistic purposes. Vasari tells us that he dissected "many bodies in order to study the action of the muscles." While this statement is open to considerable doubt, it is nevertheless true that this engraving, together with his paintings of subjects such as Hercules, with their exaggerated musculature, had an enormous influence in extending the study of anatomy in Italy and, later, in the rest of Europe.

PLATE 108. Jacopo de'Barbari, *c.* 1450 - before 1516. "Apollo and Diana." An engraver of German origin, Jacob Walch was called de'Barbari by the Italians. He is of interest to us, at least partly, because of his influence upon Dürer whom he knew in Venice. Dürer's studies of perfect human proportions stem from his enthusiasm for the works of Barbari.

PLATES 109 to 123. Leonardo da Vinci, 1452-1518. It is difficult, if not impossible, properly to evaluate Leonardo's contributions to the study of anatomy without taking into consideration the transitional state of this science in his time and the difficulty of obtaining exact information and of dealing with it in any systematic manner. Leonardo had few opportunities to study cadavers at first hand, and it is not known whether he ever had access to a complete skeleton. Yet, his drawings are remarkably accurate despite minor errors — errors which tend to become more numerous when he deals with deep dissections and inner organs. His far-ranging interests enabled him to understand and to present his material in a new light. His knowledge of mechanics, for example, was applied to the skeleton and the muscles, and he was able to demonstrate many of the facts of their operation. The plates which follow deal entirely with the osteological and myological systems and are arranged in a chronological order, so far as it can be ascertained.

PLATE 109. Bones of the foot. Upper center: Dissection of the shoulder joint.

PLATE 110. Muscles of the trunk.

PLATE 111. Muscles of the trunk and leg. The smaller sketches near the top of the plate, from left to right: thoracic and abdominal cavities, subcostal muscles in respiration, two sketches of the tensor fasciae latae muscle, three small sketches showing nerve and vascular connections, and two more sketches of the tensor fasciae latae muscle.

PLATE 112. Bones of the foot. Muscles of the head, neck, and shoulder.

PLATE 113. Surface anatomy of the shoulder.

PLATES 114-116. Muscles of the shoulder.

PLATE 117. Surface anatomy of the lower extremity. Bottom, center to right: comparative study of the hind leg of a horse with the corresponding region in man.

PLATES 118-123. Muscles of the lower extremity.

PLATE 124. Albrecht Dürer, 1471-1528. "Adam and Eve." Dürer's interests, like those of Leonardo, ranged over a wide variety of subjects including mathematics, chemistry, and physics. In his anatomical studies, he attempted to arrive at a set of ideal proportions for the human figure based upon Italian art. The results were incorporated into four books in which he considered the body to be divided into seven parts, then into eight parts. By manipulating the various proportions according to rules, he produced examples of deformed and distorted figures. The studies were often criticized as being too contrived and academic. The "Adam and Eve" is probably the most successful attempt along these lines. Compare Adam's legs with those of Barbari's Apollo, Plate 108.

PLATES 125-138. Michelangelo Buonarroti, 1474-1564. Michelangelo's efforts to achieve a high degree of realism led him to an extremely careful study of whatever he was about to execute. He is said to have bought fish in the market in order to study their scales. It is also said that he studied cadavers in preparation for a sculpture of Christ. In any event, his interest was centered on the human body to an extent which often excluded other things — landscape and portraiture among others. His anatomical studies cover a period of about twelve years in Florence and Rome. The drawings which follow were done in pen and ink, black and red chalk, and charcoal. The arrangement is chronological. See also Plate 86 for a study of proportions based upon Michelangelo.

PLATE 125. Sketch for the bronze David and arm study for the marble David.

PLATE 126. Studies of the nude.

PLATE 127. Nude seen from the back. Lower right: sketches in black chalk.

PLATE 128. Sketch for the Bruges Madonna. Three nude men.

PLATE 129. Studies for Ignudi of the Sistine Chapel ceiling.

PLATE 130. Studies for the Libyan Sibyl.

PLATE 131. Sketches for the Sistine Chapel ceiling and the tomb of Julius.

PLATE 132. Study for a Pietà.

PLATE 133. Study for a recumbent figure in the Medici Chapel.

PLATE 134. Arm and torso study for a Pietà.

PLATE 135. Study of heads for the Leda.

PLATE 136. Study for a background figure for the "Risen Christ."

PLATE 137. Study for the "Last Judgment." Detail. The spot on the lower head is oil.

PLATE 138. Crucifixion for Vittoria Colonna.

PLATE 139. Domenico del Barbiere (Fiorentino), *c.* 1506 - after 1565. Design from an anatomy book.

Barbiere was associated at Fontainebleau with Rosso de' Rossi, who was at work on an anatomy book for Francis I. This engraving presumably reproduces one of Rossi's drawings from the book which was never finished. It was, at one time, believed to have been by Michelangelo.

PLATES 140-147. Andreas Vesalius, 1514-1564. Vesalius came from a family of physicians and himself studied medicine at Louvain, Montpellier, and Paris. His later career was divided between teaching at various centers such as Louvain, Padua, Bologna, and Pisa and positions at the courts of Charles V of France and Philip II of Spain. His researches were based upon direct observation of the human body and aimed to counteract the Galenic methods based upon studies of animal anatomy. The direction of his work made him naturally opposed to such anatomists as Eustachio (see Plate 148). He is said to have kept a careful check on his assistants, and the accuracy of his plates would be sufficient to account for their revolutionizing influence if they were not equally prized today for their artistic qualities. The plates reproduced here appeared in Basel in 1543.

PLATES 140 and 141. Male and female nudes from the *Epitome*.

PLATES 142-144. Views of the skeleton from the first book of the *De Humani Corporis Fabrica.*

PLATES 145-147. Three plates showing muscles from the second book of the *De Humani Corporis Fabrica.*

PLATE 148. Bartolomeo Eustachio, (?)-1574. From the *Tabulae Anatomicae*, Rome, 1728. Eustachio's drawings, like those of Vesalius, were based upon direct observation and are characterized by a high degree of accuracy although his Galenic methods place him into a movement in anatomical studies that has been superseded. He is said to have introduced post-mortem examinations into Roman hospitals. While his drawings are exact, they are stiff and uninteresting as compared with those of Vesalius. The parts of the drawing are not labelled directly but may be referred to by the coördinates along the sides and top of the plate.

PLATE 149. Melchior Meier. "Apollo Flaying Marsyas." Not much is known about Meier except that he flourished about 1581. Subjects such as Apollo flaying Marsyas and the martyrdom of St. Sebastian became increasingly popular after the appearance of Vesalius' book of 1543 since they afforded artists the opportunity of combining a display of their anatomical knowledge with a standard artistic subject.

PLATE 150. Orazio Borgioni, died 1616. "The Dead Christ." Following Mantegna's example in his "Dead Christ," Baroque painters perfected the techniques of extreme foreshortening, a skill particularly useful in painted ceilings. Mantegna's innovation was brought to the attention of Rembrandt by Borgioni's work reproduced here.

PLATE 151. Hendrik Goltzius, 1558-1616. The "Farnese" Hercules. Just as Baroque writers derived many of the laws of their craft from a study of classical literature, so Baroque painters and sculptors went to classical models for instruction and inspiration. The "Farnese" Hercules in Rome was a favorite subject. Goltzius particularly enjoyed exaggerated muscular development and unusual attitudes and often caricatured Michelangelo in his attempt to emulate his work.

PLATE 152. Lodovico Cardi (Cigoli), 1559-1613. Both Cardi and his teacher Alessandro Allori produced brilliant examples of anatomical sculpture. The *écorché* reproduced here was a particular favorite of Italians in their studies of anatomy.

PLATE 153. Peter Paul Rubens, 1577-1640. "Studies of Venus," *recto.* Rubens' chief anatomical work is a collection of drawings published after his death. They may well have been intended as part of a larger work on anatomy. As it stands, there are forty-four copperplates together with a Latin text on anatomy and some reflections upon astrology and alchemy. The French translation, which omits the non-anatomical passages, appeared in Paris in 1773 under the title of *Théorie de la Figure humaine, considérée dans ses principes, soit en repos ou en mouvement.*

PLATE 154. William Camper, 1666-1709. This study of the hand is reproduced from Camper's book of 1697, the title of which is self-explanatory: *The Anatomy of Humane Bodies, with figures drawn after the life by some of the best masters in Europe, and curiously engraven in one hundred and fourteen copper plates, illustrated with large explications, containing many new anatomical discoveries, and chirurgical observations: to which is added an introduction explaining the animal oeconomy, with a copious index.*

PLATE 155. Francisco de Goya y Lucientes, 1746-1828. "The Giant." This aquatint is one of several which Goya made during the last years of his life. The method is particularly exacting since the picture can be lightened (by scraping) but not darkened.

PLATE 156. John Flaxman, 1755-1826. Views of the thoracic basket. The anatomical studies left by Flaxman were not intended as a textbook for artist, nor were they published during his lifetime. Nineteen plates, together with two by Robertson, were published in London in 1833 under the title of *Anatomical studies of the bones and muscles, for the use of artists, from drawings by the late John Flaxman, engraved by Henry Landseer: with two additional plates and explanatory notes by William Robertson.*

PLATES 157-166. Jules Cloquet. These plates are reproduced from his *Anatomie de l'Homme.* Paris, C. de Lasteyrie. 1821-1832.

PLATE 157. Skeleton of infant.

PLATE 158. Musculature of torso and neck, lateral view.

PLATE 159. Musculature of torso and neck, anterior view.

PLATE 160. Female pelvis, anterior view.

PLATE 161. Female pelvis, anterior view from below.

PLATE 162. Skull, anterior view.

PLATE 163. Bones of the skull, anterior view.

PLATE 164. Lateral view with sagittal section of skull.

PLATE 165. Bones of the hand.

PLATE 166. Bones of the foot, lateral view.

PLATE 167. Jean Auguste Dominique Ingres, 1780-1867. "Studies for the Dead Body of Acron." Classical themes became a favorite subject in the early nineteenth century in painting and sculpture, particularly in France. The emphasis on line rather than on color as the means of creating the illusion of volume, form, and motion led to an extremely high degree of skill in David and in Ingres, who continued the *méthode David.* Huneker spoke of Ingres as "the greatest master of line who ever lived."

PLATES 168-170. William Rimmer, 1816-1879. Rimmer was a physician who gave up his practice in order to teach art—at the Cooper Institute in New York and at the Art Museum in Boston. In 1876, he drew a series of plates in pencil on white paper, published in 1877 as *Art Anatomy.*

PLATE 168. Muscles of the Back.

PLATE 169. Muscles of the torso, solid and unalterable sections.

PLATE 170. The triceps and muscles of the arm.

PLATES 171-176. Jeno Barcsay. *Anatomy for the Artist.* Budapest, Corvina. 1956.

PLATE 171. Trunk in movement.

PLATE 172. Trunk in movement.

PLATE 173. Trunk in movement.

PLATE 174. Foreshortened body.

PLATE 175. Foreshortened body.

PLATE 176. The foot in movement.

PLATE 177. Hilaire Germain Edgar Degas, 1834-1917. "The Bath." Degas' interest in the human body, as demonstrated by his numerous canvases on ballet dancers, is often focused on considerations of balance and displacement of the center of gravity. Although his approach is often said to be intellectual rather than sensual, he compares this woman bather to a cat licking herself.

PLATES 178-189. Eadweard Muybridge. 1830-1904. These studies are reproduced from *The Human Figure in Motion,* 1913, and *Animal Locomotion,* 1887. The pioneering work of Muybridge is especially remarkable in that it was produced before the invention of the motion picture camera. The device used, the zoöpraxiscope, produced a series of transparent photographs on glass which were projected on a screen. The results of the research, carried on chiefly in California and Pennsylvania, were published in eleven volumes under the title of *Animal Locomotion.*

PLATES 178 and 179. Athlete, walking.

PLATES 180 and 181. Athlete, running.

PLATES 182 and 183. Woman, walking.

PLATES 184 and 185. Woman, throwing ball.

PLATES 186 and 187. Child, crawling.

PLATE 188. Child, walking.

PLATE 189. Child, running.

SOURCES OF ILLUSTRATIONS

PLATES 1 and 2—after Lucae, Frankfurt-an-Main.
PLATES 3 and 4—after Spalteholz.
PLATE 5—original.
PLATE 6—after Spalteholz.
PLATE 7—original.
PLATES 8-10—from life.
PLATE 11—after Spalteholz.
PLATES 14-17—from life.
PLATES 18 and 19—original.
PLATE 20—after Spalteholz and original.
PLATE 21—original.
PLATE 22—after Spalteholz.
PLATE 23—original.
PLATE 24—after Duval-Neelsen.
PLATE 25—original.
PLATE 26—after Salvage.
PLATE 27—Kollmann.
PLATE 28—from Spalteholz.
PLATE 29—after Duval-Neelsen.
PLATE 30—from life.
PLATE 31—original.
PLATES 32 and 33—after Salvage.
PLATE 34—Kollmann.
PLATES 35, 37, and 39-44—original.
PLATES 36 and 38—from photographs of van der Weyde.
PLATES 45 and 46—after Richer.
PLATE 47—after Spalteholz.
PLATES 48-52—from life and anatomical preparations.
PLATES 53 and 54—after Spalteholz.
PLATE 55—from anatomical preparation.
PLATES 56-59—from life and anatomical preparations.
PLATE 60—from anatomical preparation.
PLATES 61-63—after a model, A. B'renek in Vienna, and from life.
PLATE 64—from anatomical preparations.
PLATES 65 and 66—after Richer.
PLATES 70-73—from life and anatomical preparations.
PLATES 74 and 75—from plaster models.
PLATE 76—after Spalteholz.
PLATE 77—original.
PLATES 78-81—from life and anatomical preparations.
PLATE 82—after Neelsen and anatomical preparations.
PLATES 83 and 84—after Richer.
PLATE 85—Kollmann.
PLATE 86—after a sketch by Michelangelo.
PLATES 87-189—see Introduction.

PLATES

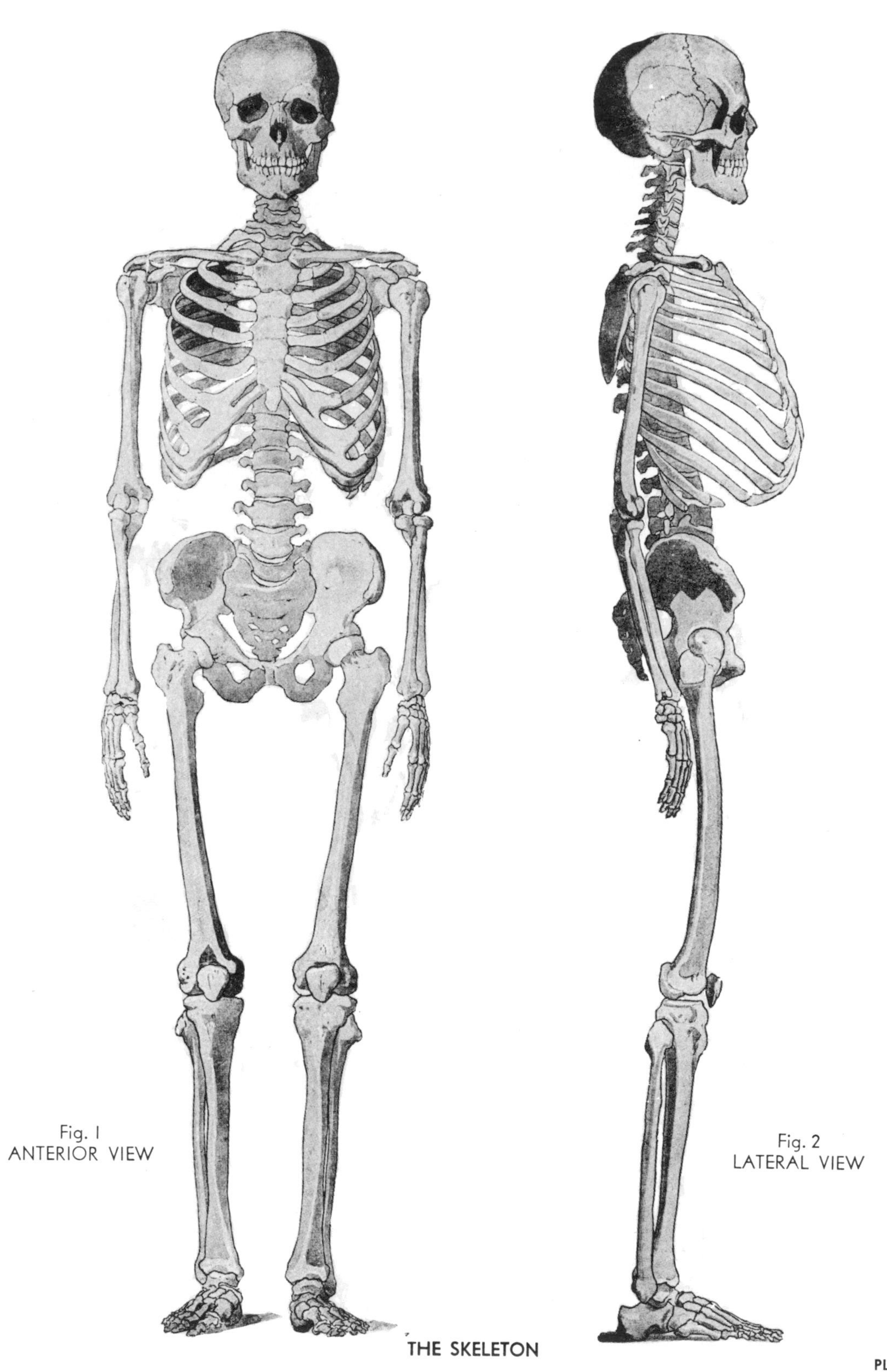

THE SKELETON

PLATE 1

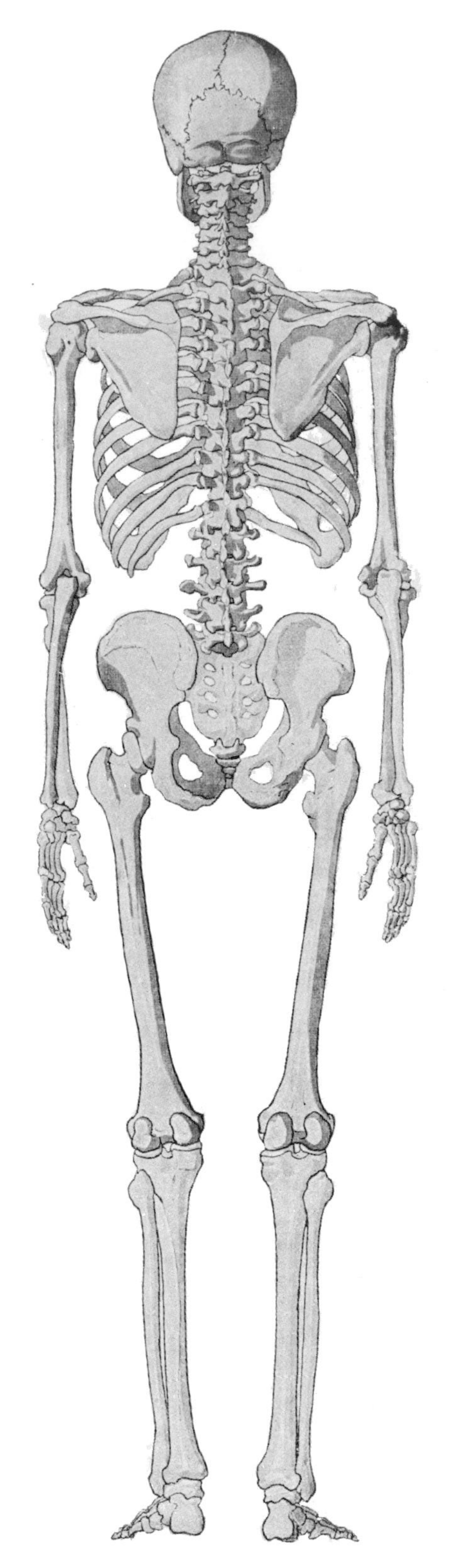

PLATE 2 POSTERIOR VIEW OF THE SKELETON

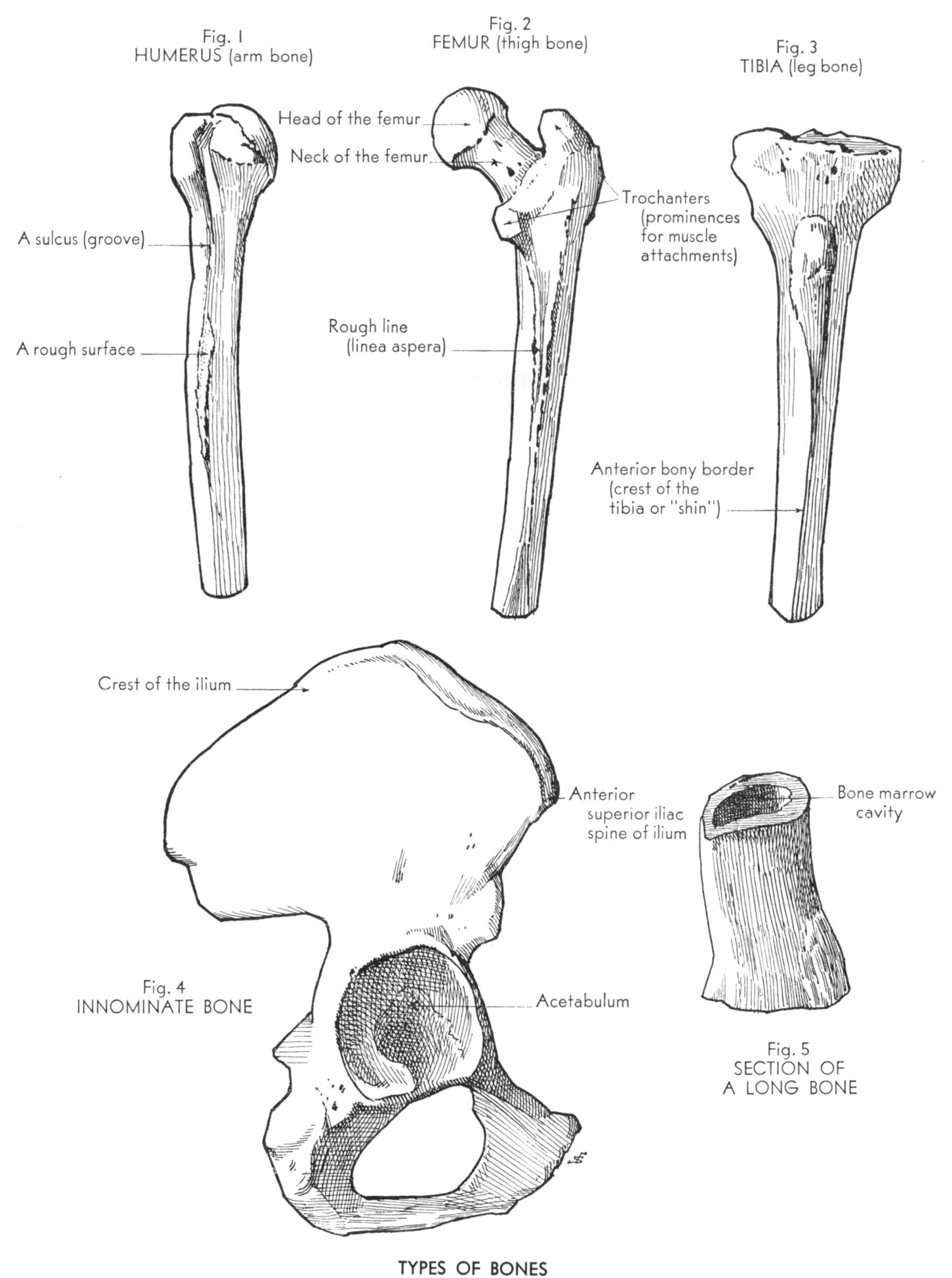

TYPES OF BONES

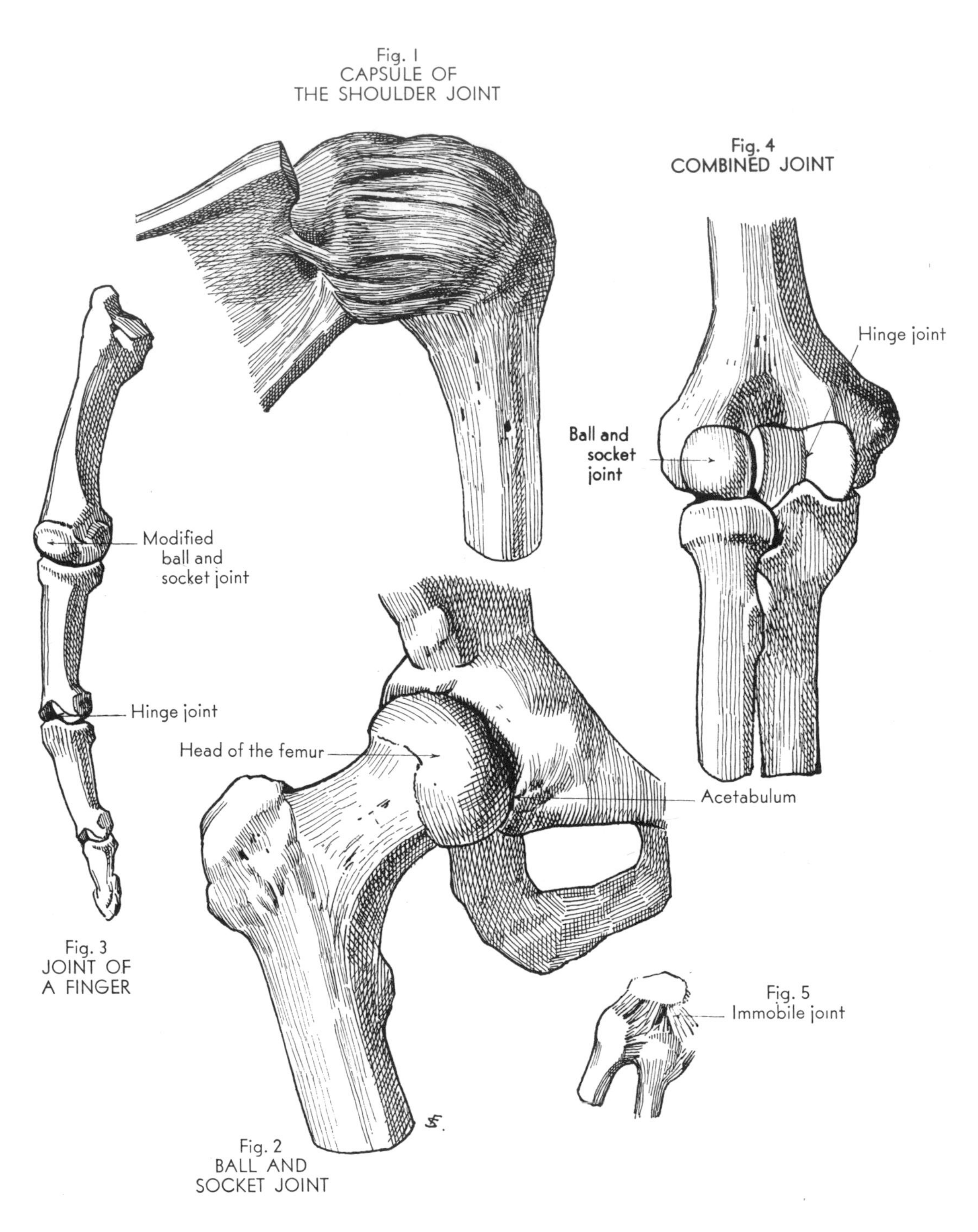

TYPES OF JOINTS

PLATE 4

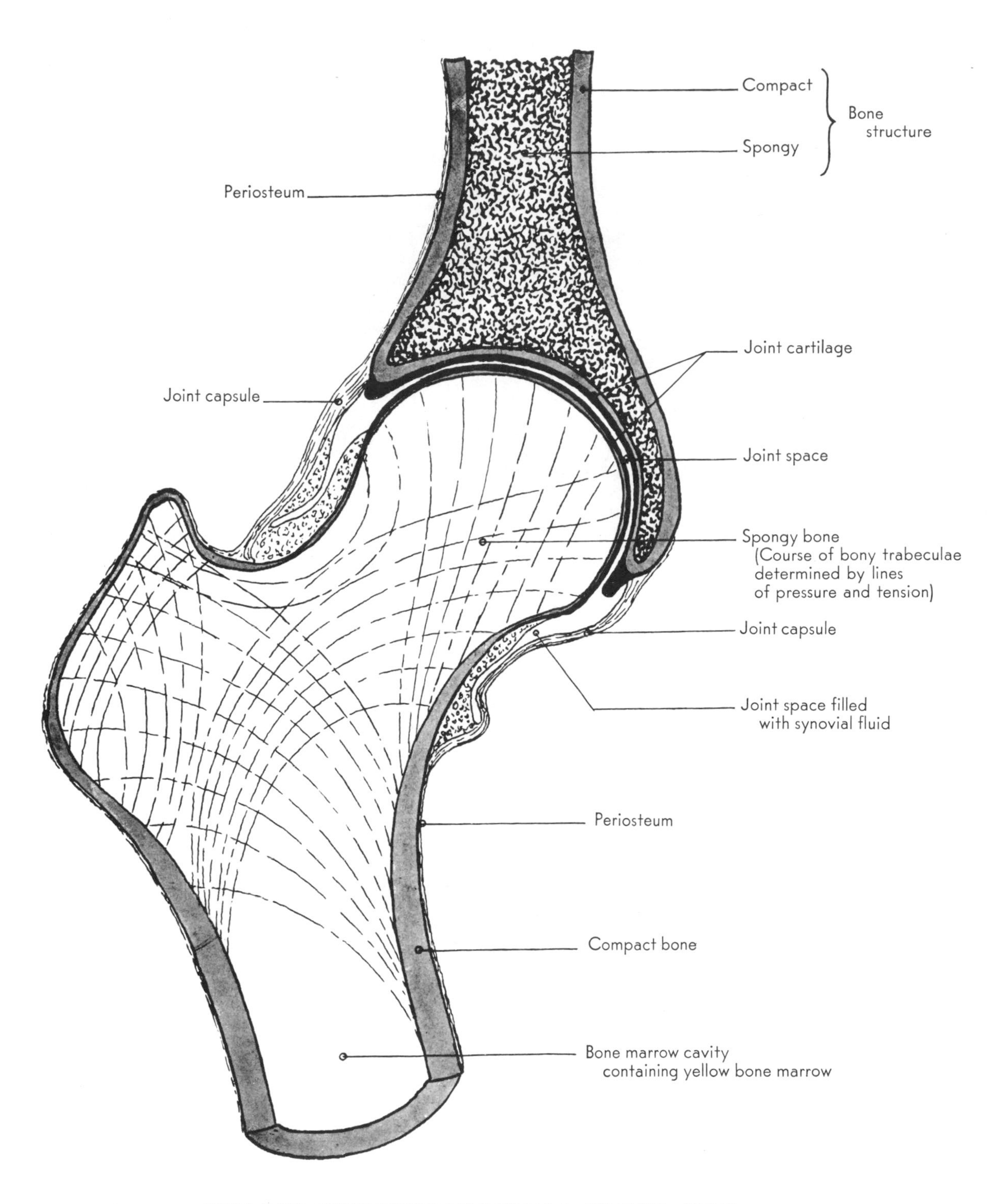

SCHEMATIC CROSS-SECTION THROUGH A JOINT (HIP JOINT)

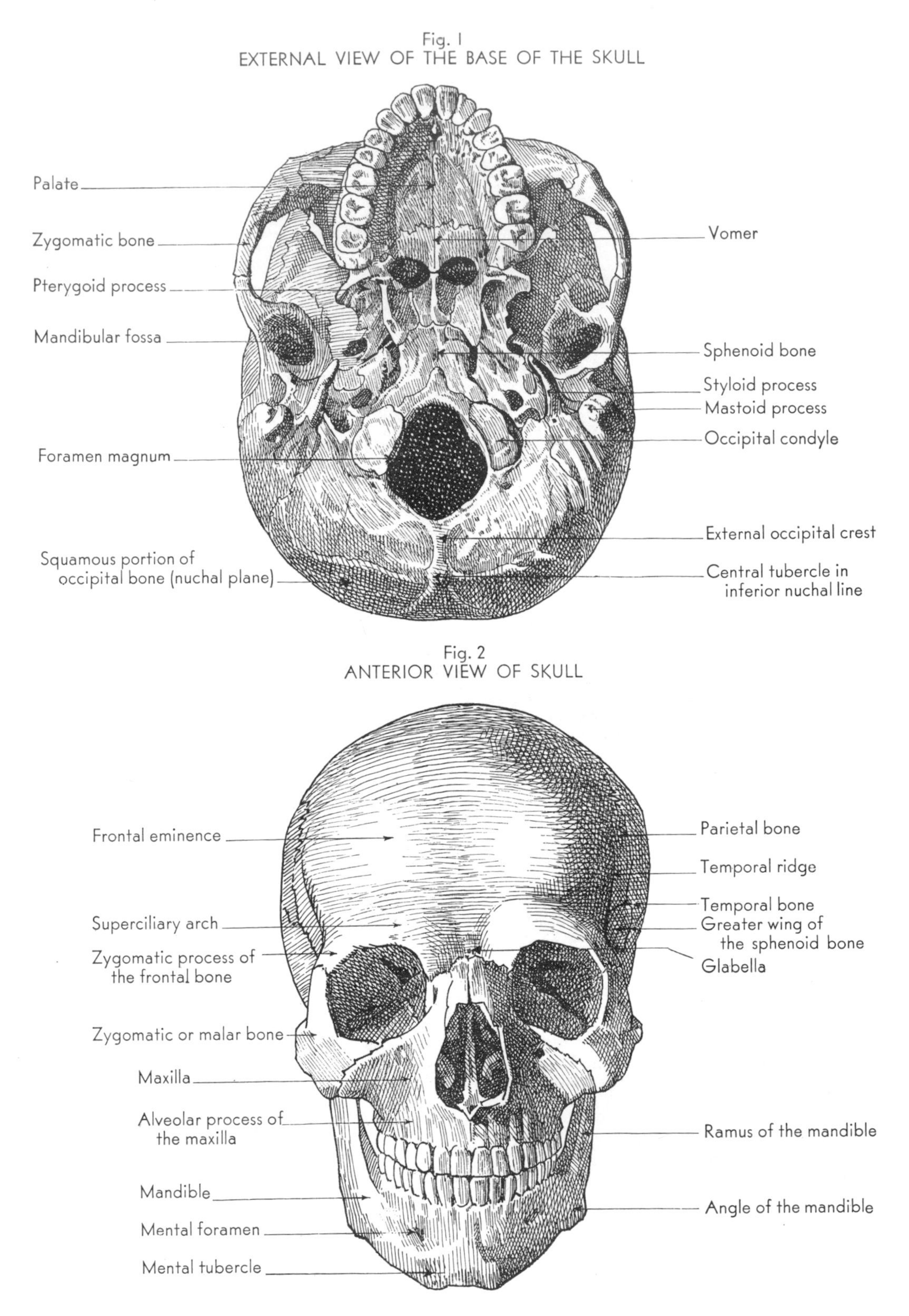

PLATE 6

THE SKULL

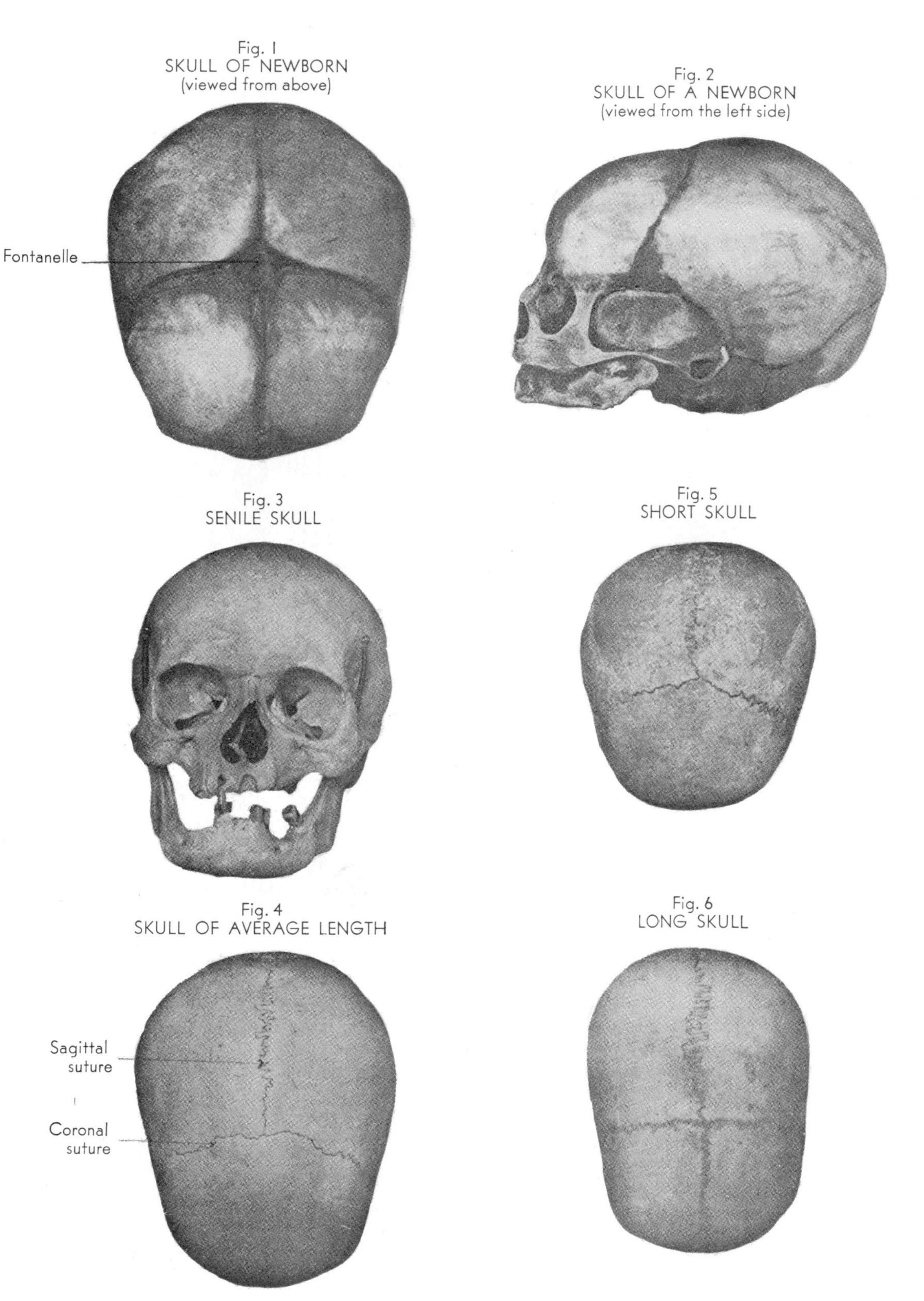

TYPES OF SKULLS

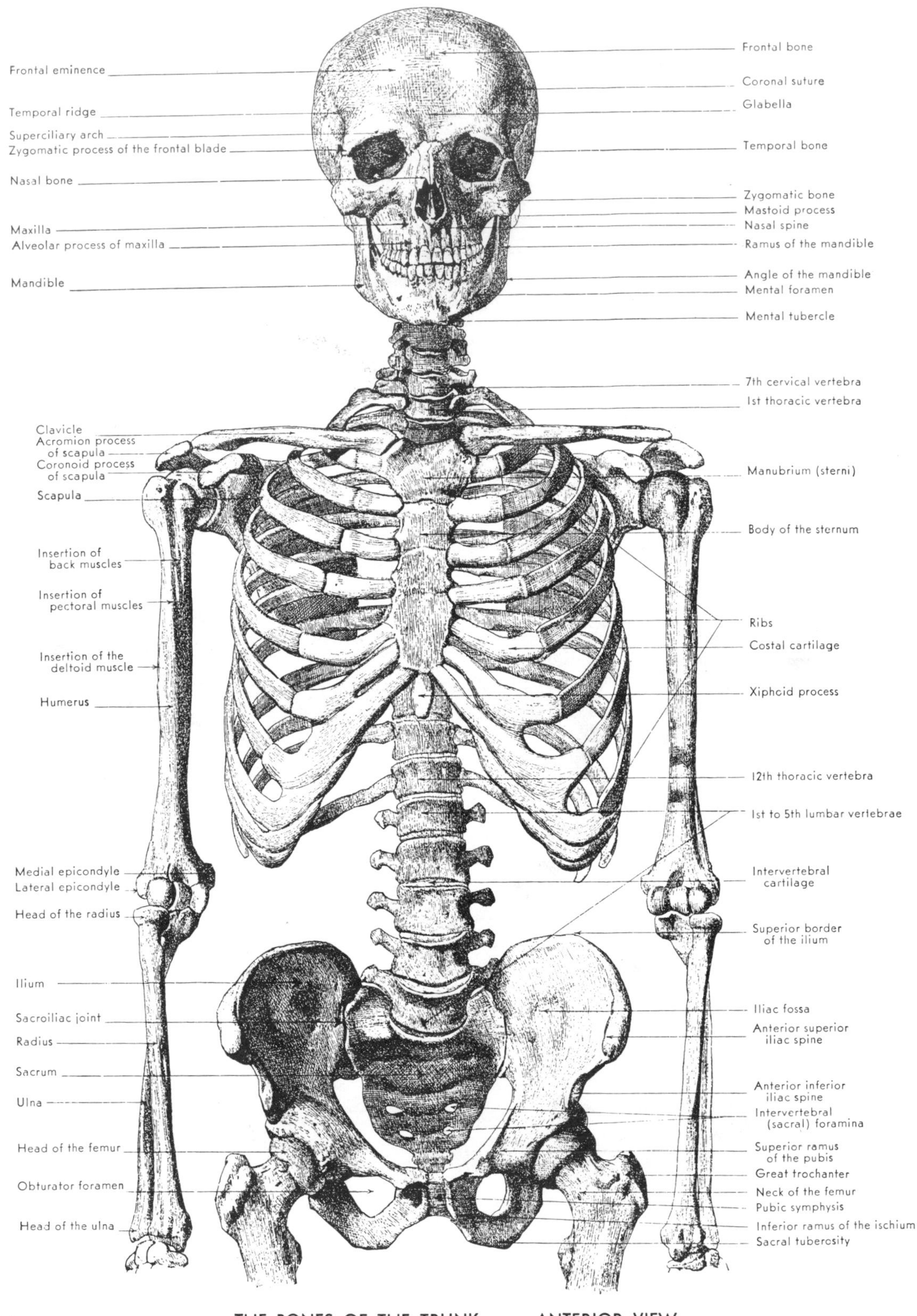

PLATE 8 THE BONES OF THE TRUNK ANTERIOR VIEW

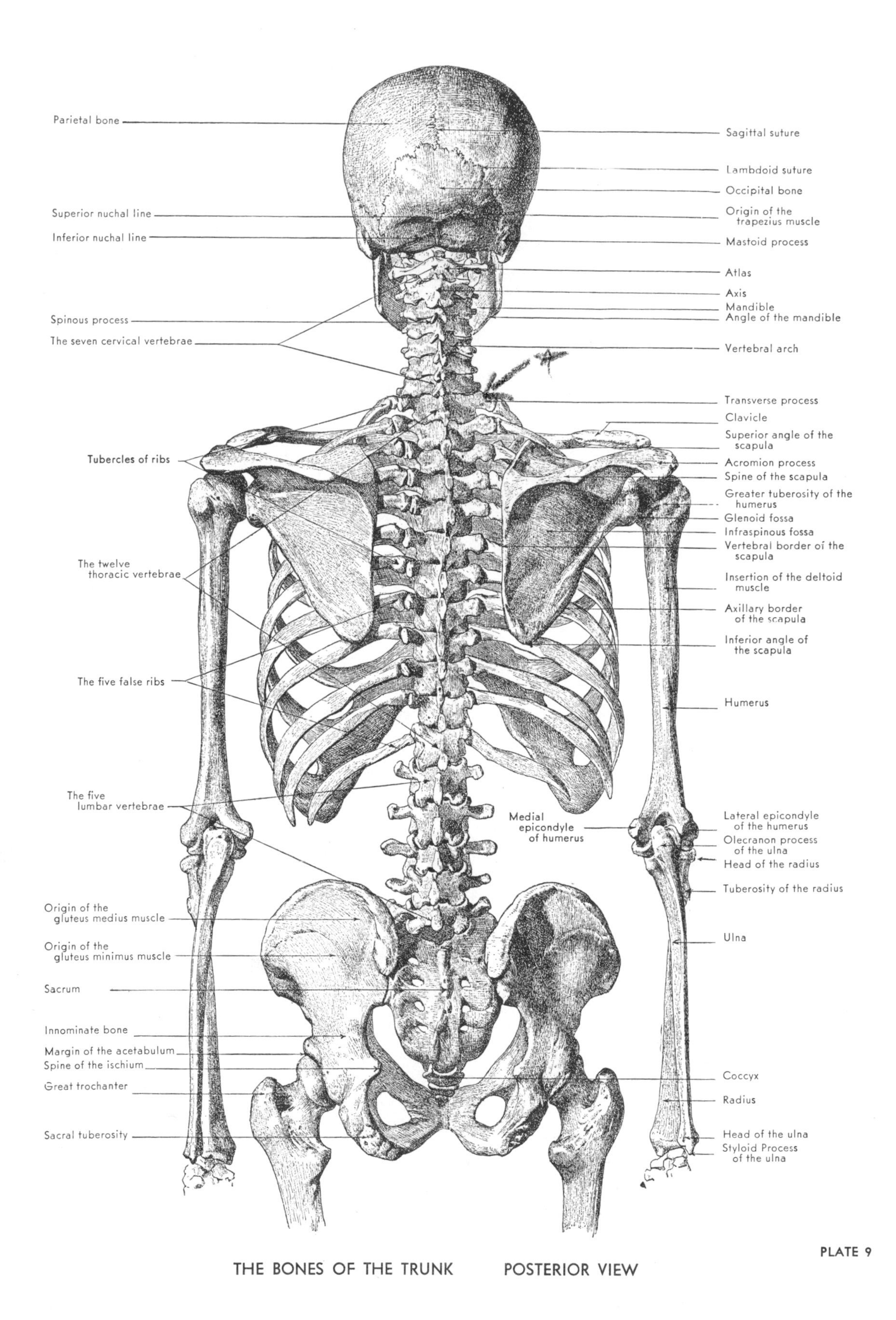

PLATE 9

THE BONES OF THE TRUNK POSTERIOR VIEW

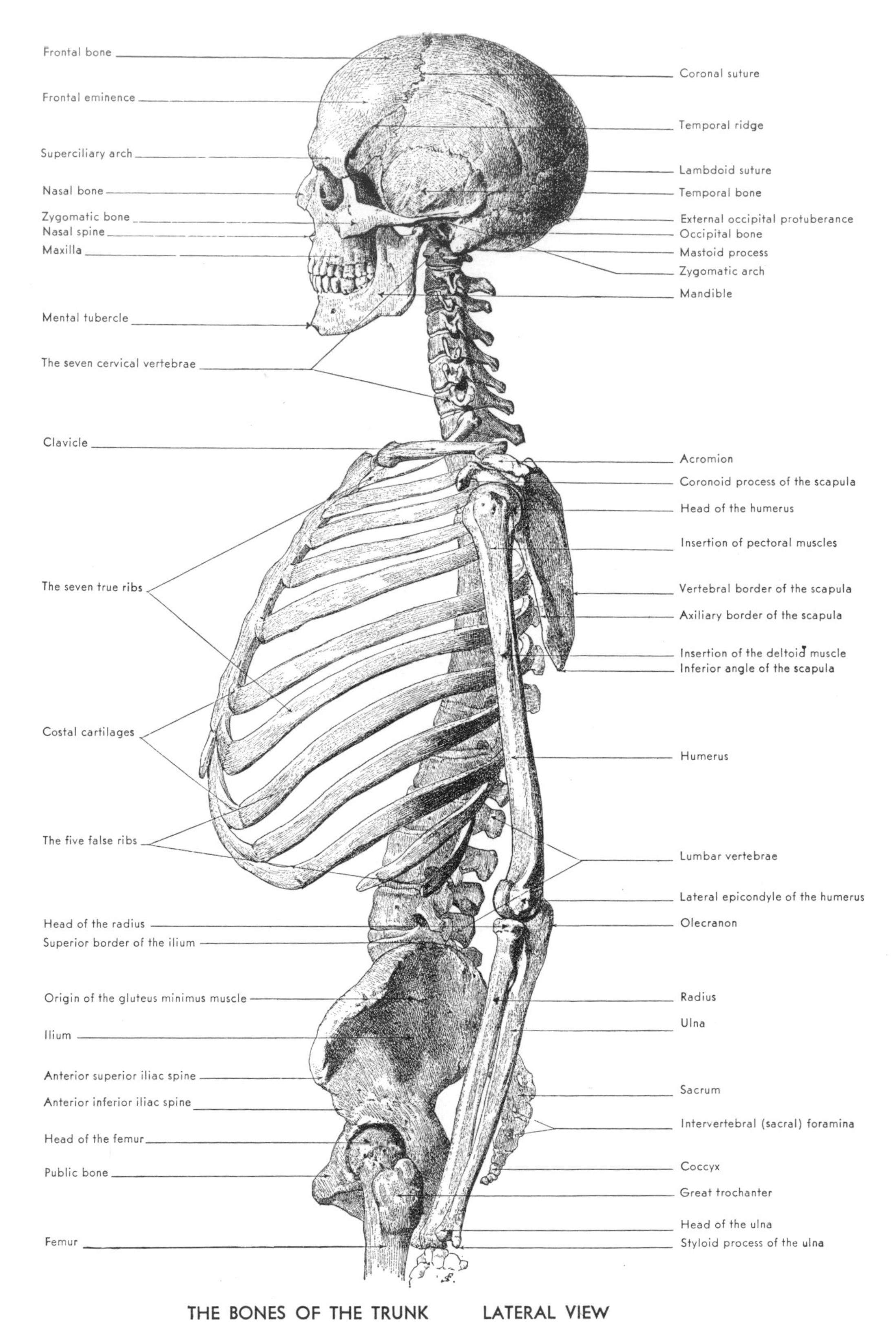

PLATE 10 THE BONES OF THE TRUNK LATERAL VIEW

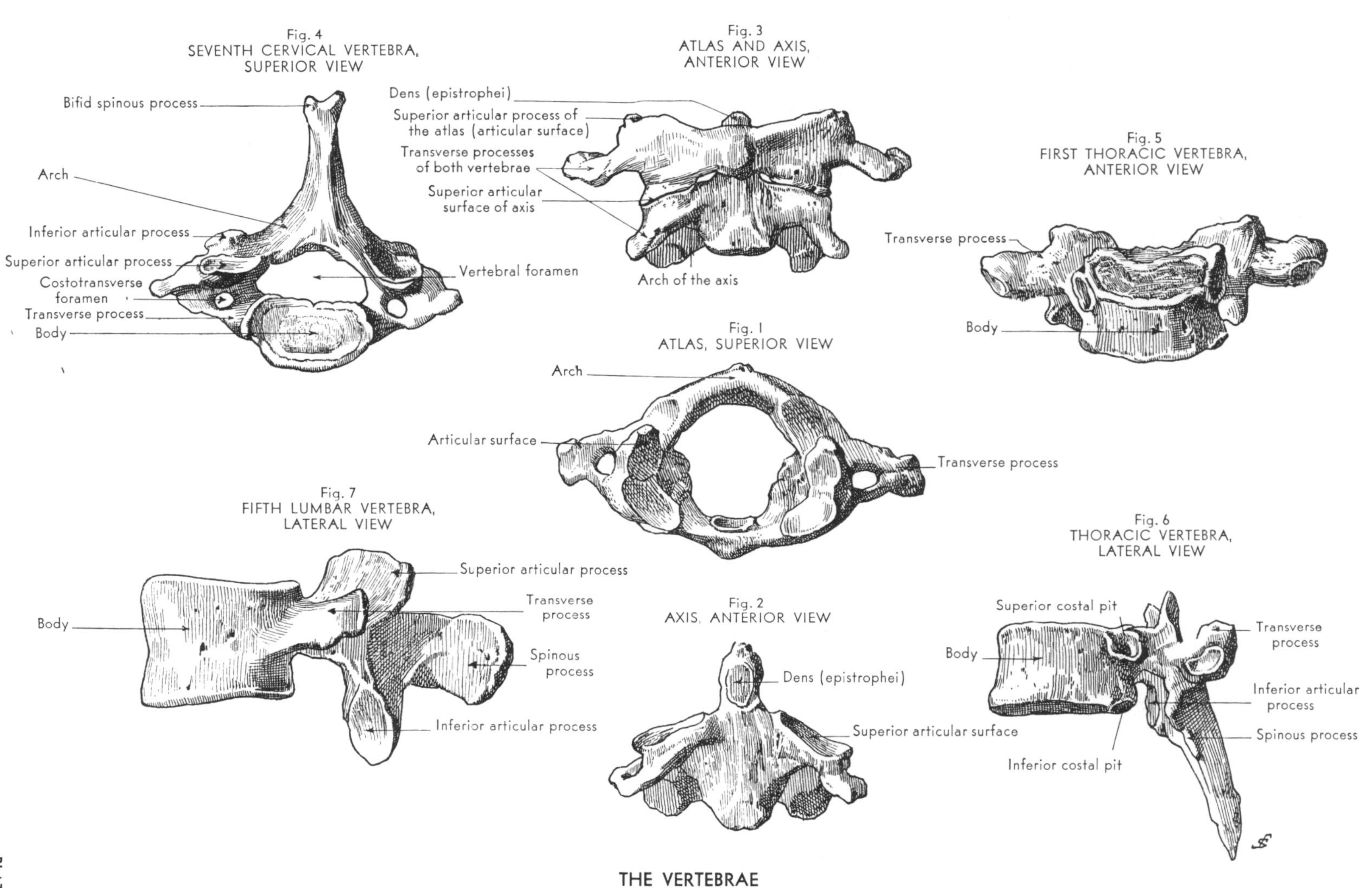

THE VERTEBRAE

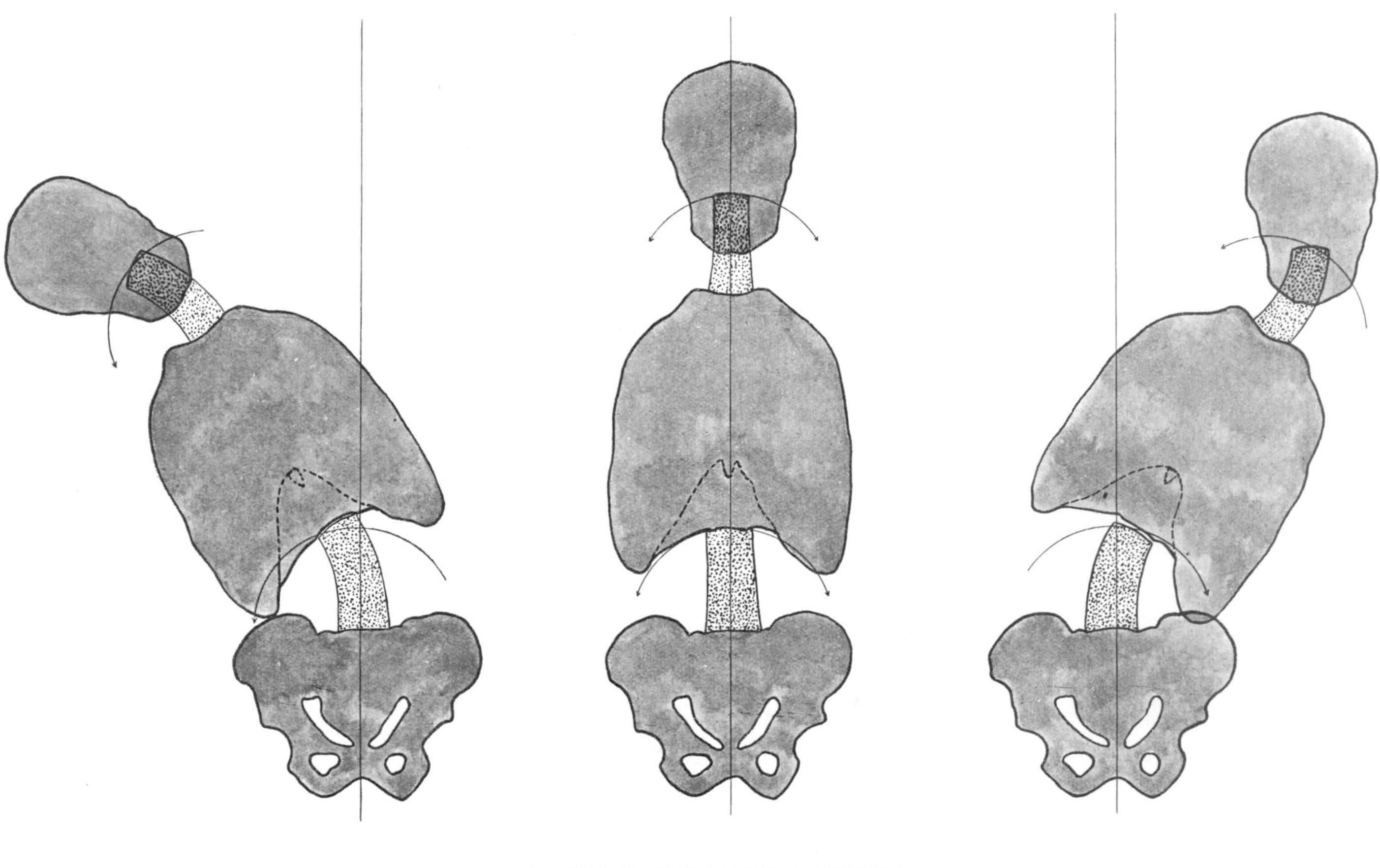

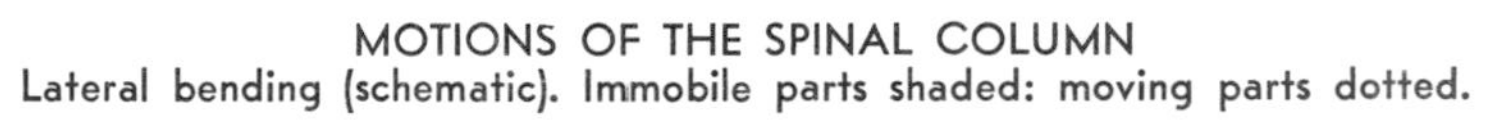

MOTIONS OF THE SPINAL COLUMN

Lateral bending (schematic). Immobile parts shaded: moving parts dotted.

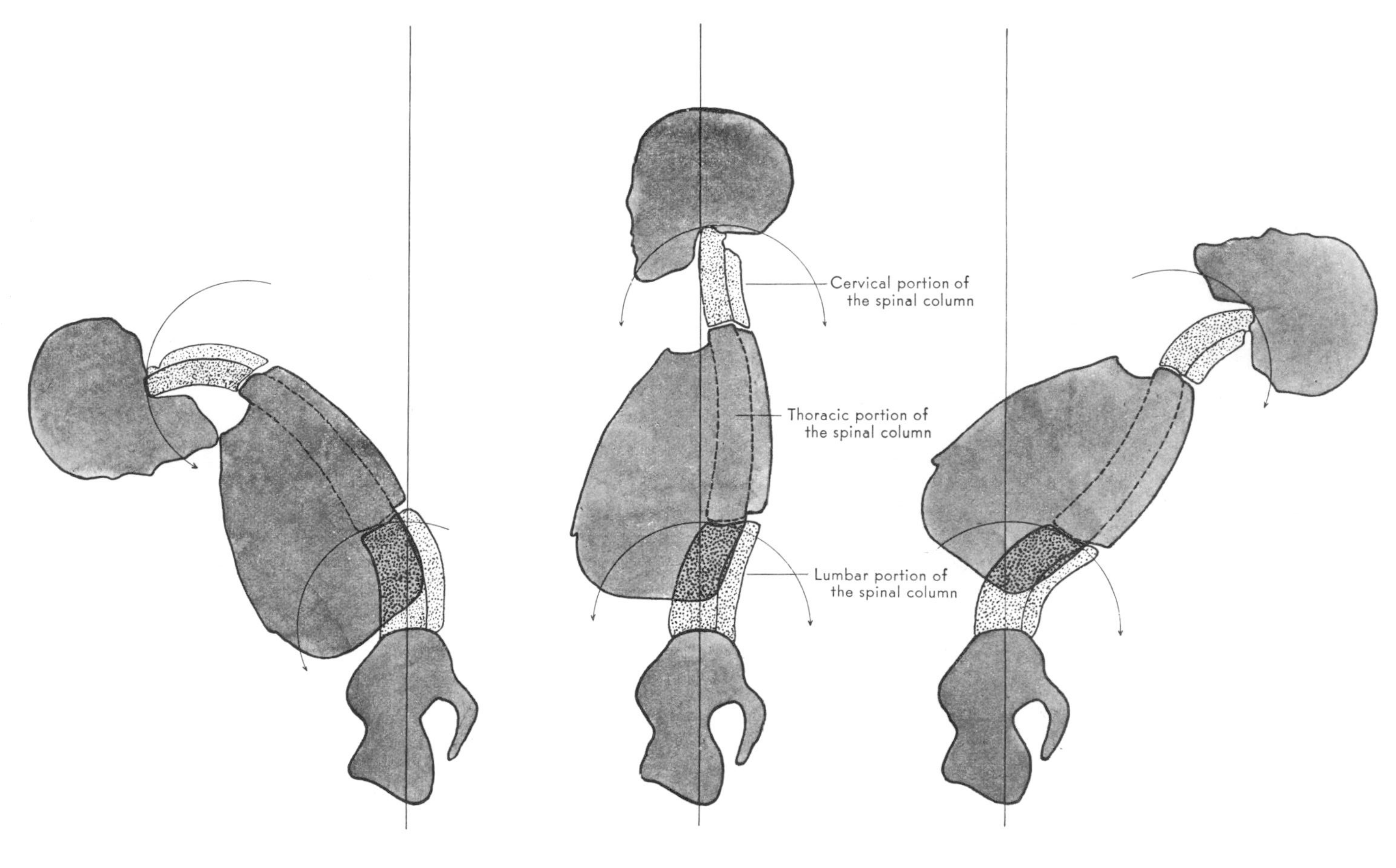

MOTIONS OF THE SPINAL COLUMN
Forward and backward bending (schematic). Immobile parts shaded: moving parts dotted.

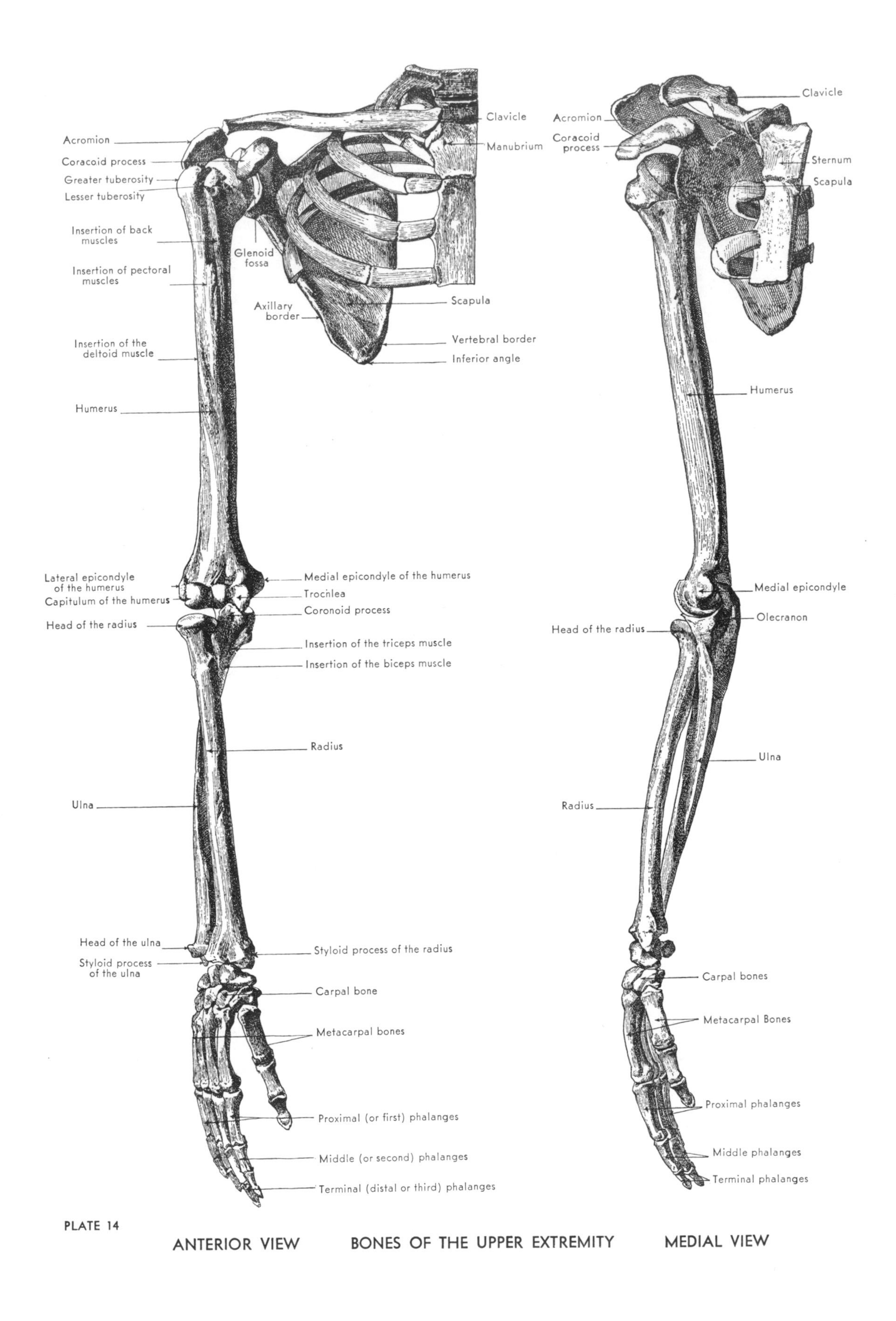

PLATE 14

ANTERIOR VIEW BONES OF THE UPPER EXTREMITY MEDIAL VIEW

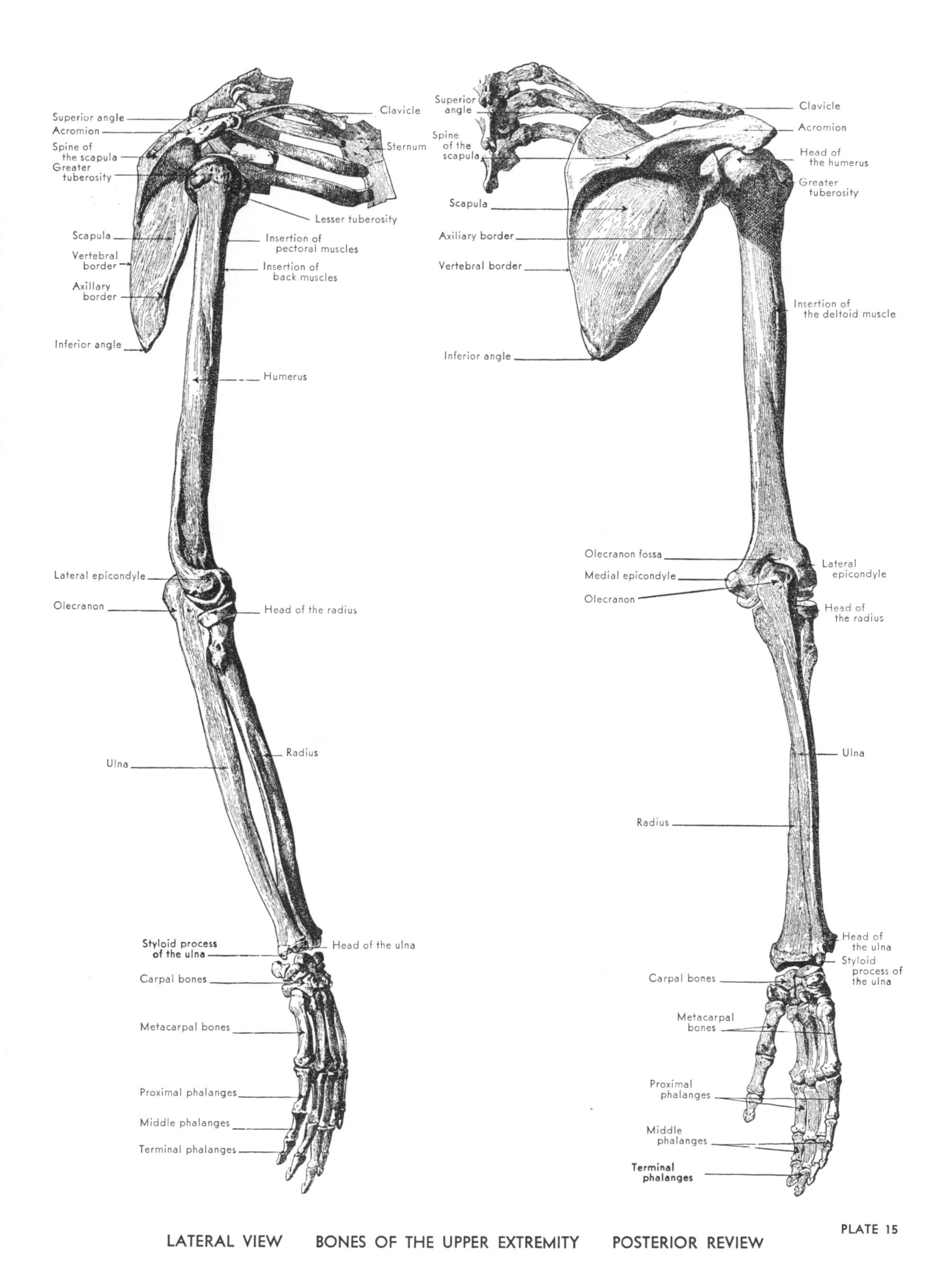

PLATE 15

LATERAL VIEW BONES OF THE UPPER EXTREMITY POSTERIOR REVIEW

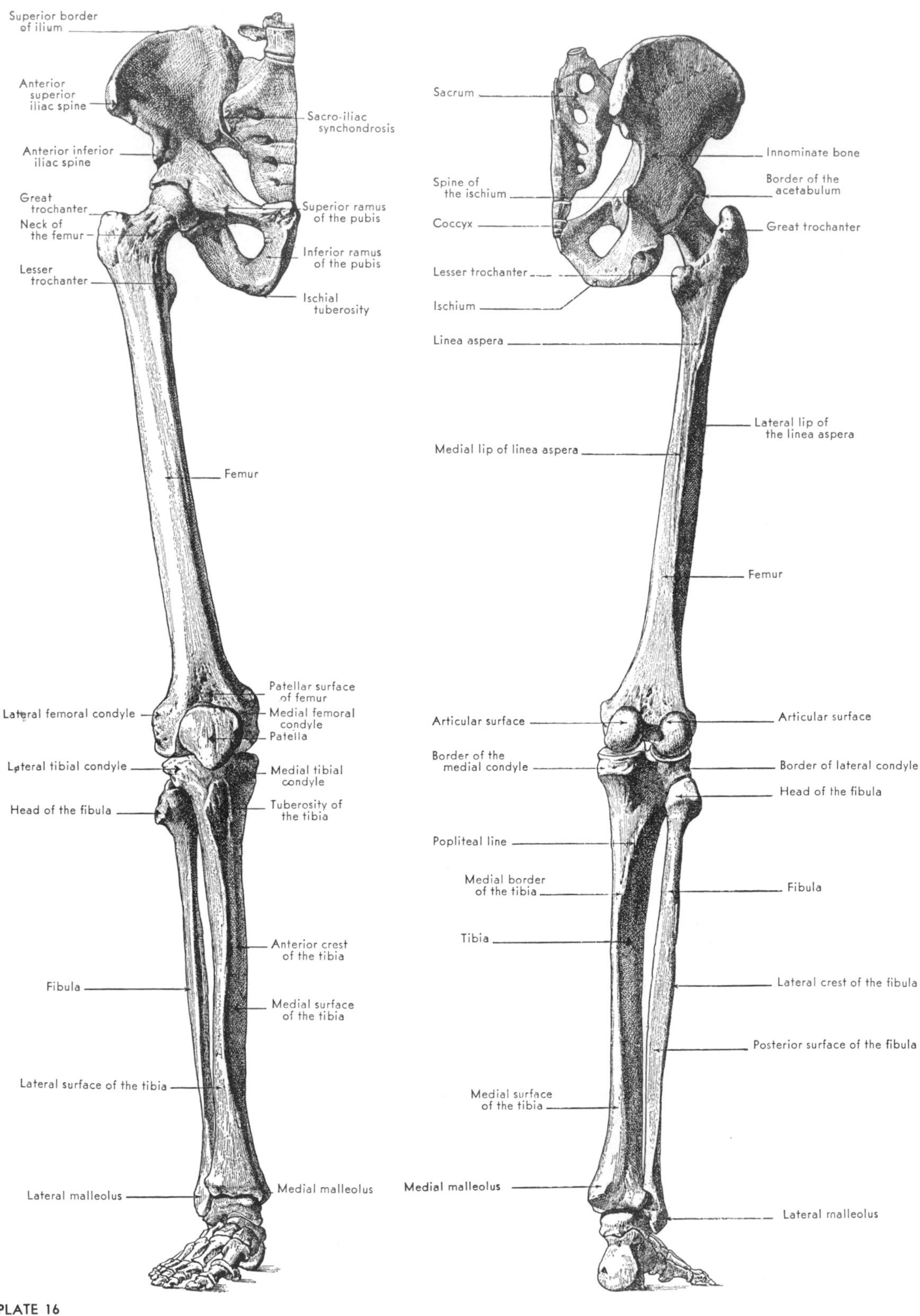

PLATE 16

THE BONES OF THE LOWER EXTREMITY ANTERIOR AND POSTERIOR VIEWS

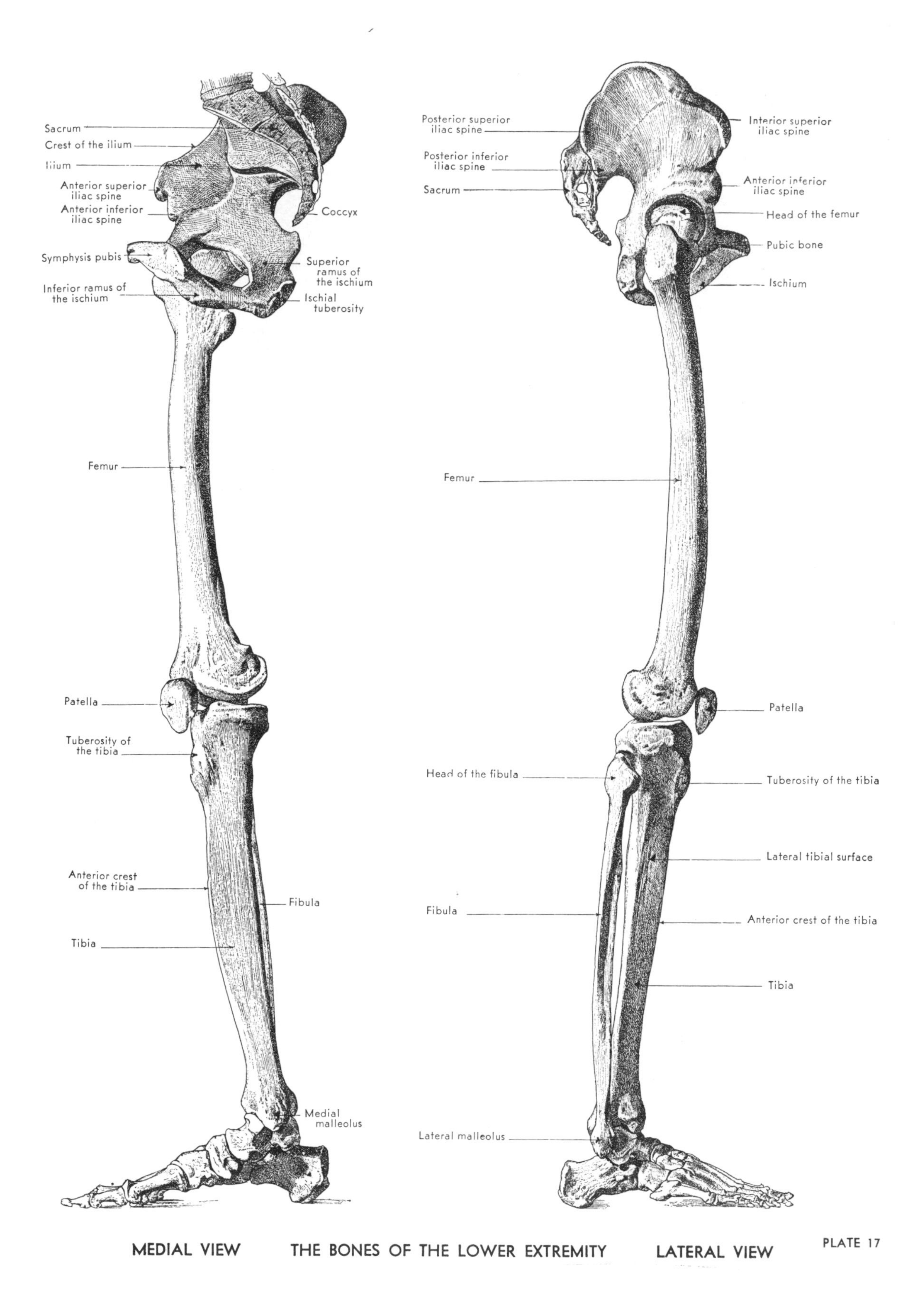

MEDIAL VIEW THE BONES OF THE LOWER EXTREMITY LATERAL VIEW

PLATE 17

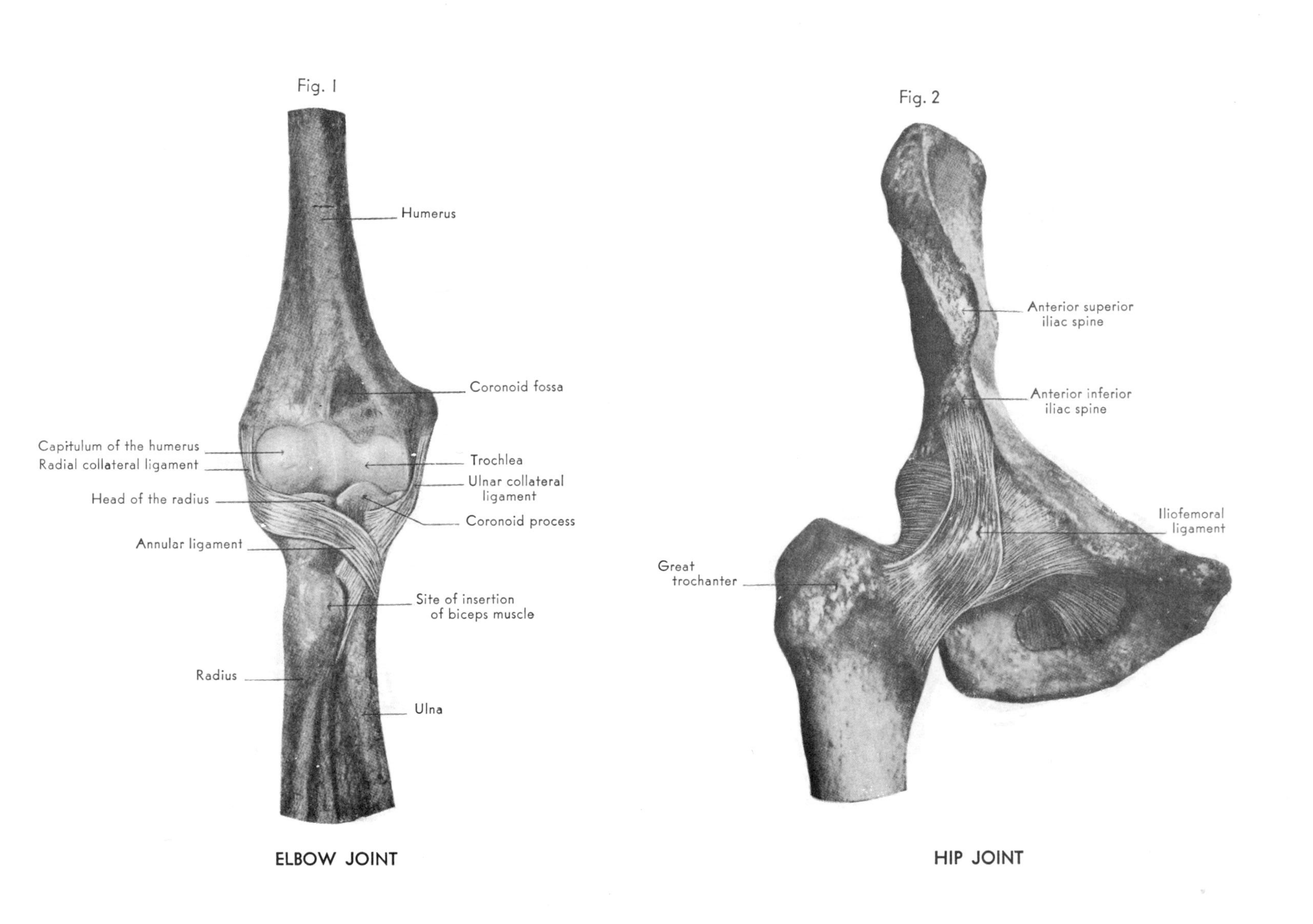

ELBOW JOINT

HIP JOINT

Fig. 1

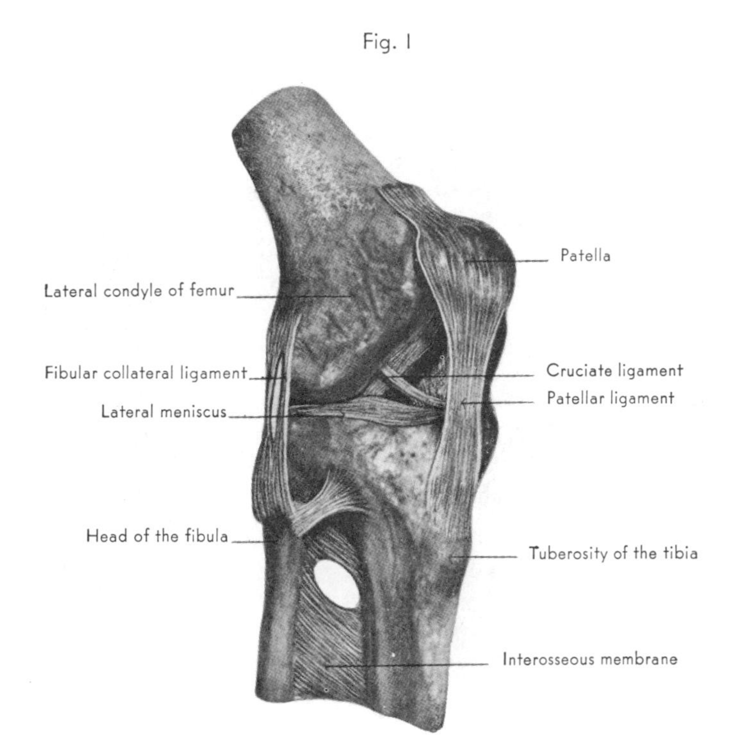

Fig. 2

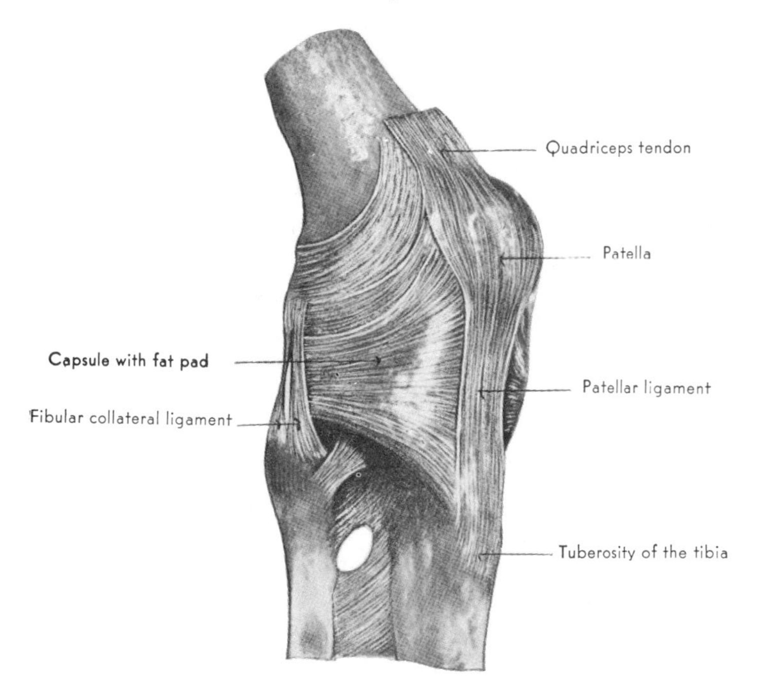

KNEE JOINT

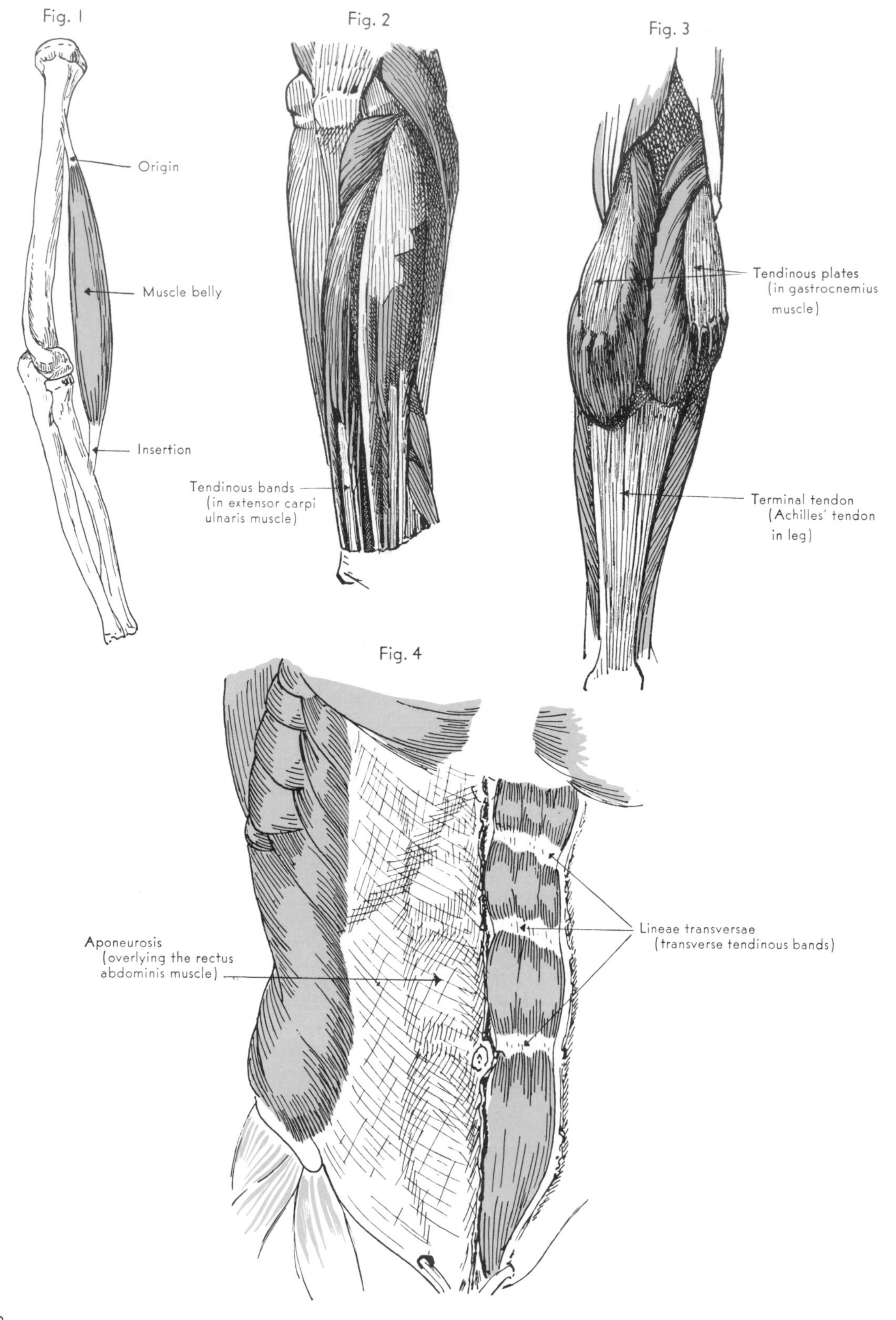

TYPES OF TENDONS AND MUSCLES

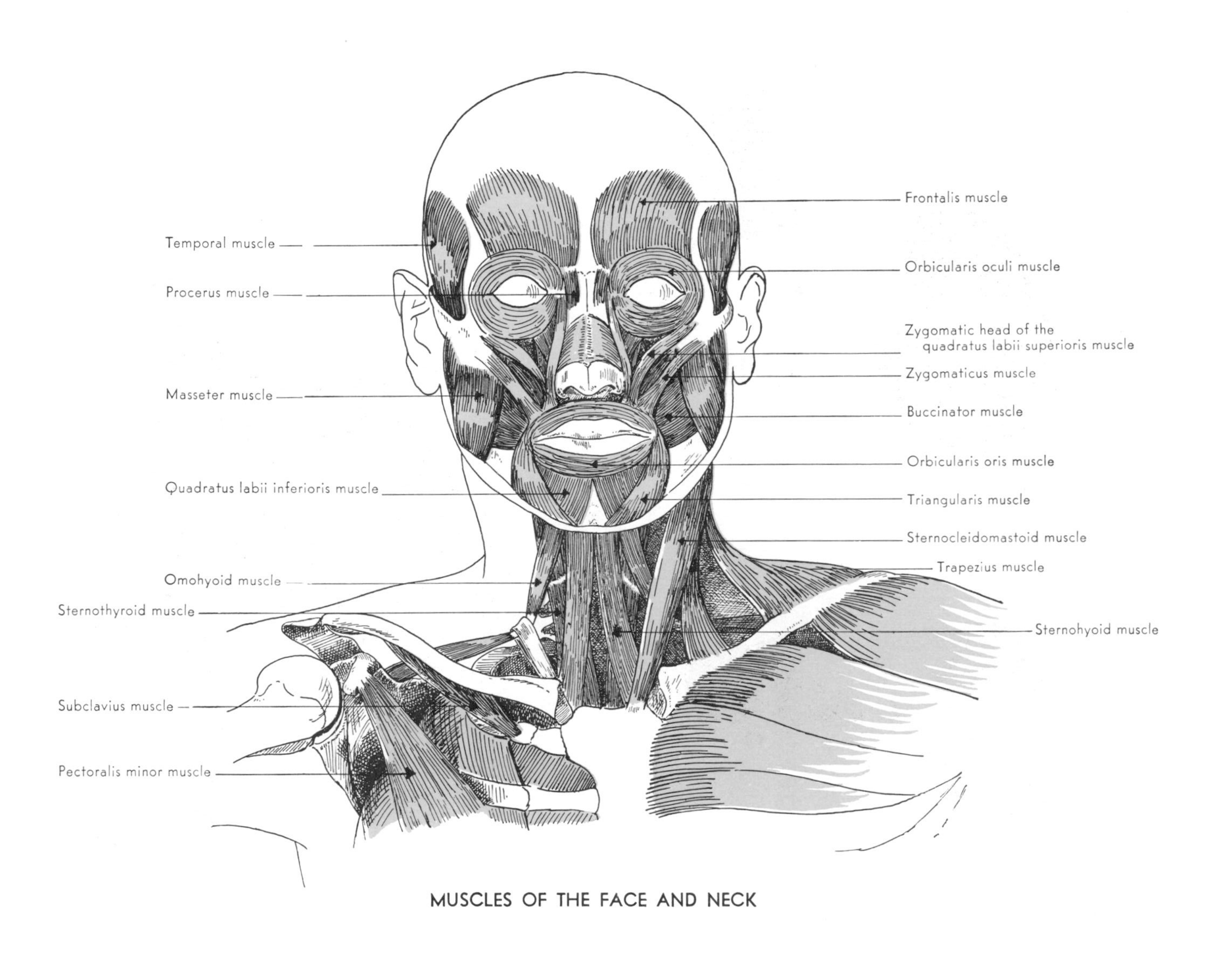

MUSCLES OF THE FACE AND NECK

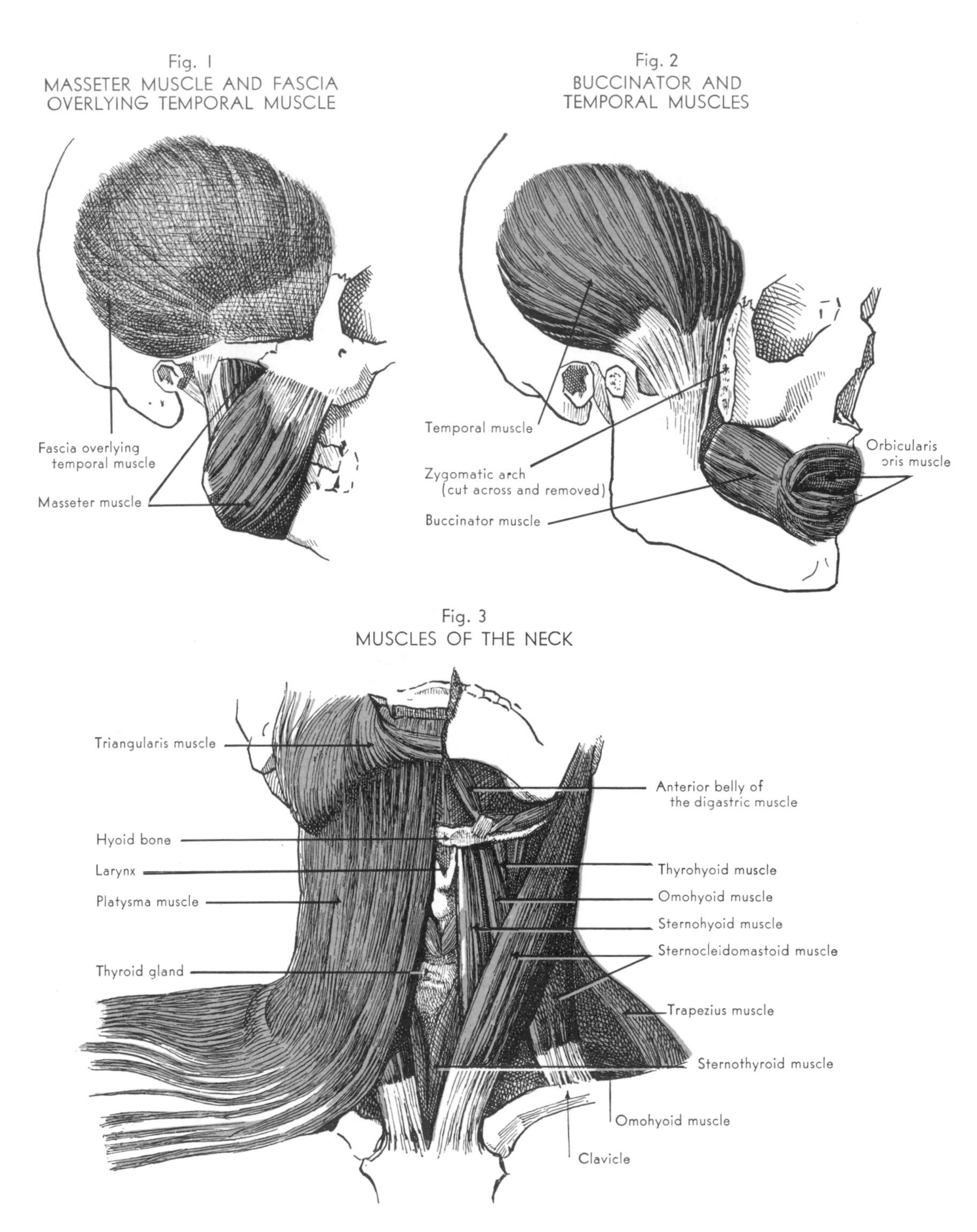

MUSCLES OF THE HEAD AND NECK

Fig. 1

Fig. 2

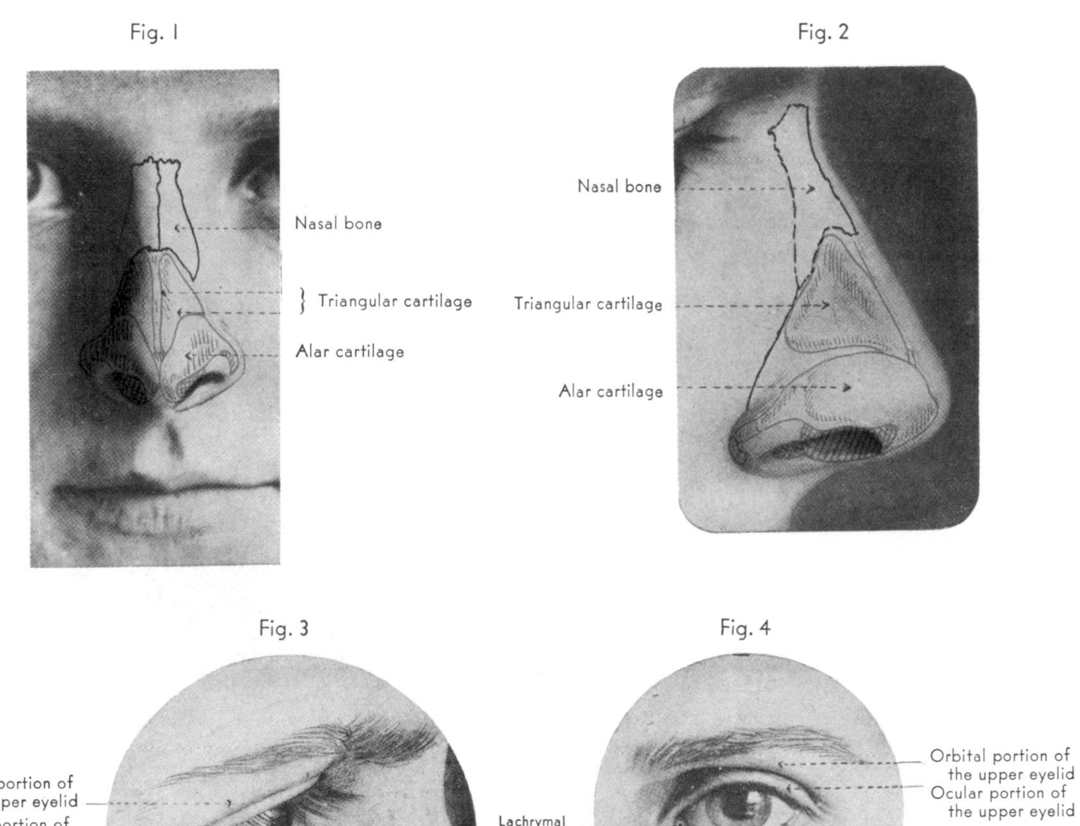

Fig. 3

Orbital portion of the upper eyelid
Ocular portion of the upper eyelid
Ocular portion of the lower eyelid
Orbital portion of the lower eyelid

Fig. 4

Lachrymal caruncle
Orbital portion of the upper eyelid
Ocular portion of the upper eyelid
Ocular portion of the lower eyelid
Orbital portion of the lower eyelid

Fig. 5

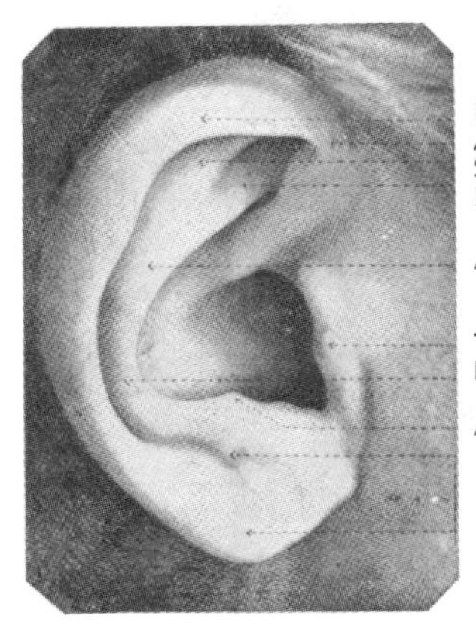

Helix of the ear
Anterior branch of the helix
Superior branch } of the anti-helix
Inferior branch
Anti-helix
Tragus
Posterior branch of the anti-helix
Anti-tragus
Lower end of the anti-helix
Lobule of the ear

Fig. 4

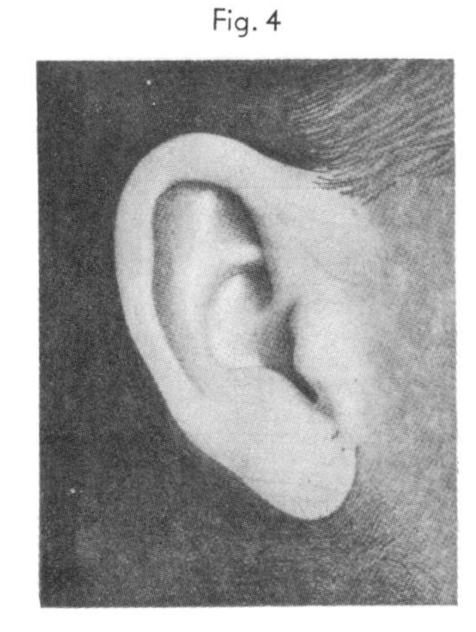

NOSE, EYE, AND EAR

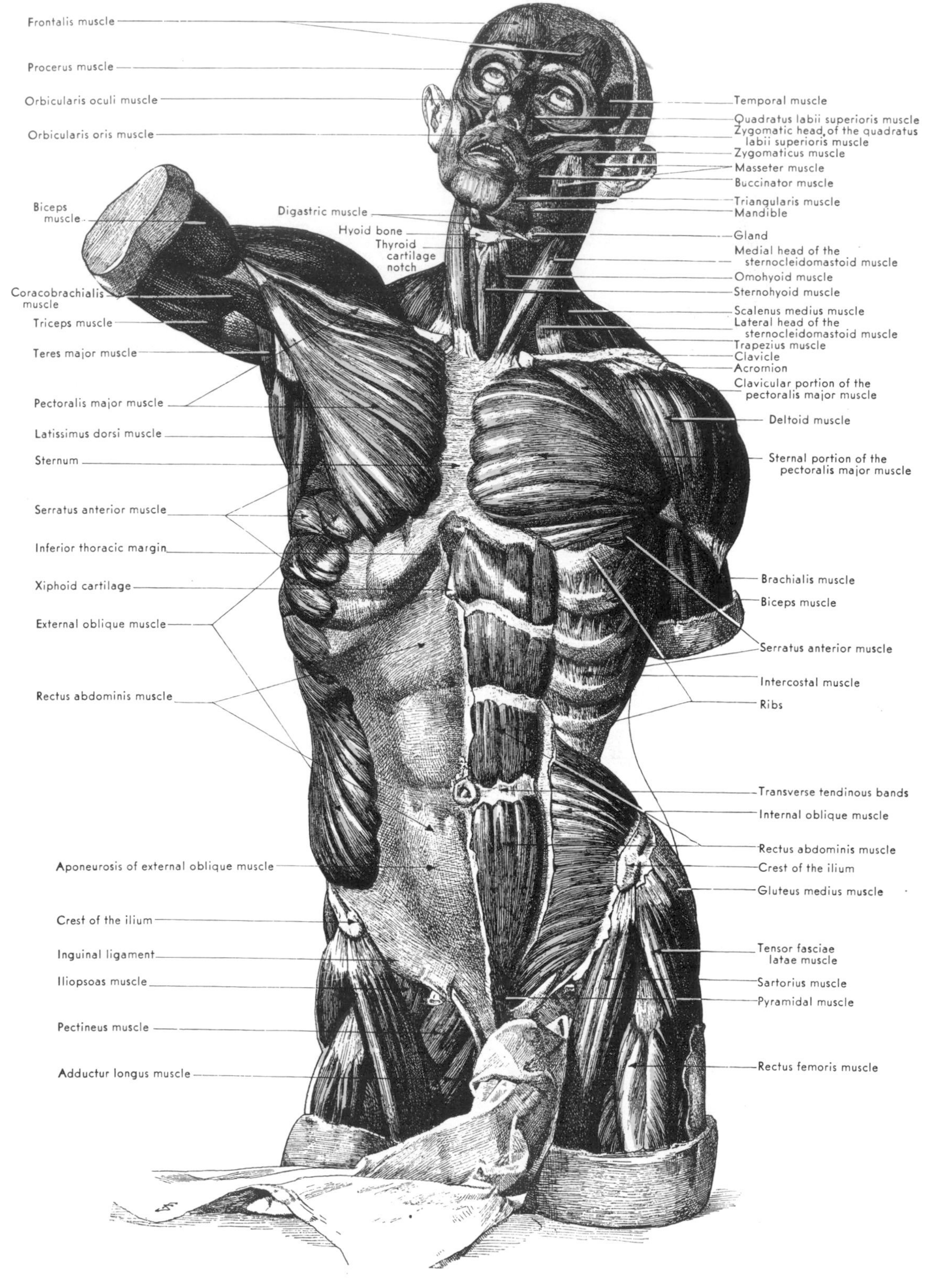

PLATE 24

MUSCLES OF THE TRUNK ANTERIOR VIEW

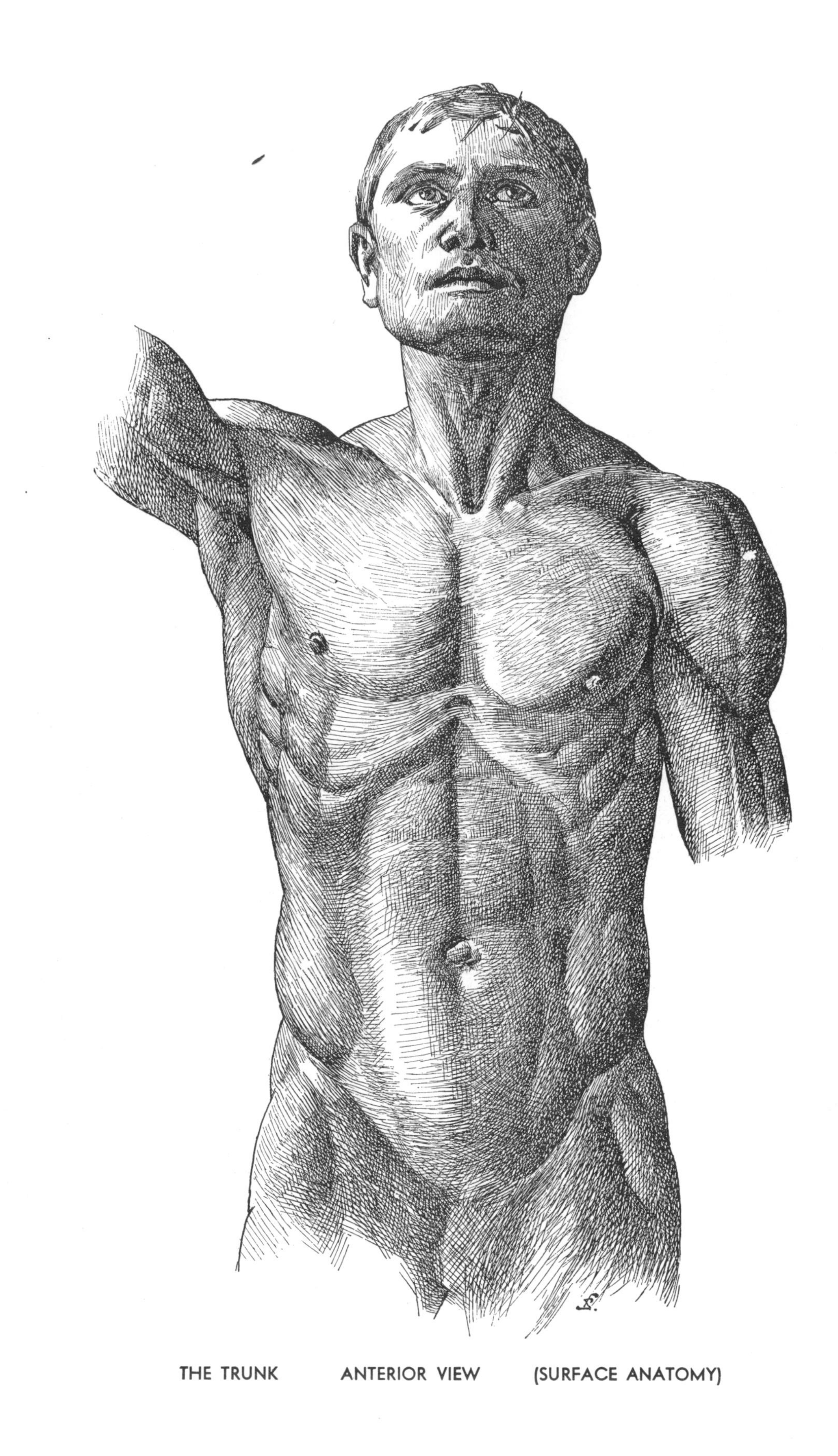

THE TRUNK ANTERIOR VIEW (SURFACE ANATOMY) PLATE 25

PLATE 26

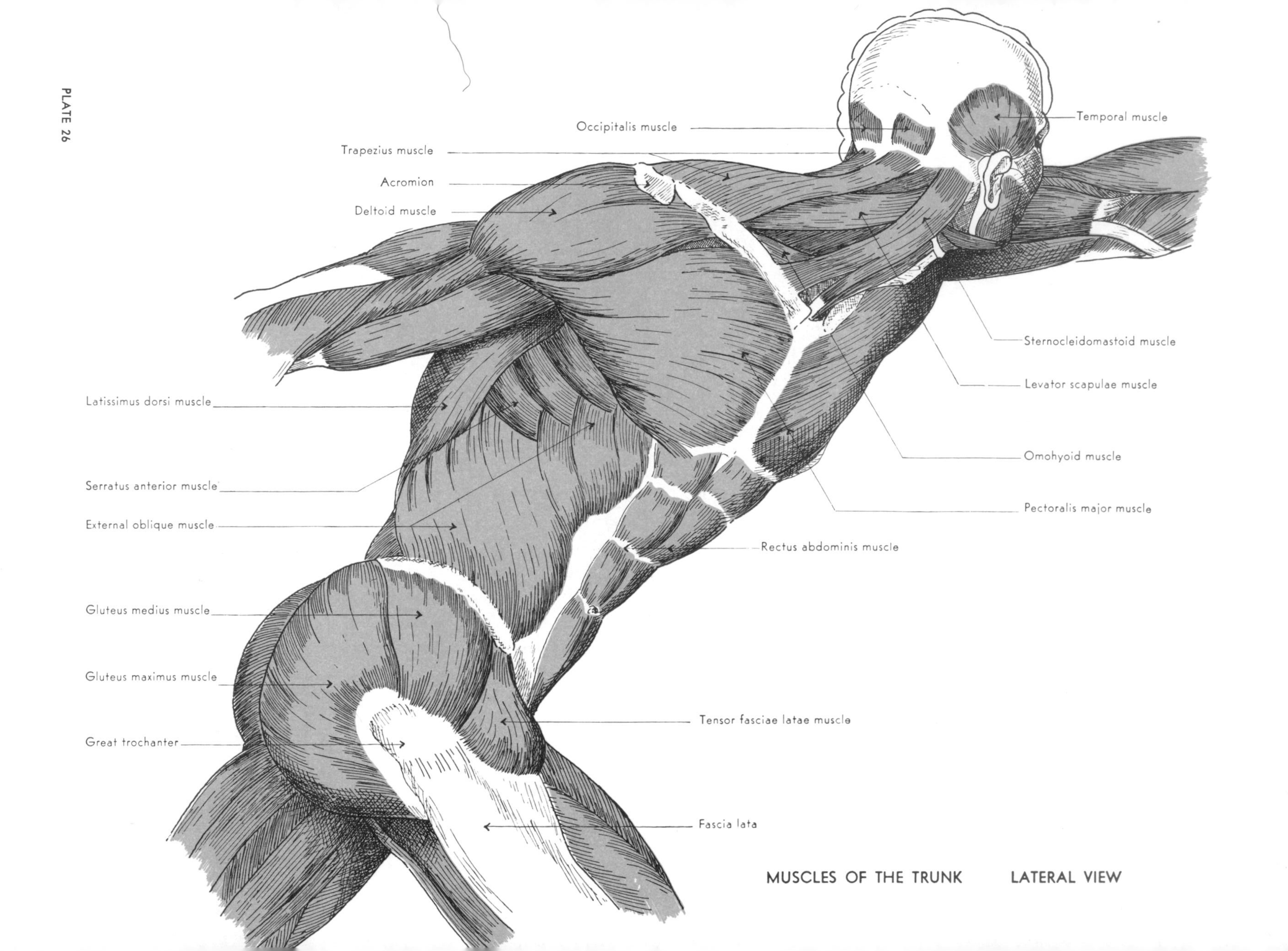

MUSCLES OF THE TRUNK LATERAL VIEW

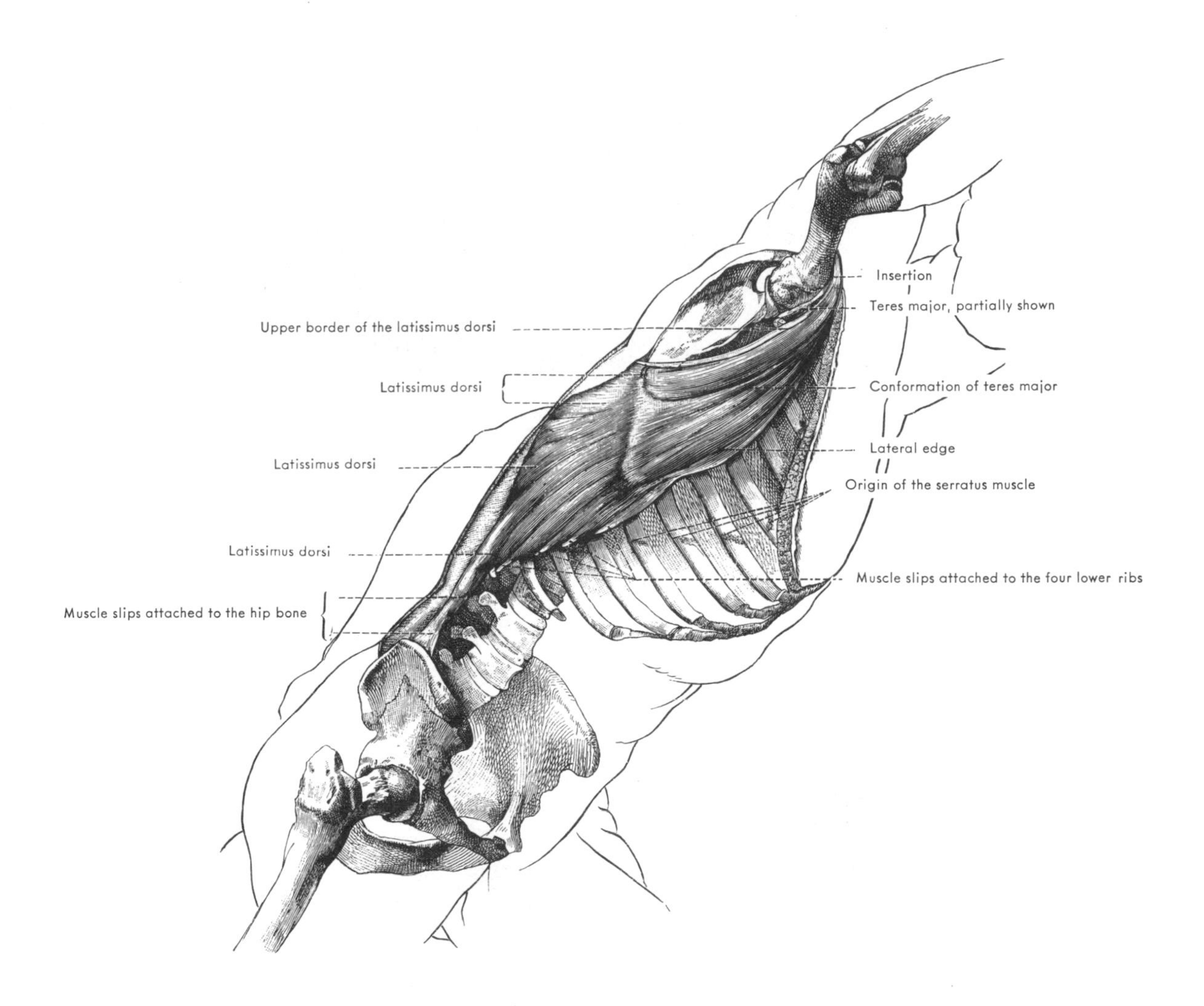

KOLLMANN, THE LATISSIMUS DORSI, WITH ARM RAISED

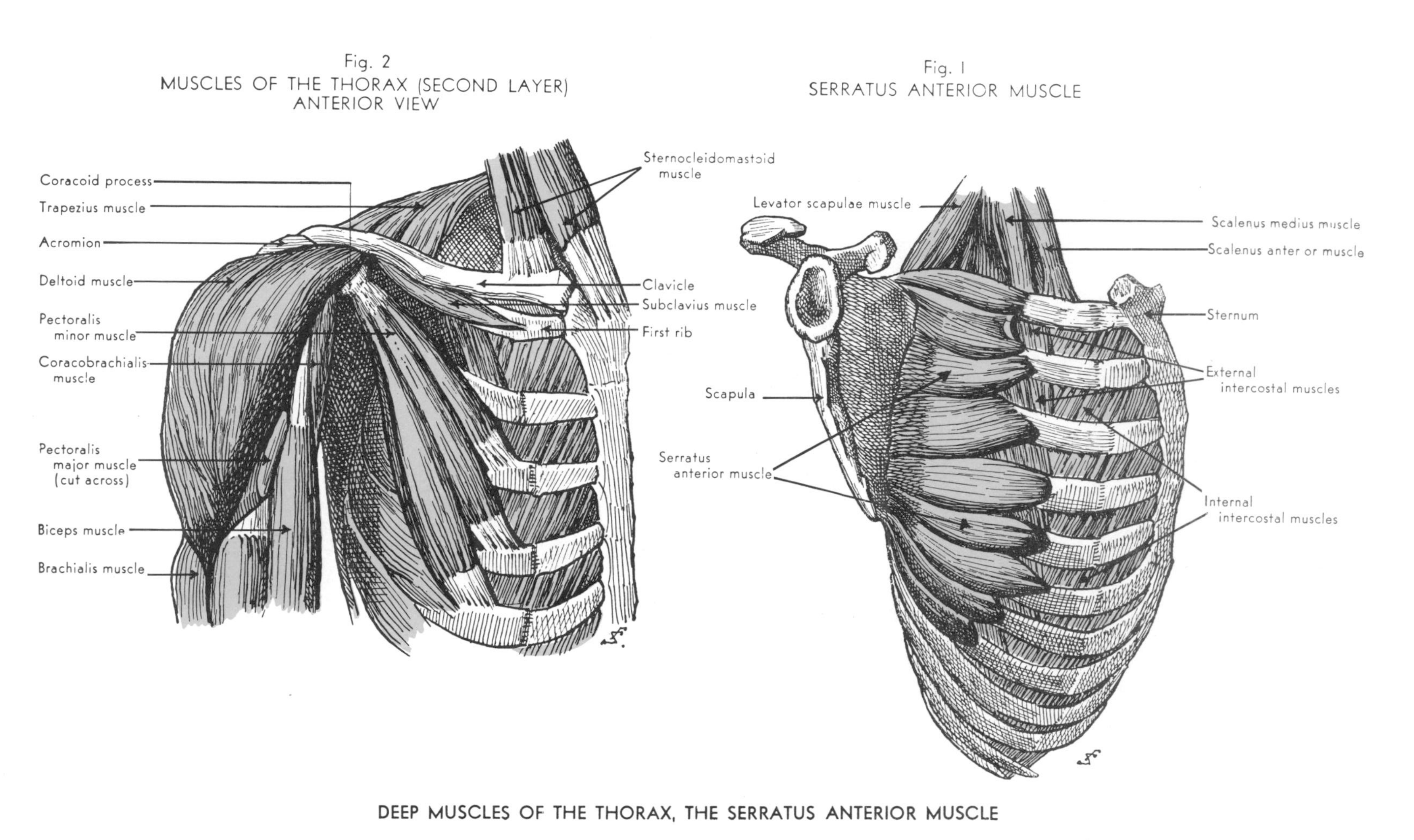

DEEP MUSCLES OF THE THORAX, THE SERRATUS ANTERIOR MUSCLE

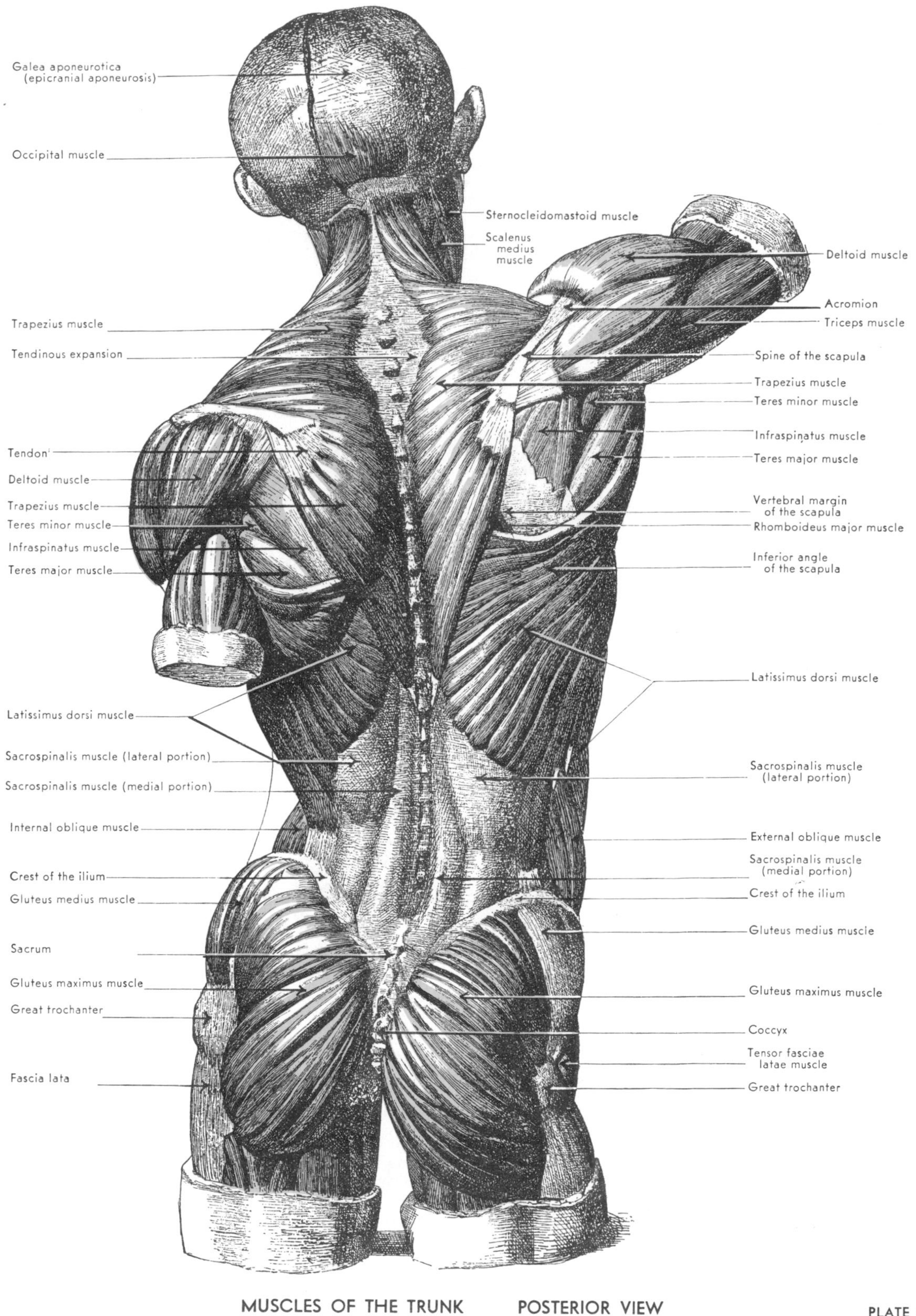

MUSCLES OF THE TRUNK POSTERIOR VIEW

PLATE 29

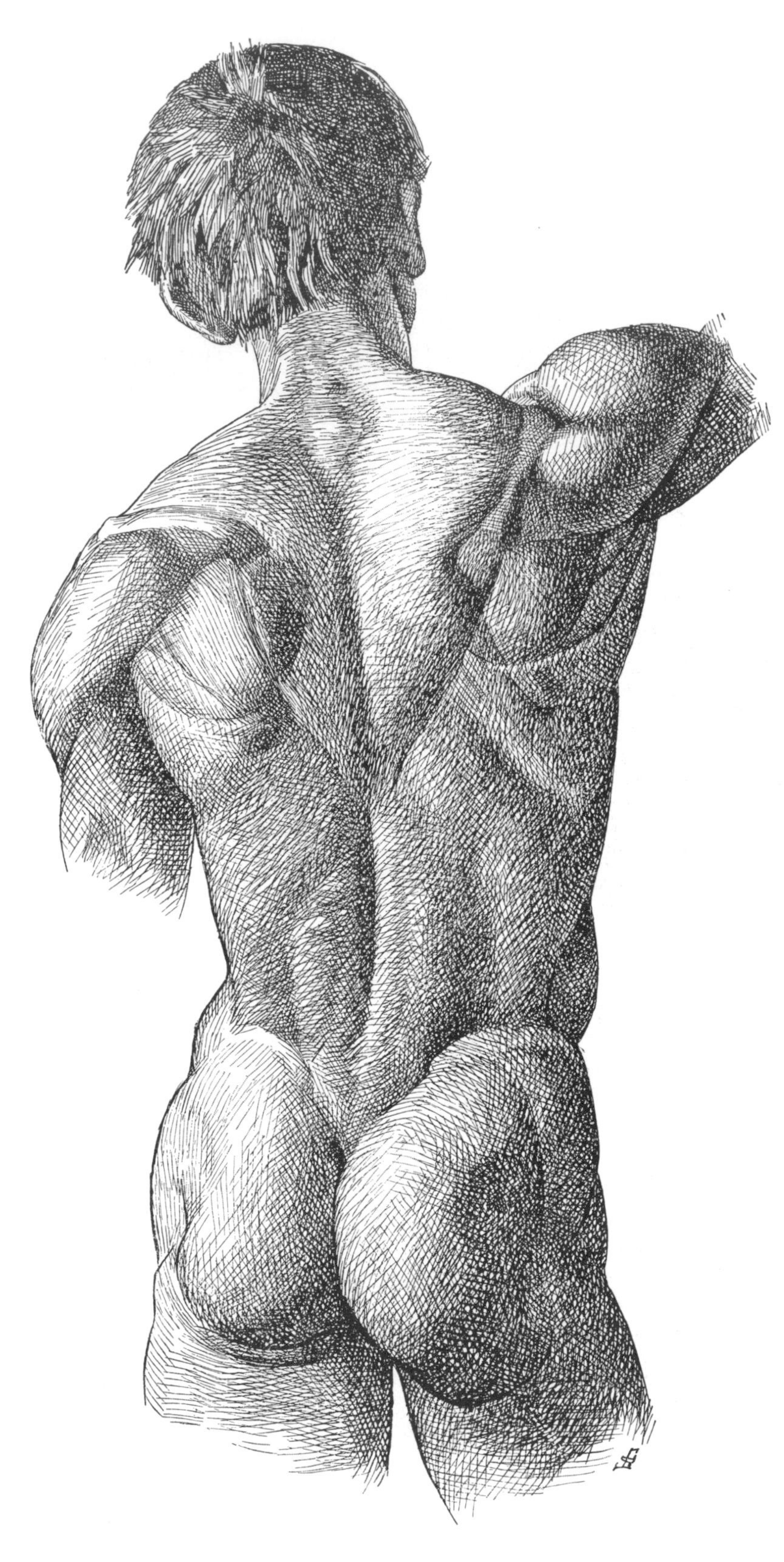

PLATE 30 THE TRUNK POSTERIOR VIEW (SURFACE ANATOMY)

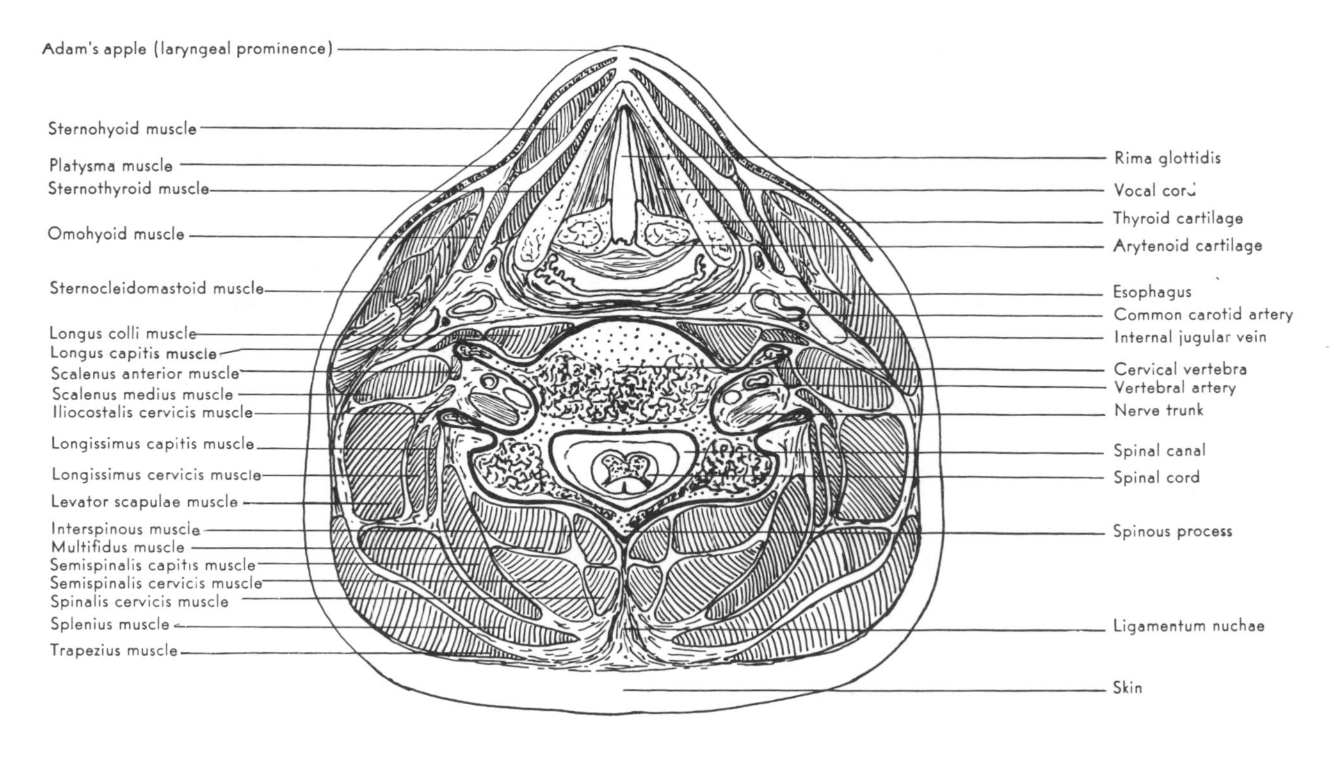

CROSS-SECTION THROUGH THE NECK OF AN ADULT

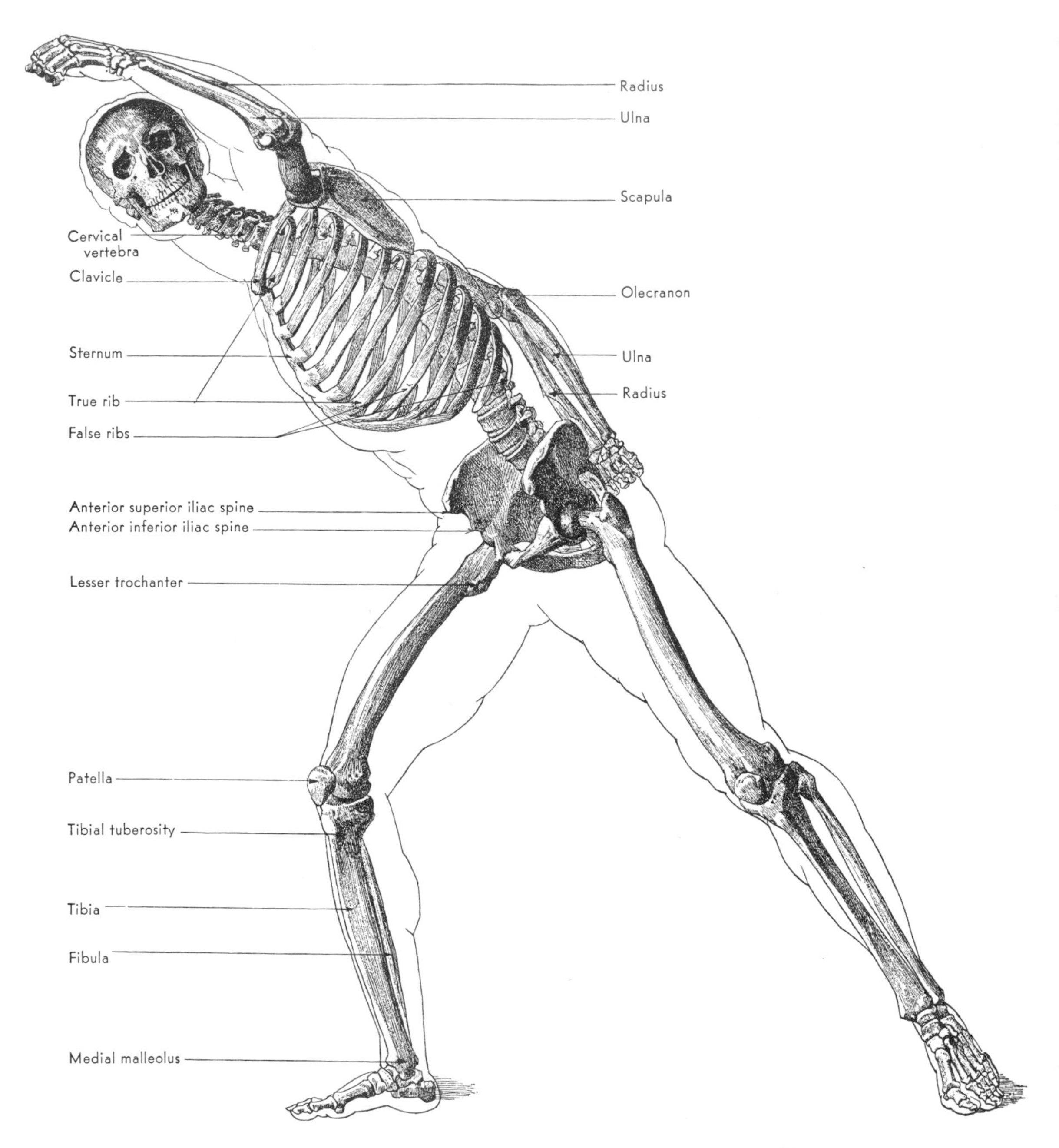

SKELETON IN POSITION OF "THE FIGHTER" BY BORGHESE

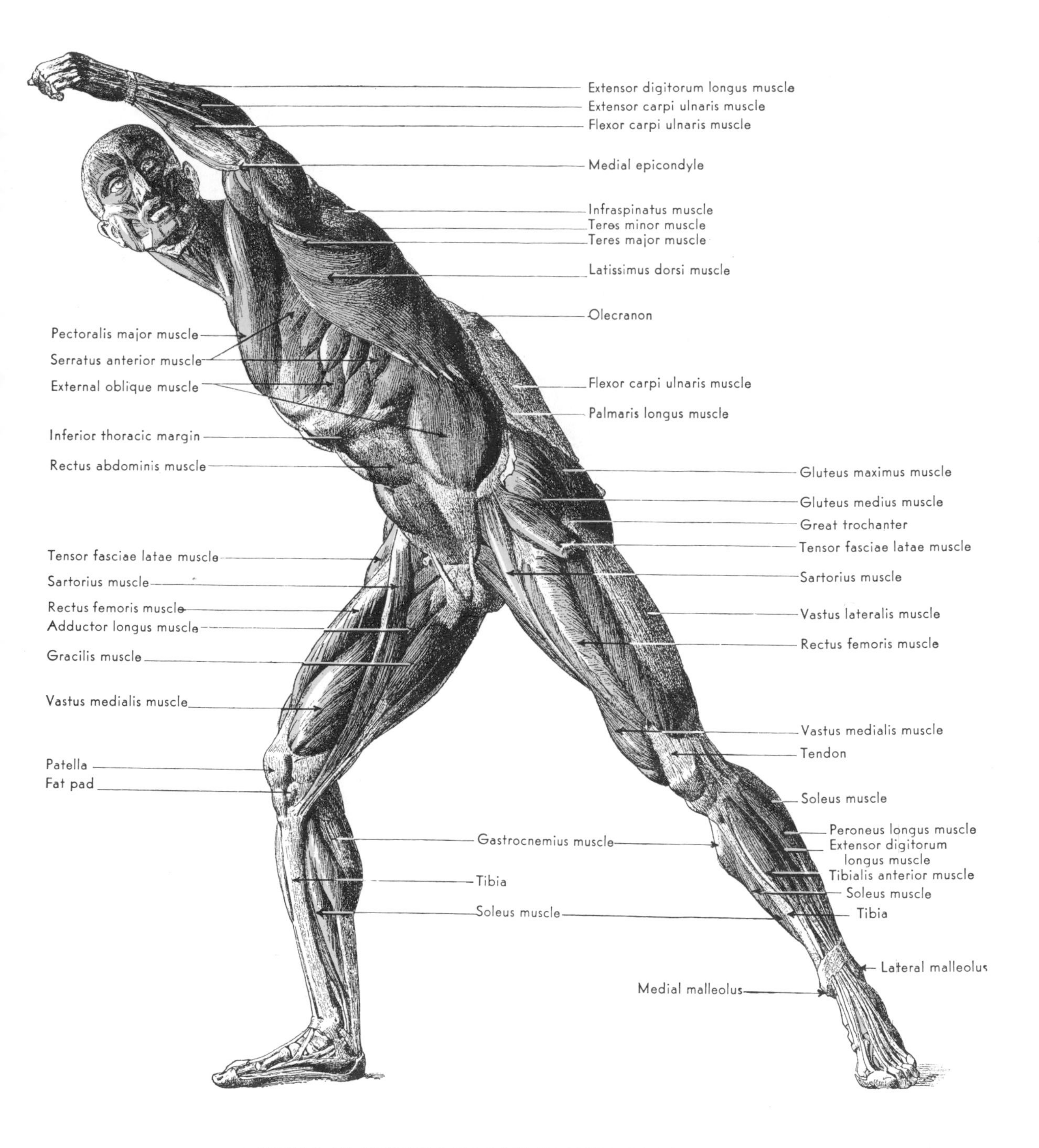

MUSCLES IN POSITION OF "THE FIGHTER" BY BORGHESE

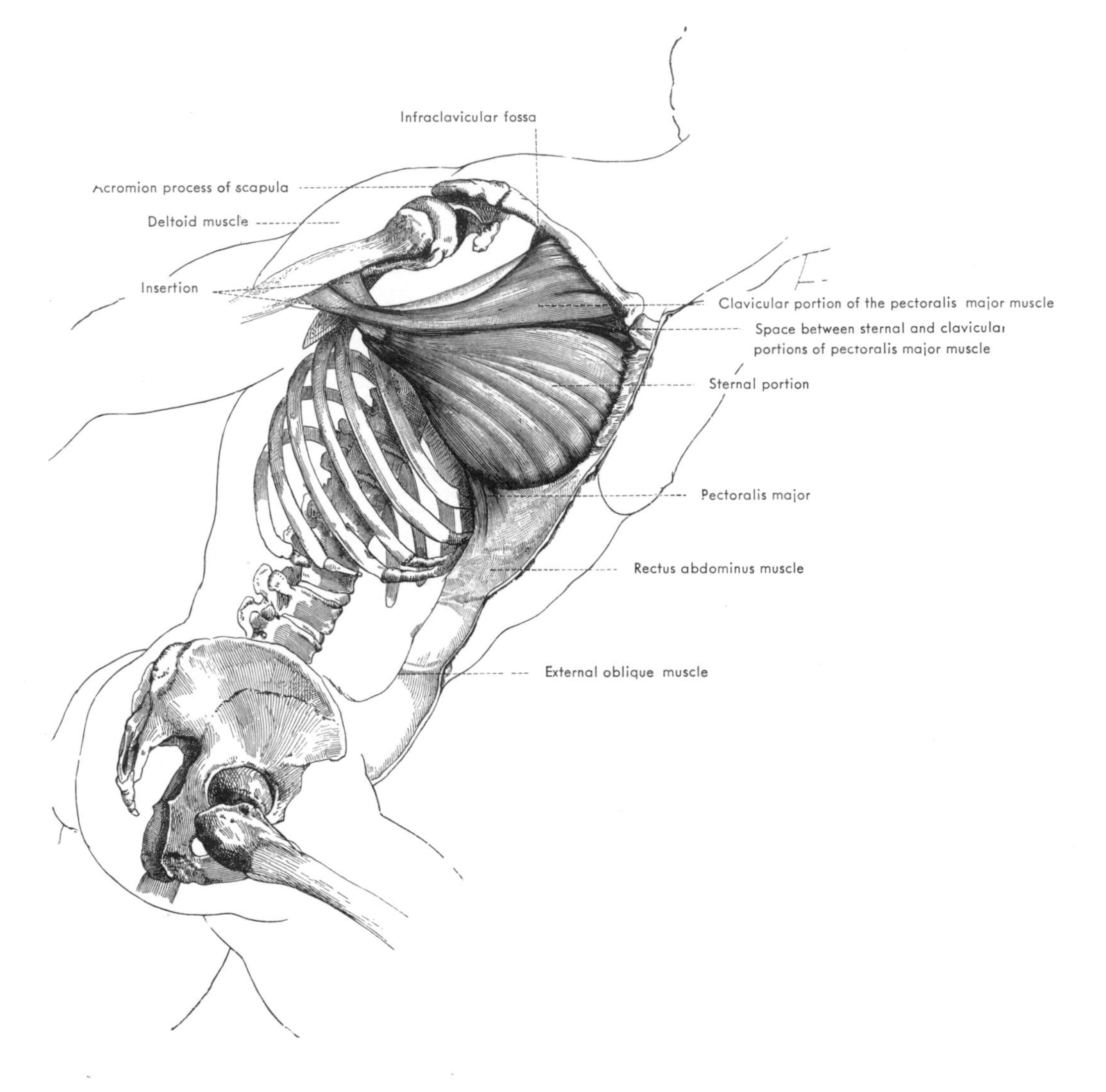

KOLLMANN, THE PECTORALIS MAJOR, BASED ON THE BORGHESE "FIGHTER"

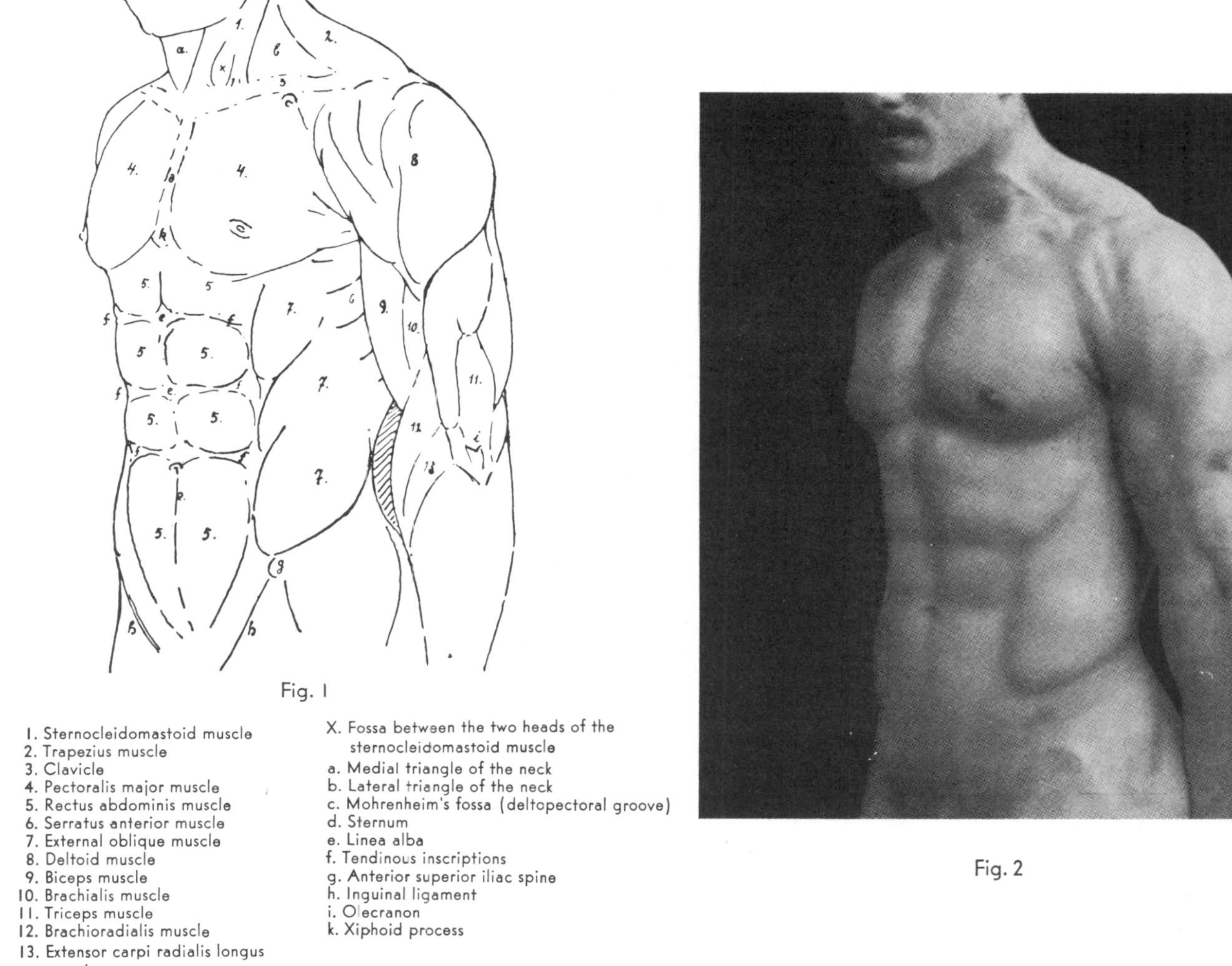

Fig. 1

1. Sternocleidomastoid muscle
2. Trapezius muscle
3. Clavicle
4. Pectoralis major muscle
5. Rectus abdominis muscle
6. Serratus anterior muscle
7. External oblique muscle
8. Deltoid muscle
9. Biceps muscle
10. Brachialis muscle
11. Triceps muscle
12. Brachioradialis muscle
13. Extensor carpi radialis longus muscle

X. Fossa between the two heads of the sternocleidomastoid muscle
a. Medial triangle of the neck
b. Lateral triangle of the neck
c. Mohrenheim's fossa (deltopectoral groove)
d. Sternum
e. Linea alba
f. Tendinous inscriptions
g. Anterior superior iliac spine
h. Inguinal ligament
i. Olecranon
k. Xiphoid process

Fig. 2

ATHLETE THORAX I

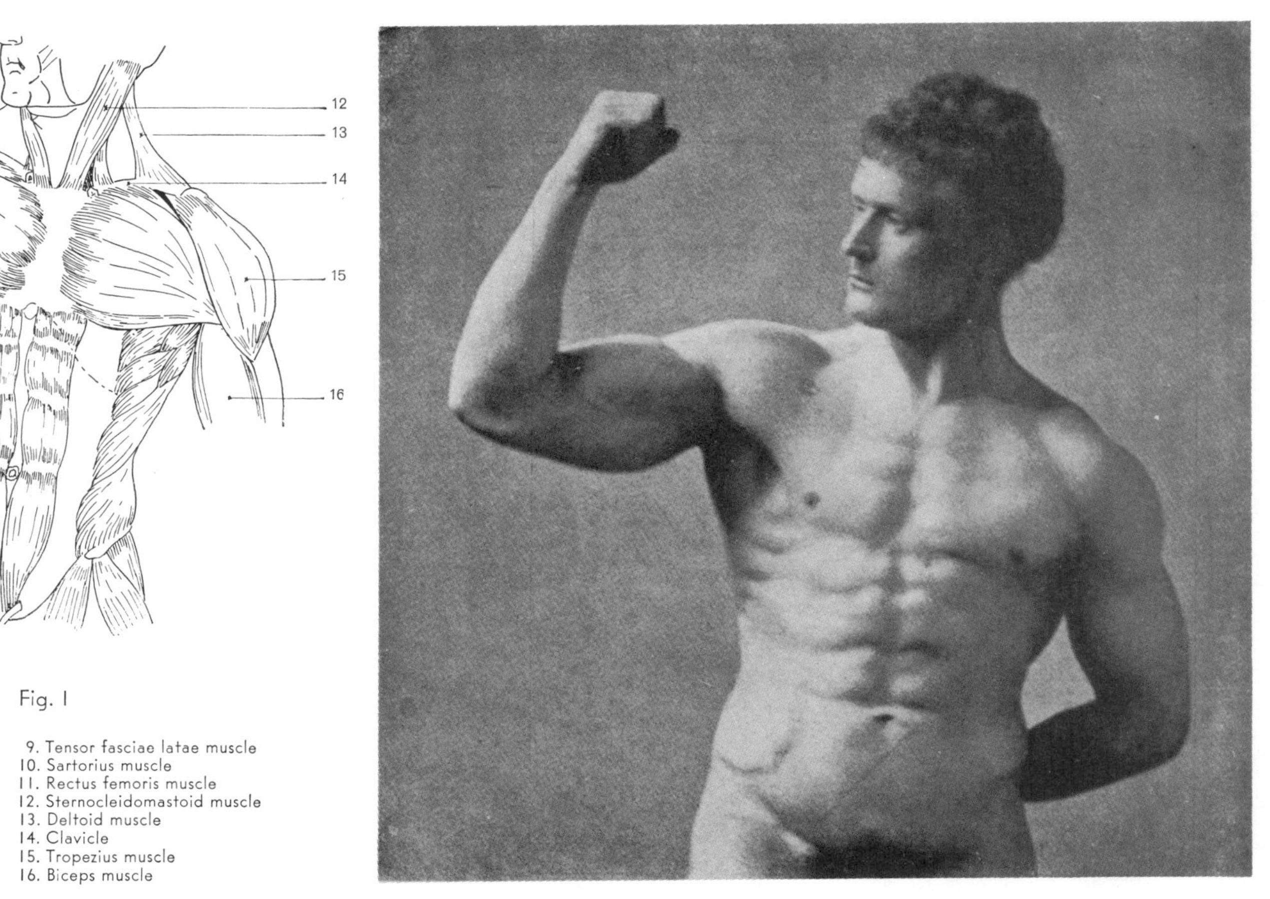

Fig. 1

1. Pectoralis major muscle
2. Latissimus dorsi muscle
3. Serratus anterior muscle
4. Inferior thoracic margin
5. External oblique muscle
6. Rectus abdominis muscle
7. Gluteus medius muscle
8. Crest of the ilium
9. Tensor fasciae latae muscle
10. Sartorius muscle
11. Rectus femoris muscle
12. Sternocleidomastoid muscle
13. Deltoid muscle
14. Clavicle
15. Tropezius muscle
16. Biceps muscle

Fig. 2

ATHLETE ANTERIOR VIEW

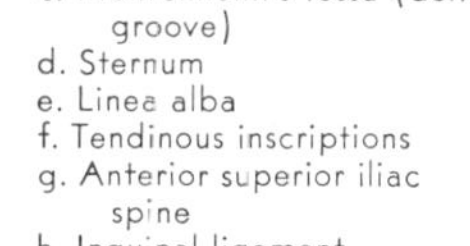

Fig. 1

1. Sternocleidomastoid muscle
2. Trapezius muscle
3. Clavicle
4. Pectoralis major muscle
5. Rectus abdominis muscle
6. Serratus anterior muscle
7. External oblique muscle
8. Deltoid muscle

X. Fossa between the two heads of the sternocleidomastoid muscle

a. Medial triangle of the neck
b. Lateral triangle of the neck
c. Mohrenheim's fossa (deltopectoral groove)
d. Sternum
e. Linea alba
f. Tendinous inscriptions
g. Anterior superior iliac spine
h. Inguinal ligament
i. Acromion
k. Xiphoid process

Fig. 2

ATHLETE THORAX II

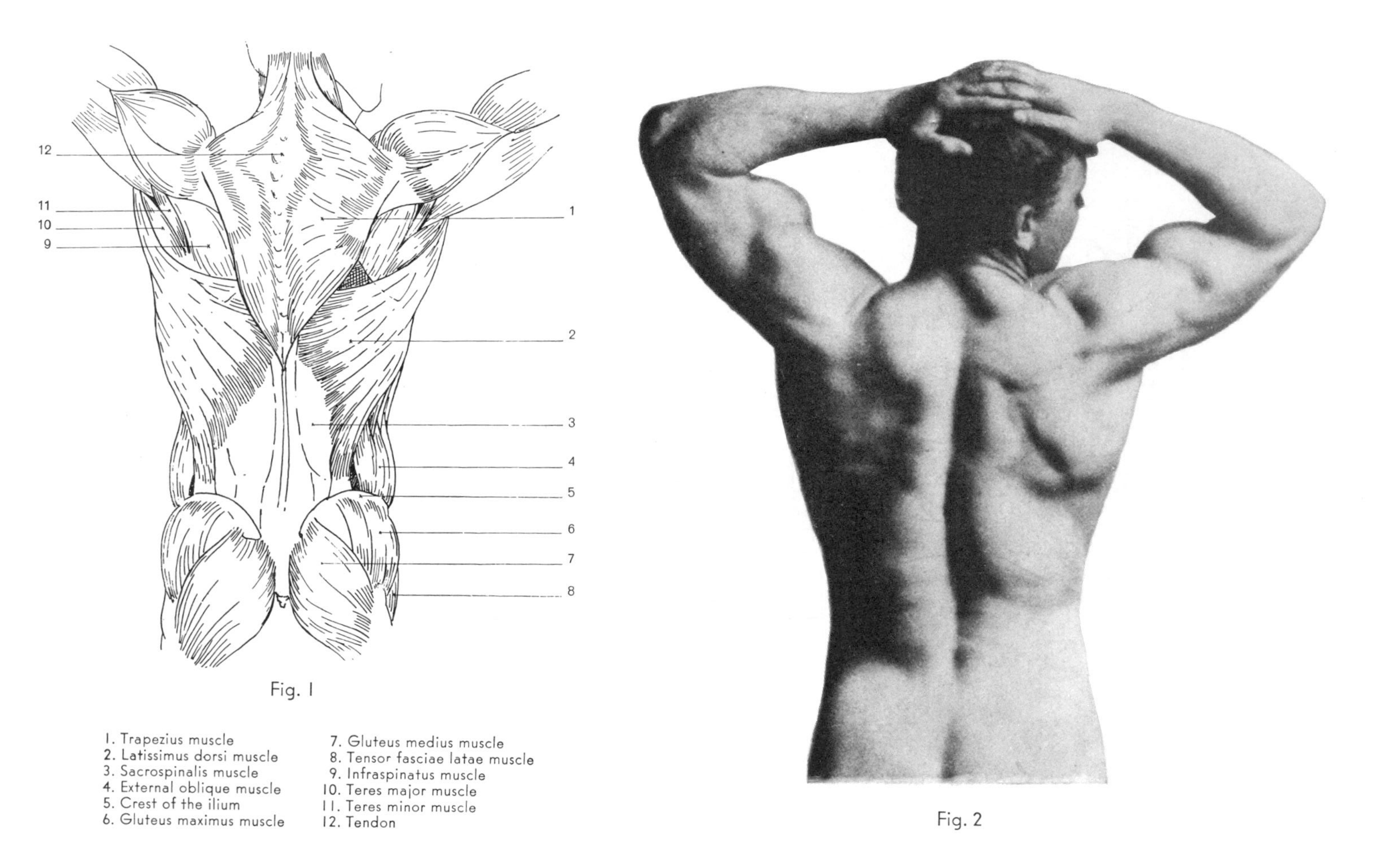

Fig. 1

1. Trapezius muscle
2. Latissimus dorsi muscle
3. Sacrospinalis muscle
4. External oblique muscle
5. Crest of the ilium
6. Gluteus maximus muscle
7. Gluteus medius muscle
8. Tensor fasciae latae muscle
9. Infraspinatus muscle
10. Teres major muscle
11. Teres minor muscle
12. Tendon

Fig. 2

ATHLETE POSTERIOR VIEW

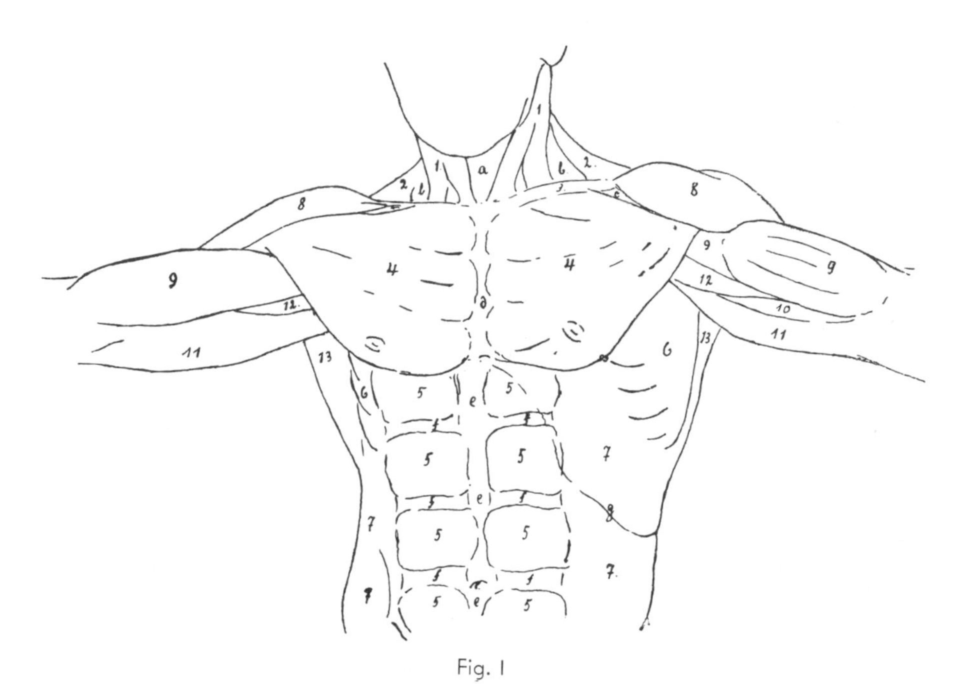

Fig. 1

1. Sternocleidomastoid muscle
2. Trapezius muscle
3. Clavicle
4. Pectoralis major muscle
5. Rectus abdominis muscle
6. Serratus anterior muscle
7. External oblique muscle
8. Deltoid muscle
9. Biceps muscle
10. Brachialis muscle
11. Triceps muscle
12. Coracobrachialis muscle
13. Latissimus dorsi muscle

X. Fossa between the two heads of the sternocleidomastoid muscle
a. Medial triangle of the neck
b. Lateral triangle of the neck
c. Mohrenheim's fossa (deltopectoral groove)
d. Sternum
e. Linea alba
f. Tendinous inscriptions
g. Inferior thoracic margin

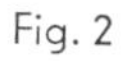

Fig. 2

ATHLETE THORAX III

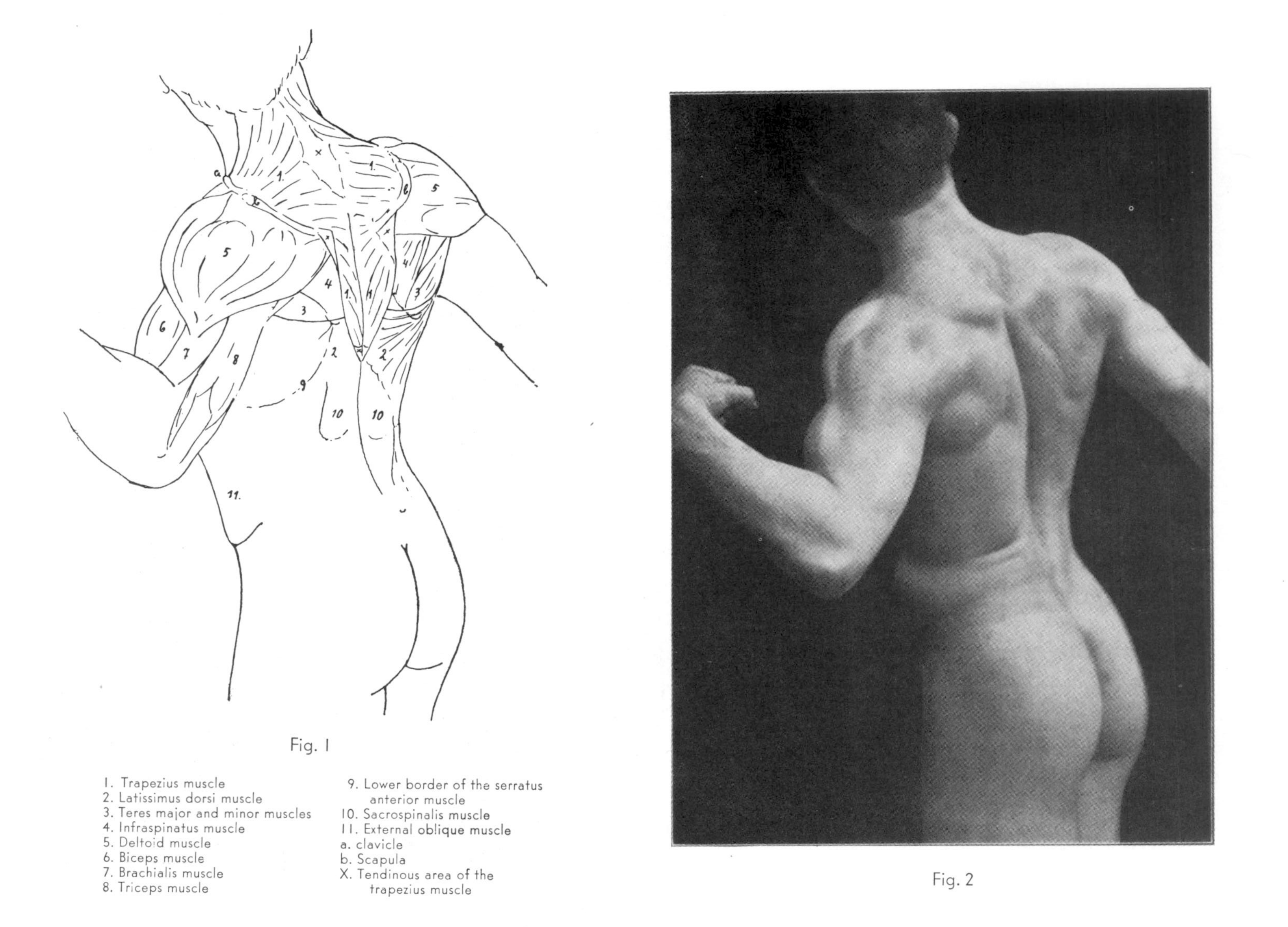

Fig. 1

1. Trapezius muscle
2. Latissimus dorsi muscle
3. Teres major and minor muscles
4. Infraspinatus muscle
5. Deltoid muscle
6. Biceps muscle
7. Brachialis muscle
8. Triceps muscle
9. Lower border of the serratus anterior muscle
10. Sacrospinalis muscle
11. External oblique muscle

a. clavicle
b. Scapula
X. Tendinous area of the trapezius muscle

Fig. 2

ATHLETE BACK I

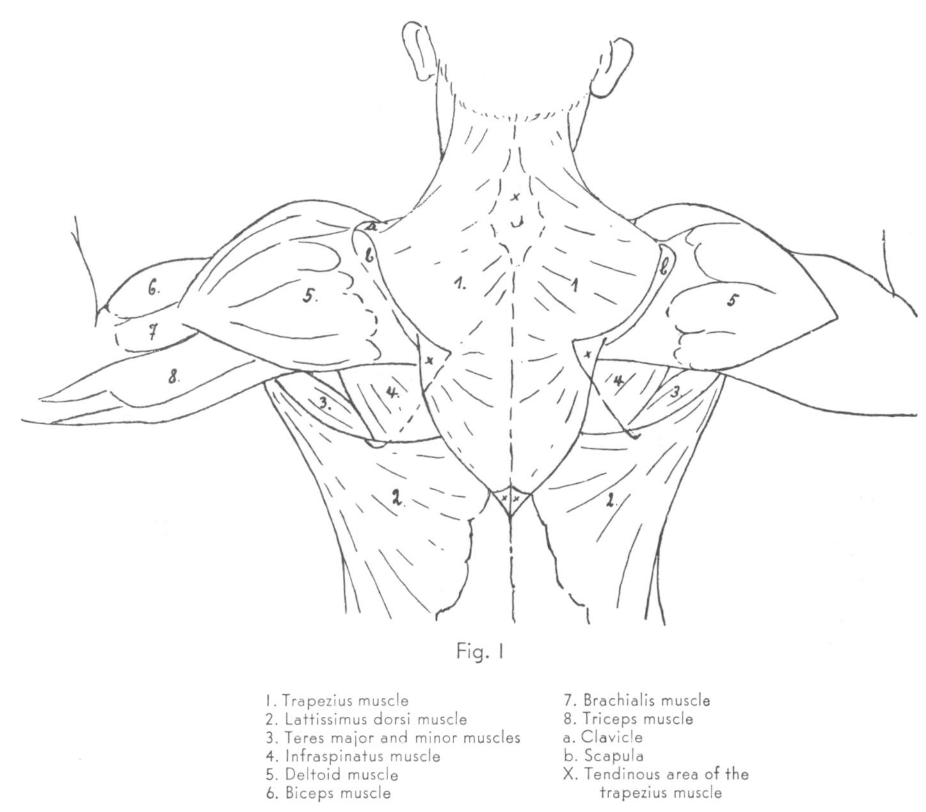

Fig. I

1. Trapezius muscle
2. Lattissimus dorsi muscle
3. Teres major and minor muscles
4. Infraspinatus muscle
5. Deltoid muscle
6. Biceps muscle
7. Brachialis muscle
8. Triceps muscle

a. Clavicle
b. Scapula
X. Tendinous area of the trapezius muscle

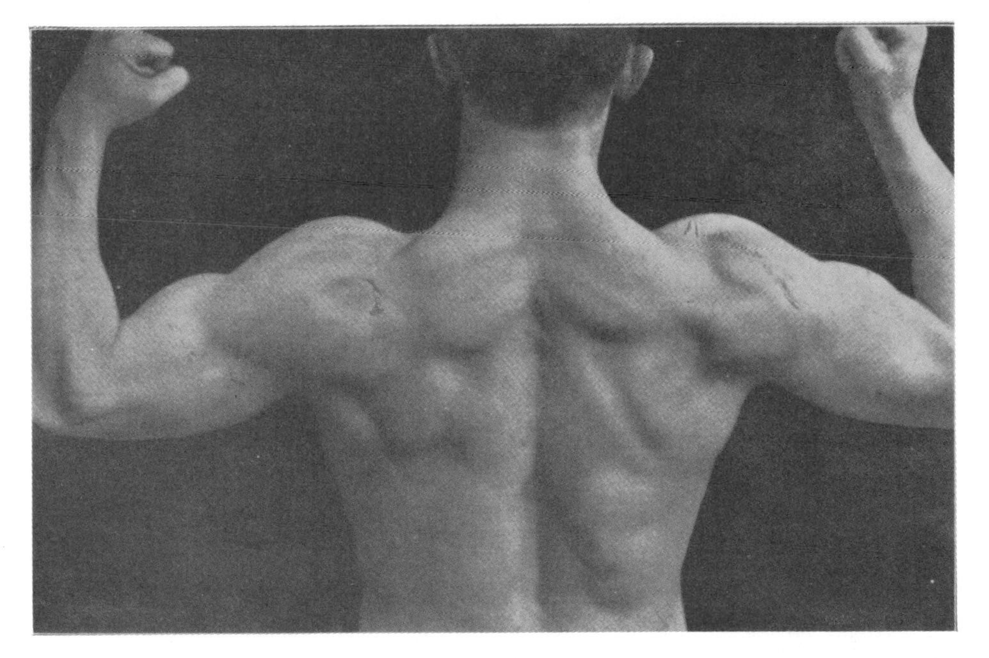

Fig. 2

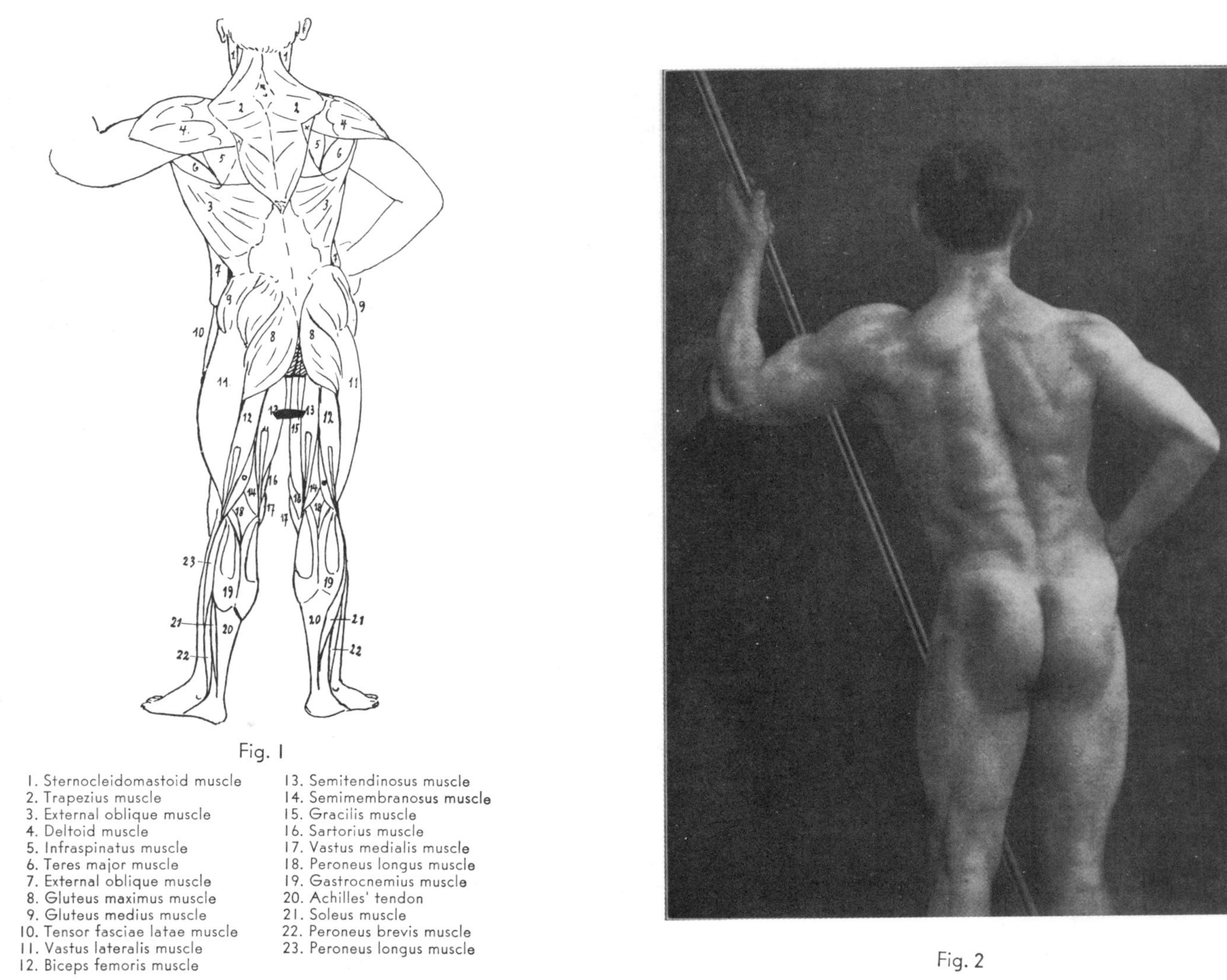

Fig. 1

1. Sternocleidomastoid muscle
2. Trapezius muscle
3. External oblique muscle
4. Deltoid muscle
5. Infraspinatus muscle
6. Teres major muscle
7. External oblique muscle
8. Gluteus maximus muscle
9. Gluteus medius muscle
10. Tensor fasciae latae muscle
11. Vastus lateralis muscle
12. Biceps femoris muscle
13. Semitendinosus muscle
14. Semimembranosus muscle
15. Gracilis muscle
16. Sartorius muscle
17. Vastus medialis muscle
18. Peroneus longus muscle
19. Gastrocnemius muscle
20. Achilles' tendon
21. Soleus muscle
22. Peroneus brevis muscle
23. Peroneus longus muscle

Fig. 2

ATHLETE BACK III

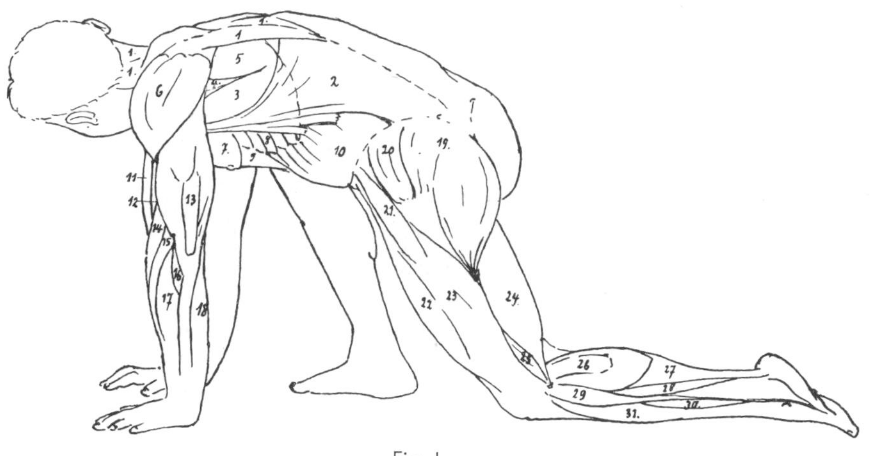

Fig. 1

1. Trapzius muscle
2. Latissimus dorsi muscle
3. Teres major muscle
4. Teres minor muscle
5. Infraspinatus muscle
6. Deltoid muscle
7. Pectoralis major muscle
8. Serratus anterior muscle
9. Rectus abdominis muscle
10. External oblique muscle
11. Biceps muscle
12. Brachialis muscle
13. Triceps muscle
14. Brachioradialis muscle
15. Extensor carpi radialis longus muscle
16. Anconeus muscle
17. Extensor carpi radialis brevis muscle
18. Extensor carpi ulnaris muscle
19. Gluteus maximus muscle
20. Gluteus medius muscle
21. Tensor fasciae latae muscle
22. Rectus femoris muscle
23. Vastus lateralis muscle
24. Biceps femoris muscle, long head
25. Biceps femoris muscle, short head
26. Gastrocnemius muscle
27. Achilles' tendon
28. Soleus muscle
29. Peroneus longus muscle
30. Peroneus brevis muscle
31. Extensor digitorum longus muscle

Fig. 2

ATHLETE KNEELING

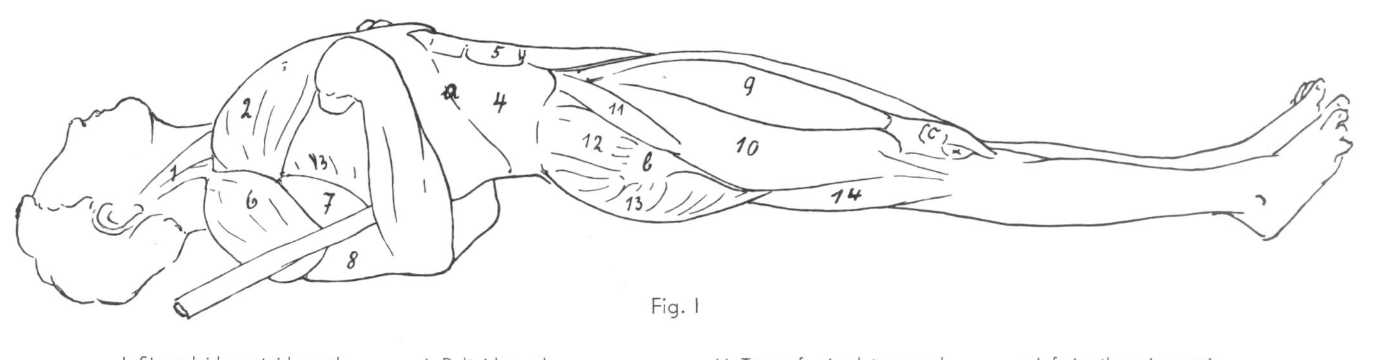

Fig. 1

1. Stenocleidomastoid muscle
2. Pectoralis major muscle
3. Serratus anterior muscle
4. External oblique muscle
5. Rectus abdominis muscle
6. Deltoid muscle
7. Biceps muscle
8. Triceps muscle
9. Rectus femoris muscle
10. Vastus lateralis muscle
11. Tensor fasciae latae muscle
12. Gluteus medius muscle
13. Gluteus maximus muscle
14. Biceps femoris muscle

a. Inferior thoracic margin
b. Tendinous insertion of the gluteus maximus muscle
c. Patella
X. Fat pad of knee joint

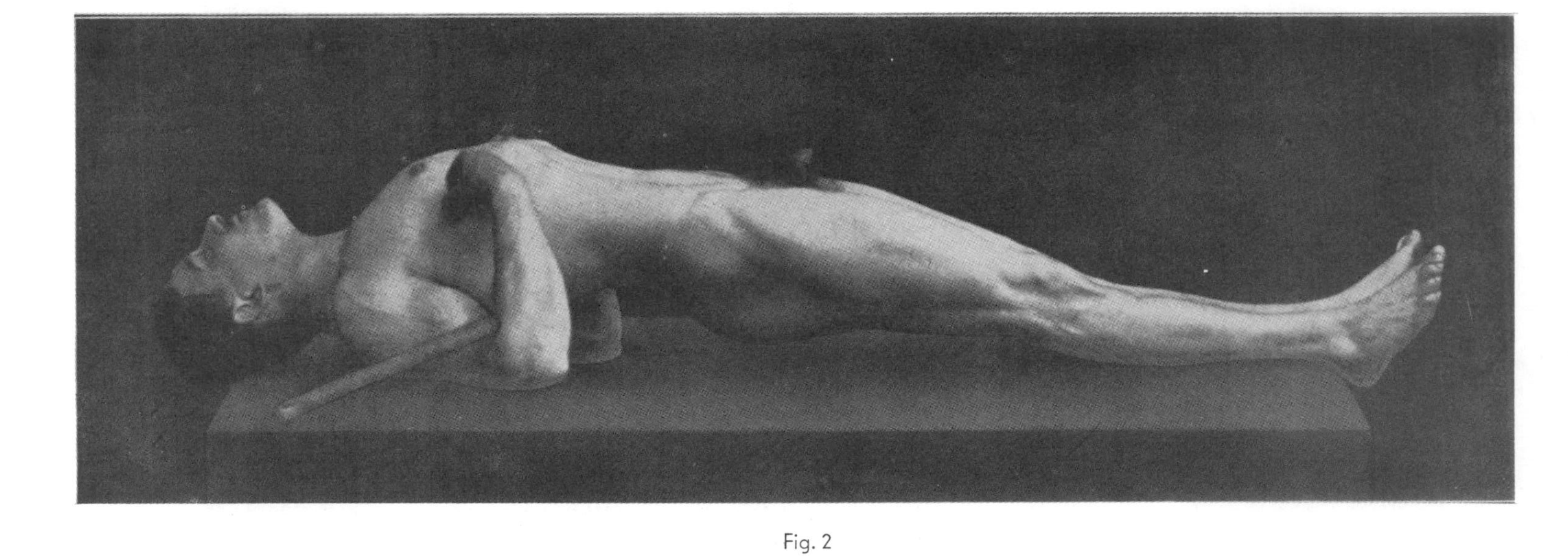

Fig. 2

ATHLETE SUPINE

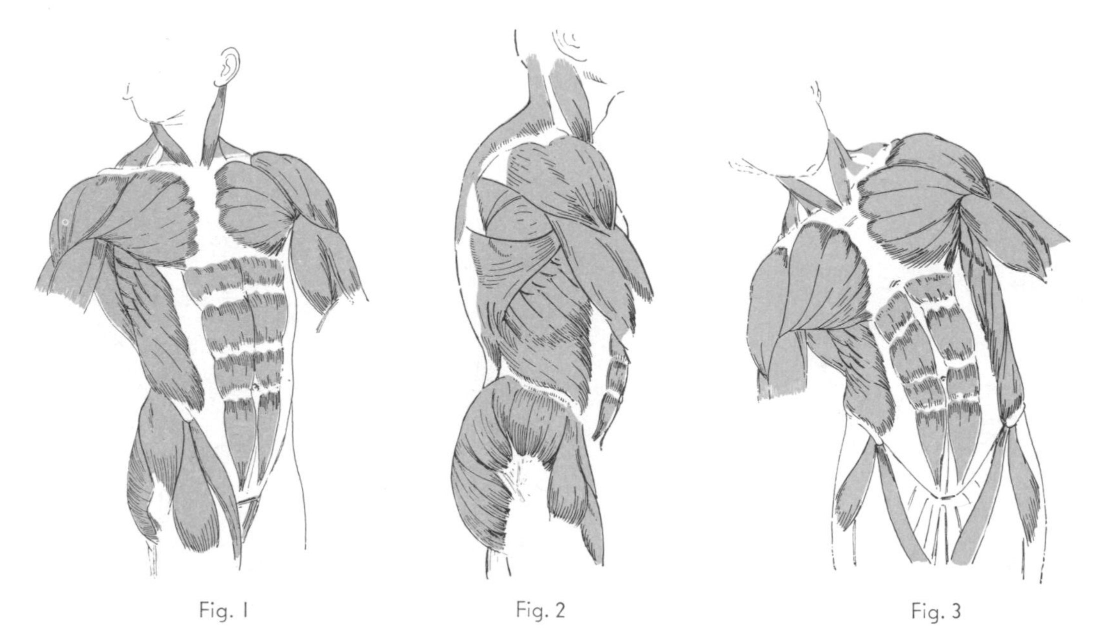

Fig. 1 Fig. 2 Fig. 3

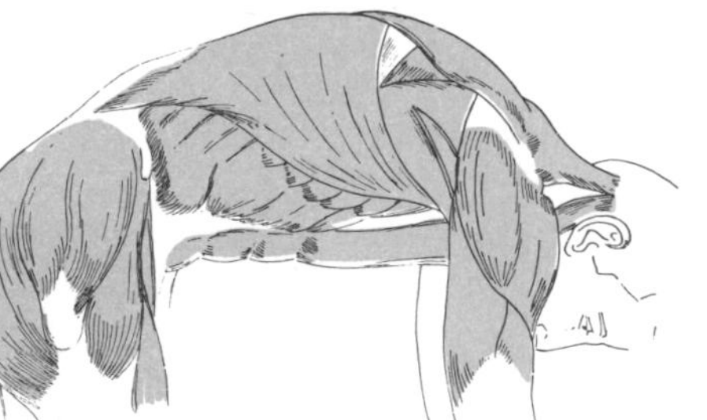

Fig. 4

Fig. 5

Fig. 6

POSITIONS OF THE TRUNK I

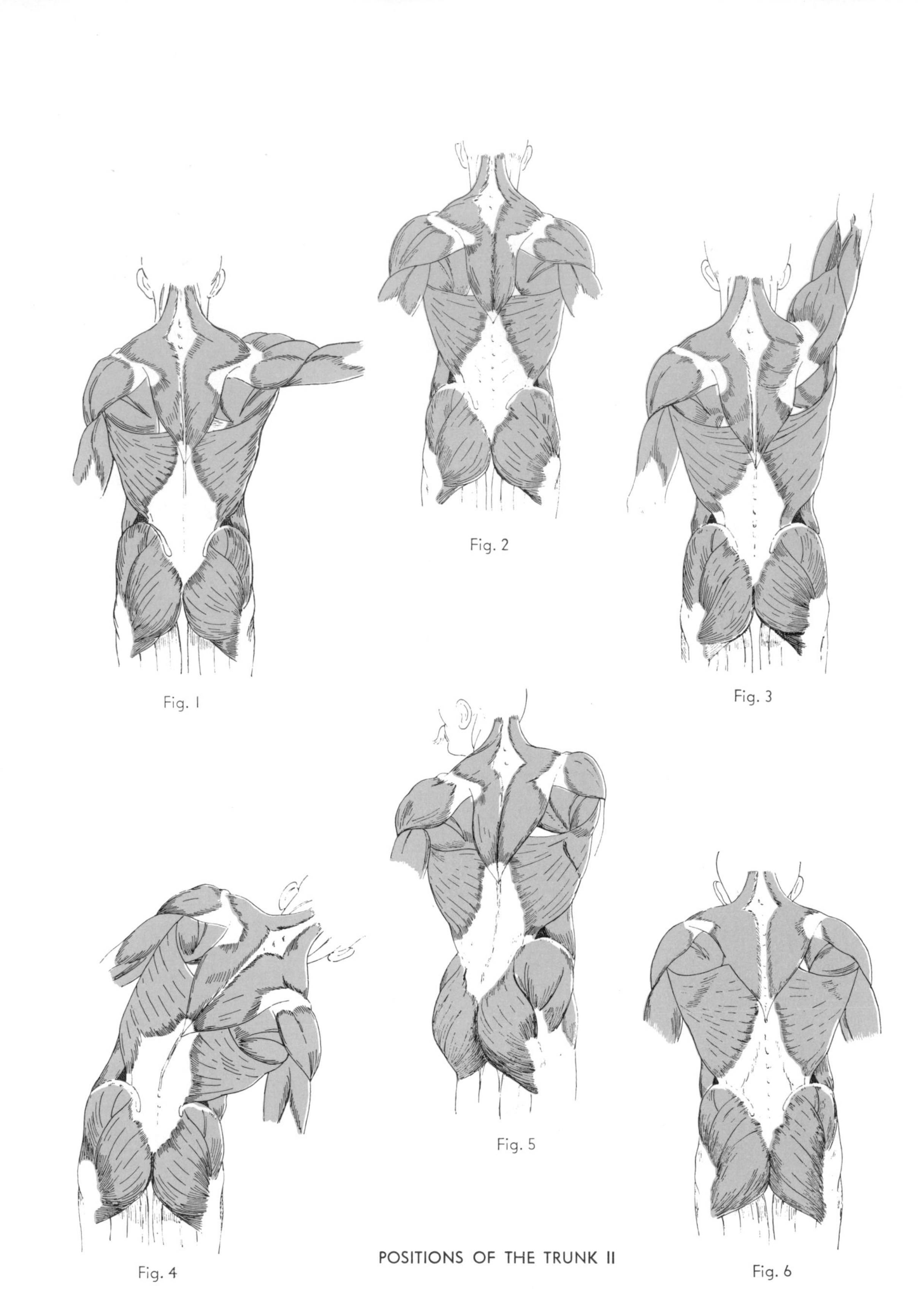
Fig. 1
Fig. 2
Fig. 3
Fig. 4
Fig. 5
Fig. 6

POSITIONS OF THE TRUNK II

Fig. 1
TRANSVERSE MUSCLES OF THE BACK (SECOND LAYER), POSTERIOR VIEW

Fig. 2
LONGITUDINAL MUSCLES OF THE BACK (SECOND LAYER), POSTERIOR VIEW

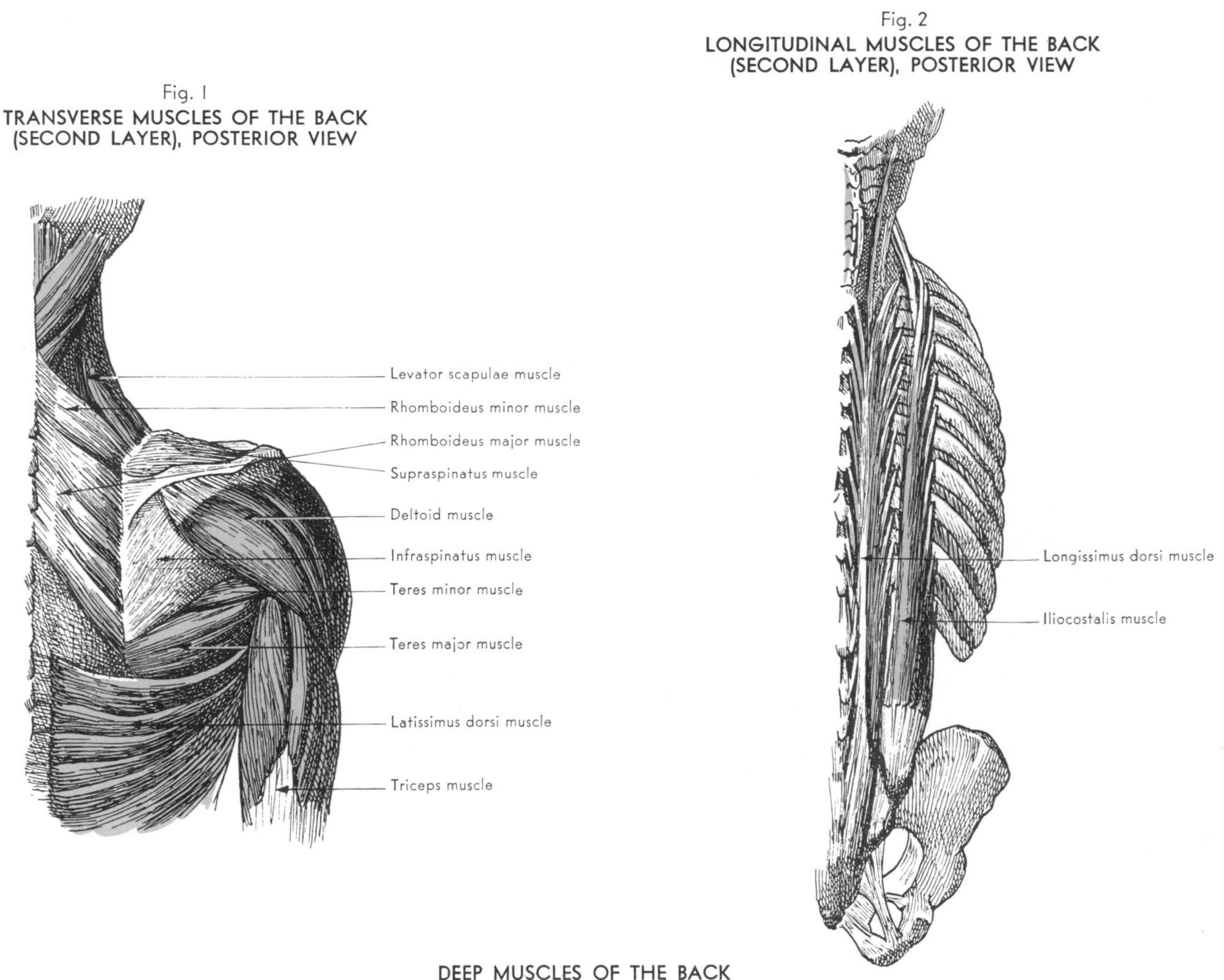

DEEP MUSCLES OF THE BACK

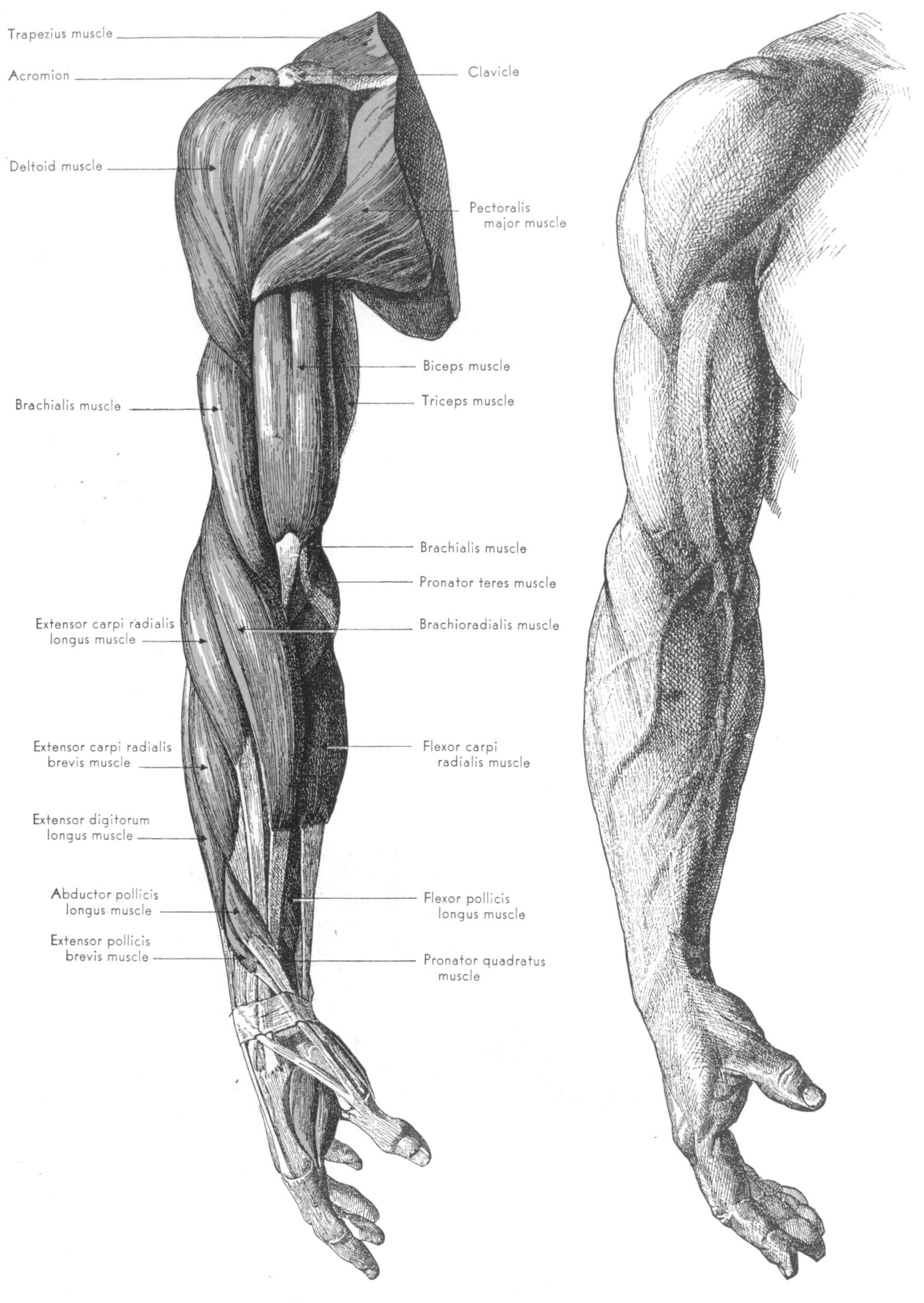

PLATE 48

ANTERIOR VIEW FOR COMPARISON WITH SURFACE ANATOMY
THE MUSCLES OF THE UPPER EXTREMITY,

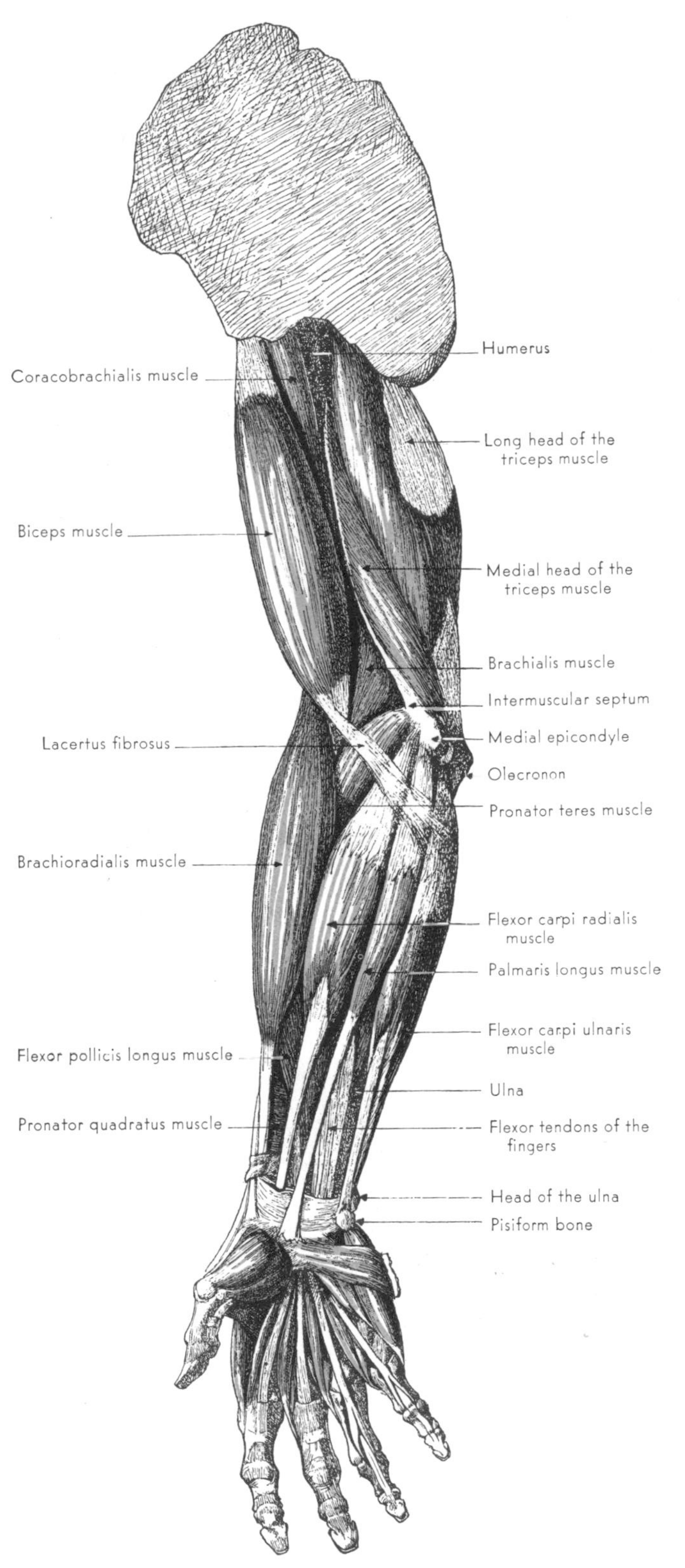

THE MUSCLES OF THE UPPER EXTREMITY, MEDIAL VIEW

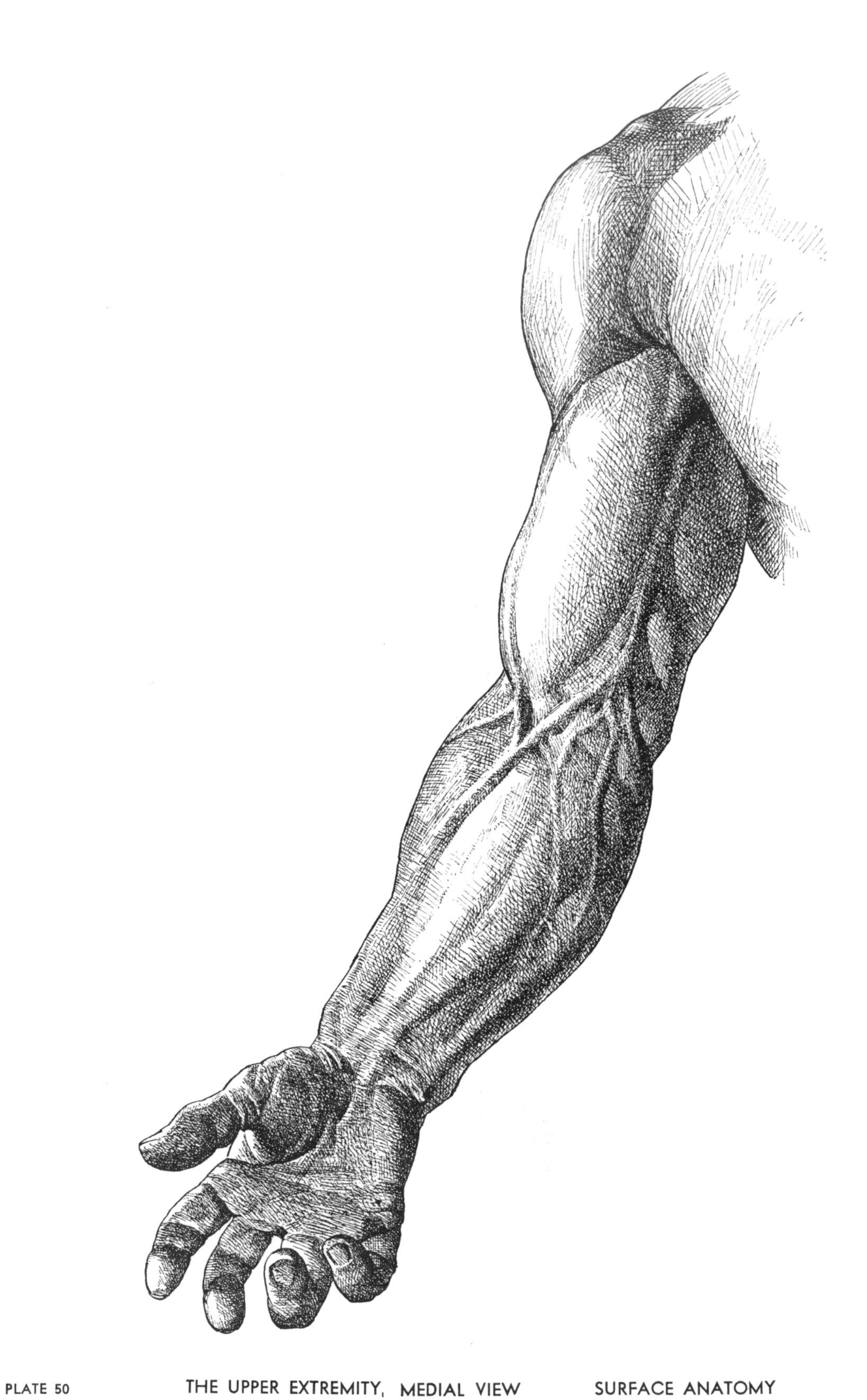

PLATE 50 THE UPPER EXTREMITY, MEDIAL VIEW SURFACE ANATOMY

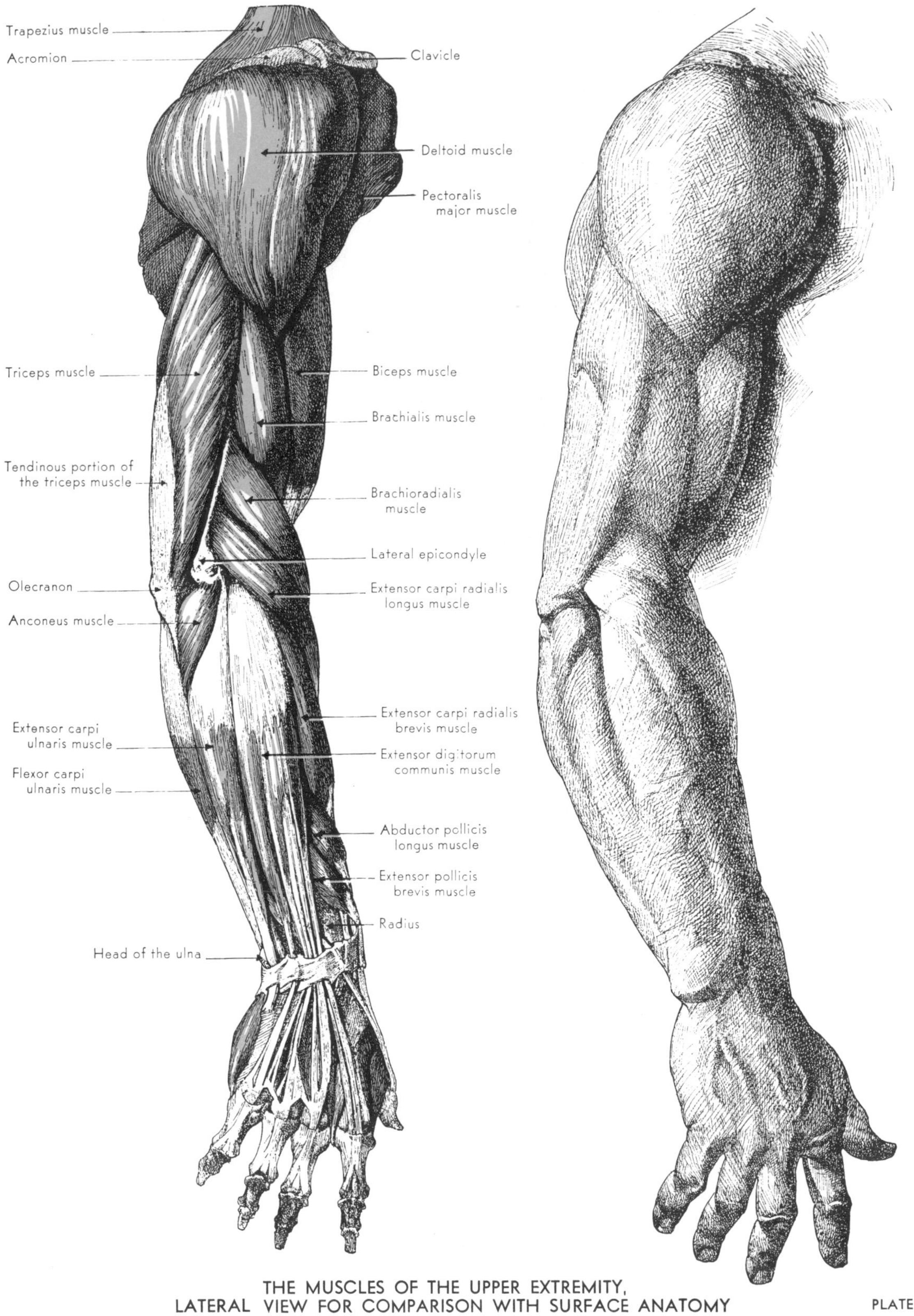

THE MUSCLES OF THE UPPER EXTREMITY,
LATERAL VIEW FOR COMPARISON WITH SURFACE ANATOMY

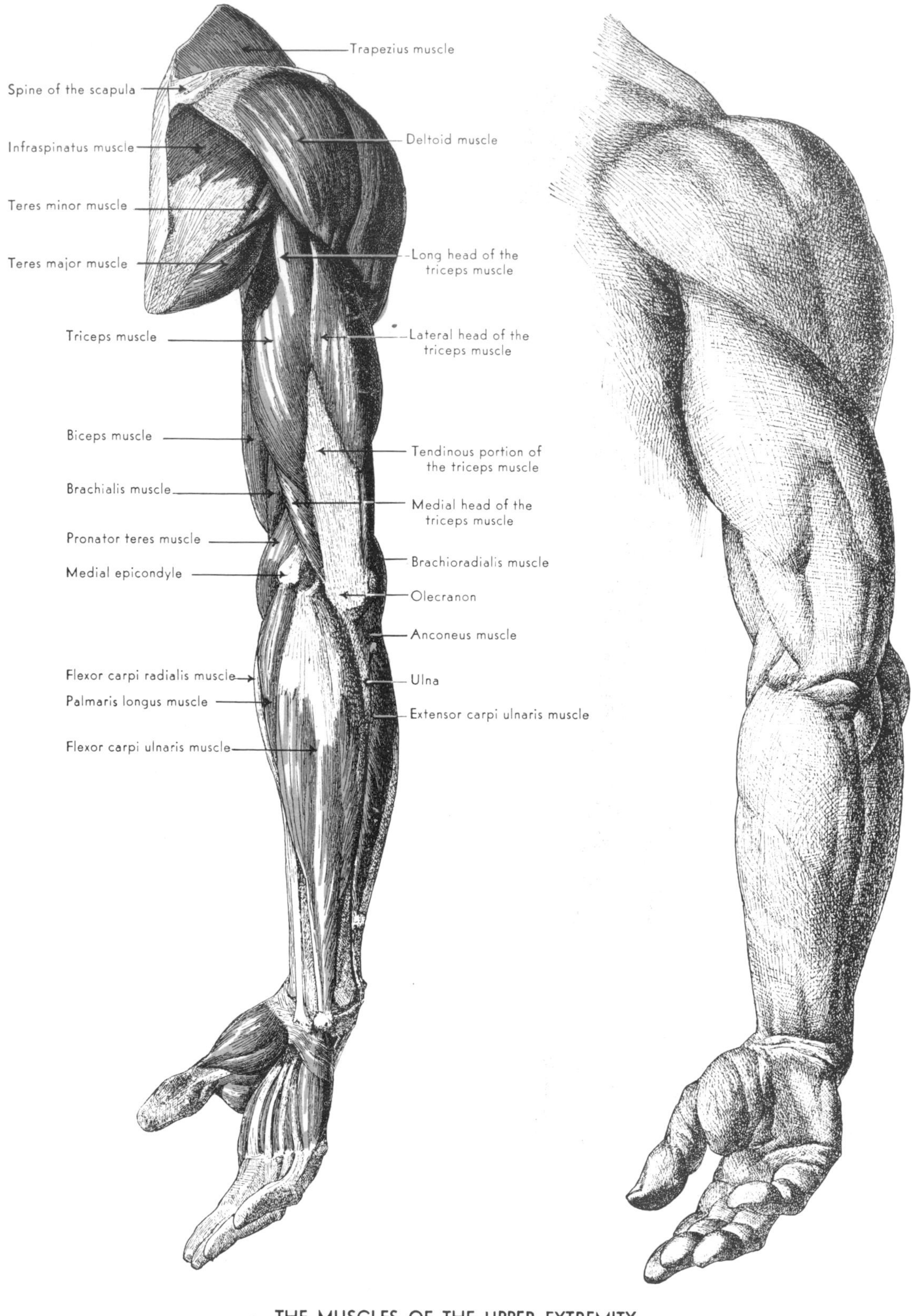

PLATE 52

THE MUSCLES OF THE UPPER EXTREMITY,
POSTERIOR VIEW FOR COMPARISON WITH SURFACE ANATOMY

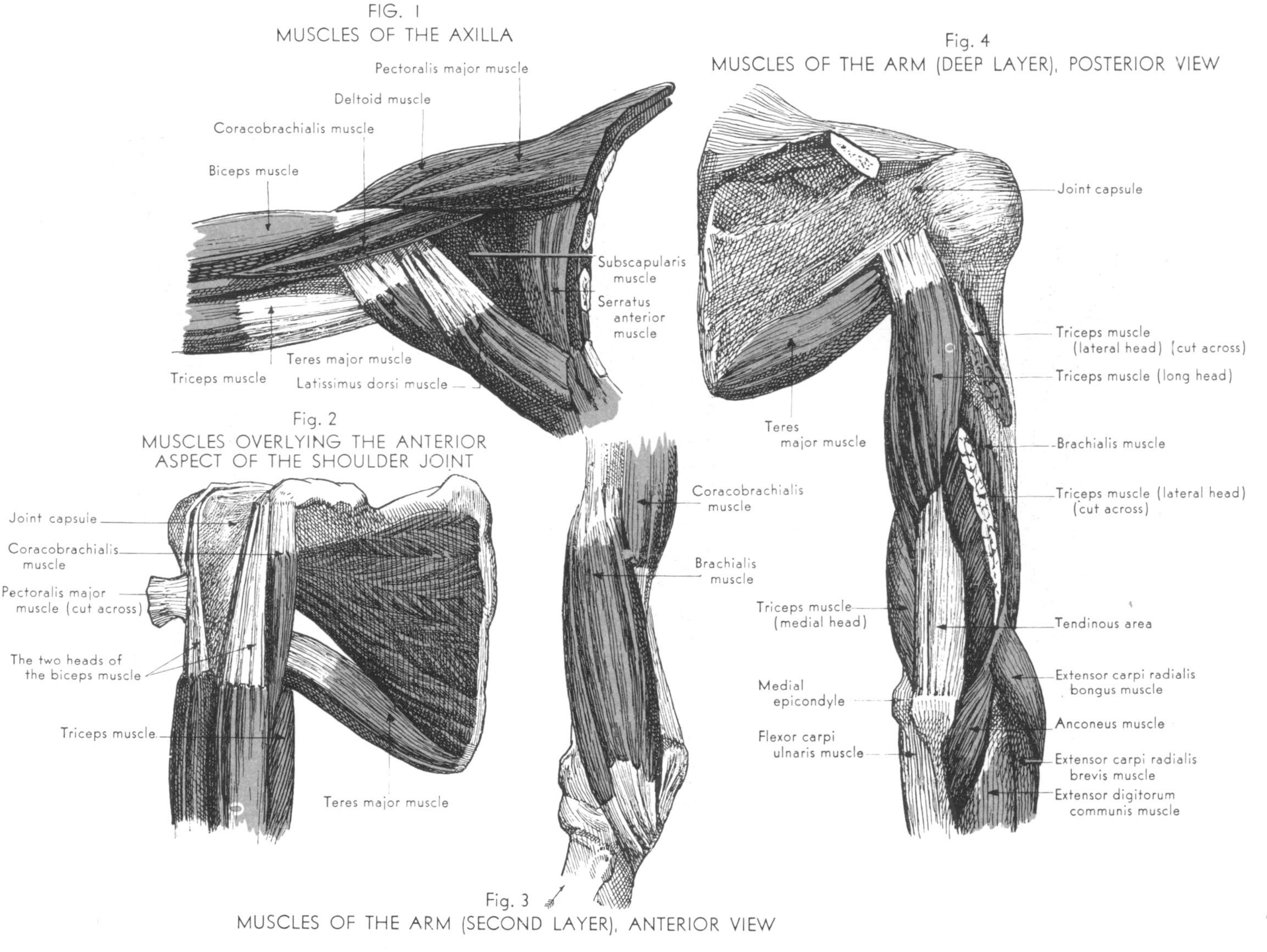

FIG. 1
MUSCLES OF THE AXILLA

Fig. 2
MUSCLES OVERLYING THE ANTERIOR ASPECT OF THE SHOULDER JOINT

Fig. 3
MUSCLES OF THE ARM (SECOND LAYER), ANTERIOR VIEW

Fig. 4
MUSCLES OF THE ARM (DEEP LAYER), POSTERIOR VIEW

THE MUSCLES OF THE ARM

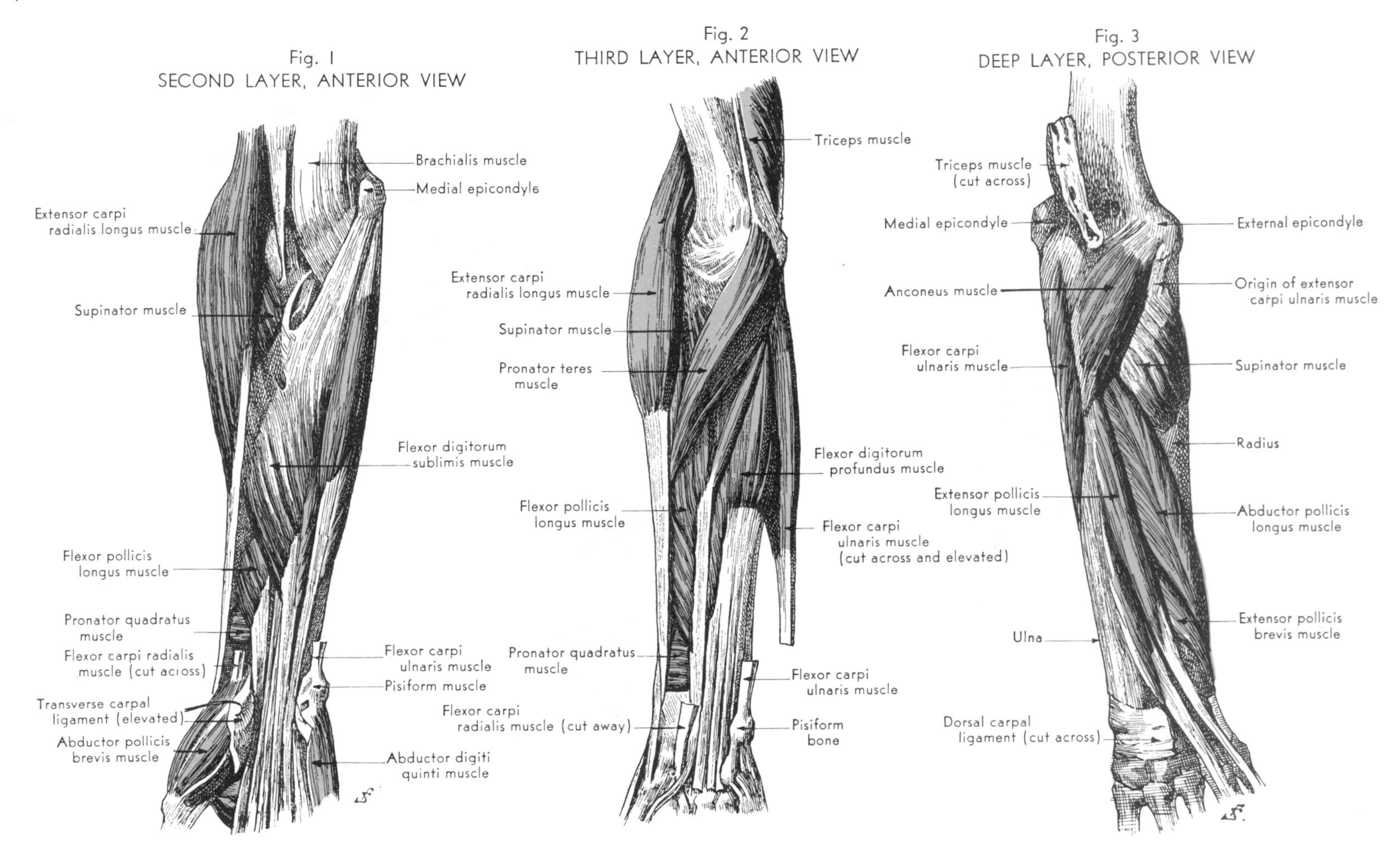

THE MUSCLES OF THE FOREARM

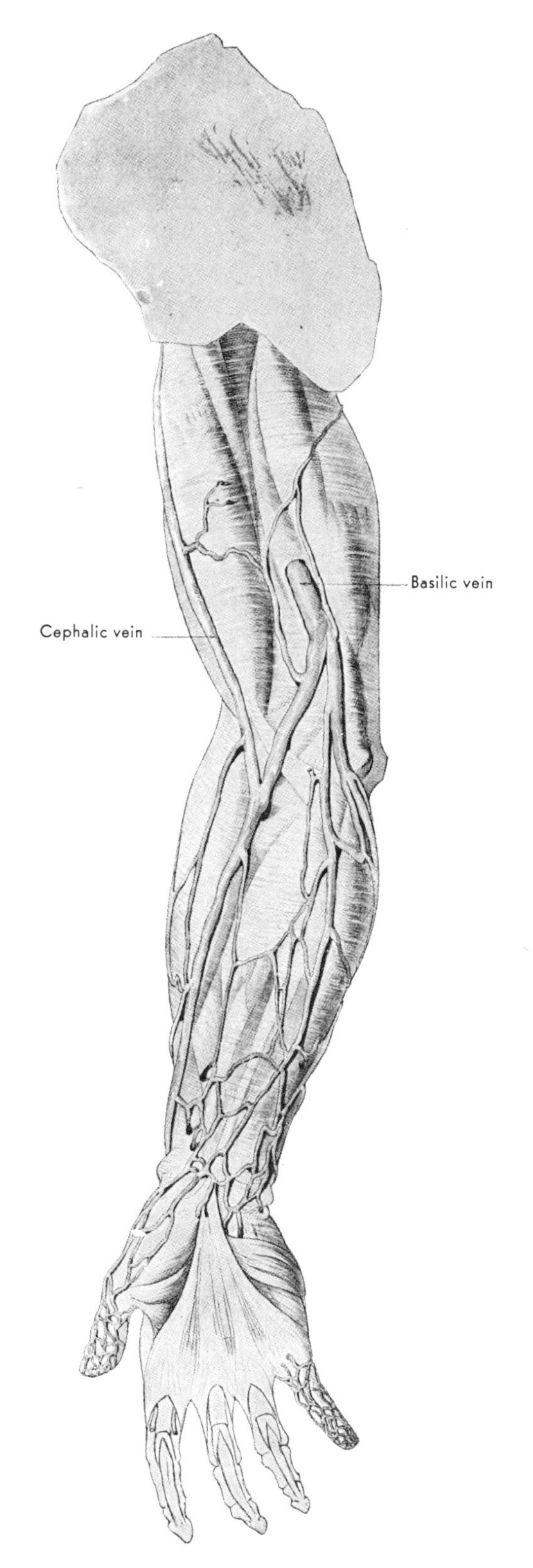

SUPERFICIAL VEINS OF THE UPPER EXTREMITY

PLATE 55

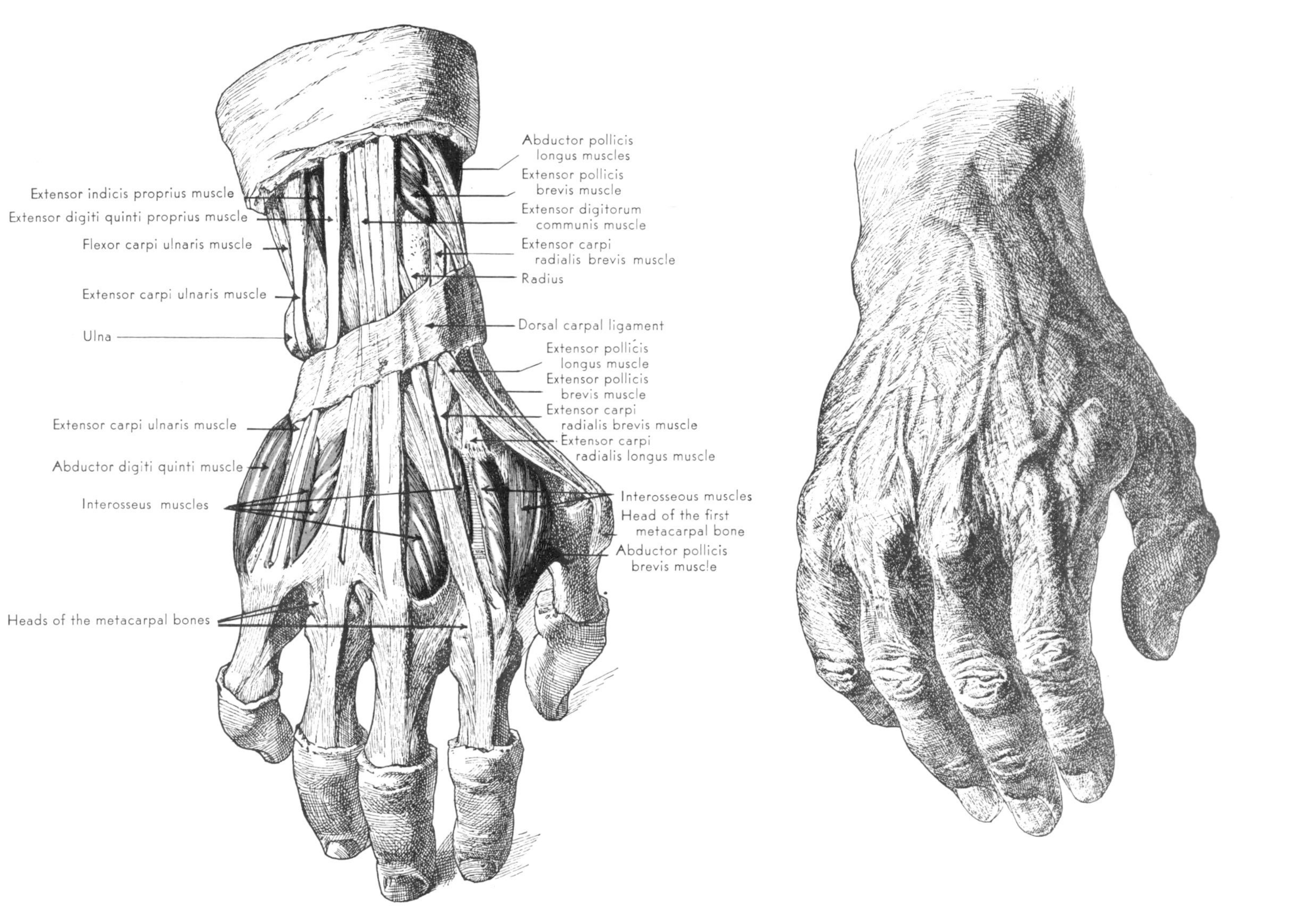

THE MUSCLES OF THE HAND, POSTERIOR VIEW WITH SURFACE ANATOMY FOR COMPARISON

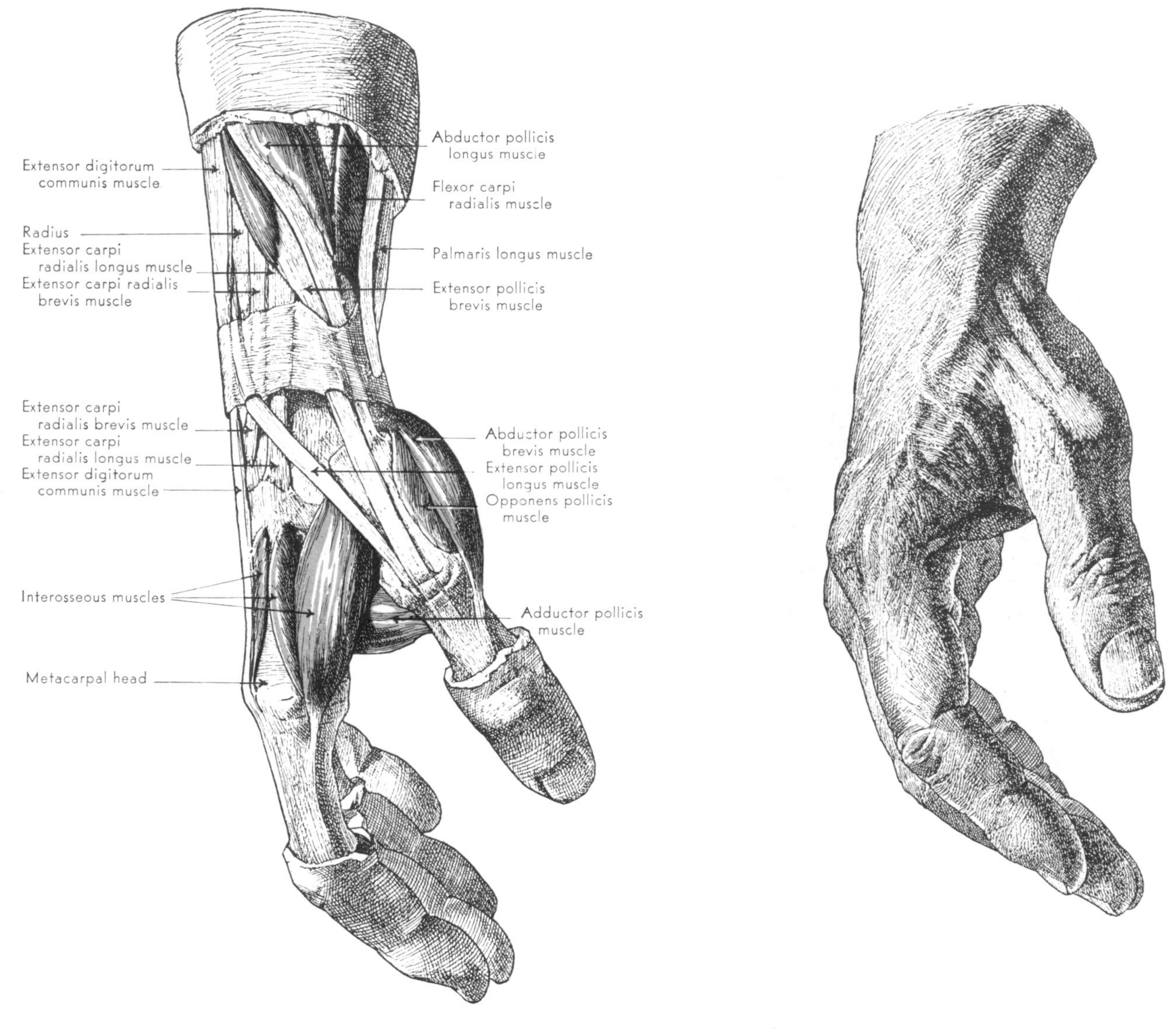

THE MUSCLES OF THE HAND, LATERAL VIEW WITH SURFACE ANATOMY FOR COMPARISON

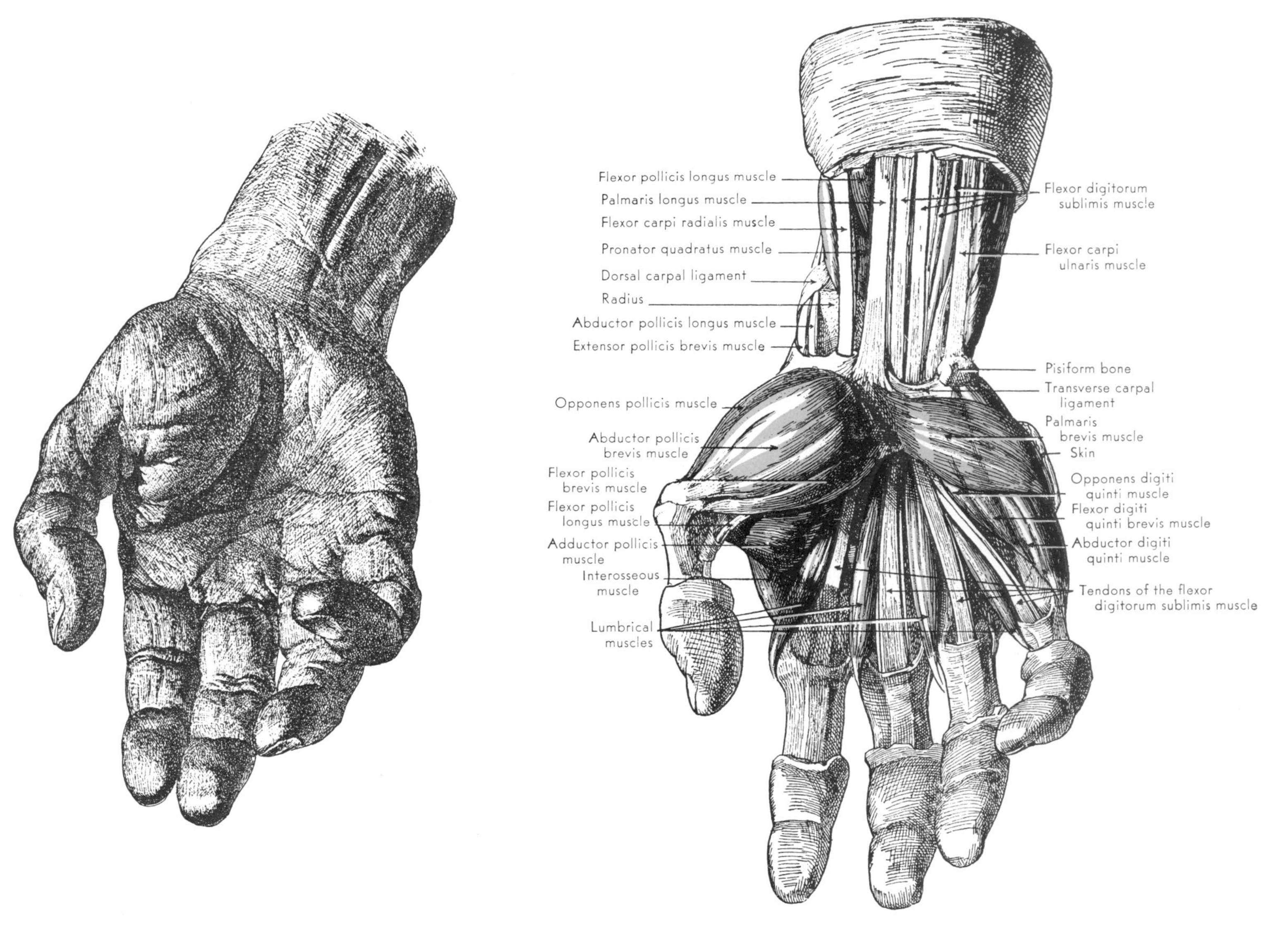

THE MUSCLES OF THE HAND, ANTERIOR VIEW WITH SURFACE ANATOMY FOR COMPARISON

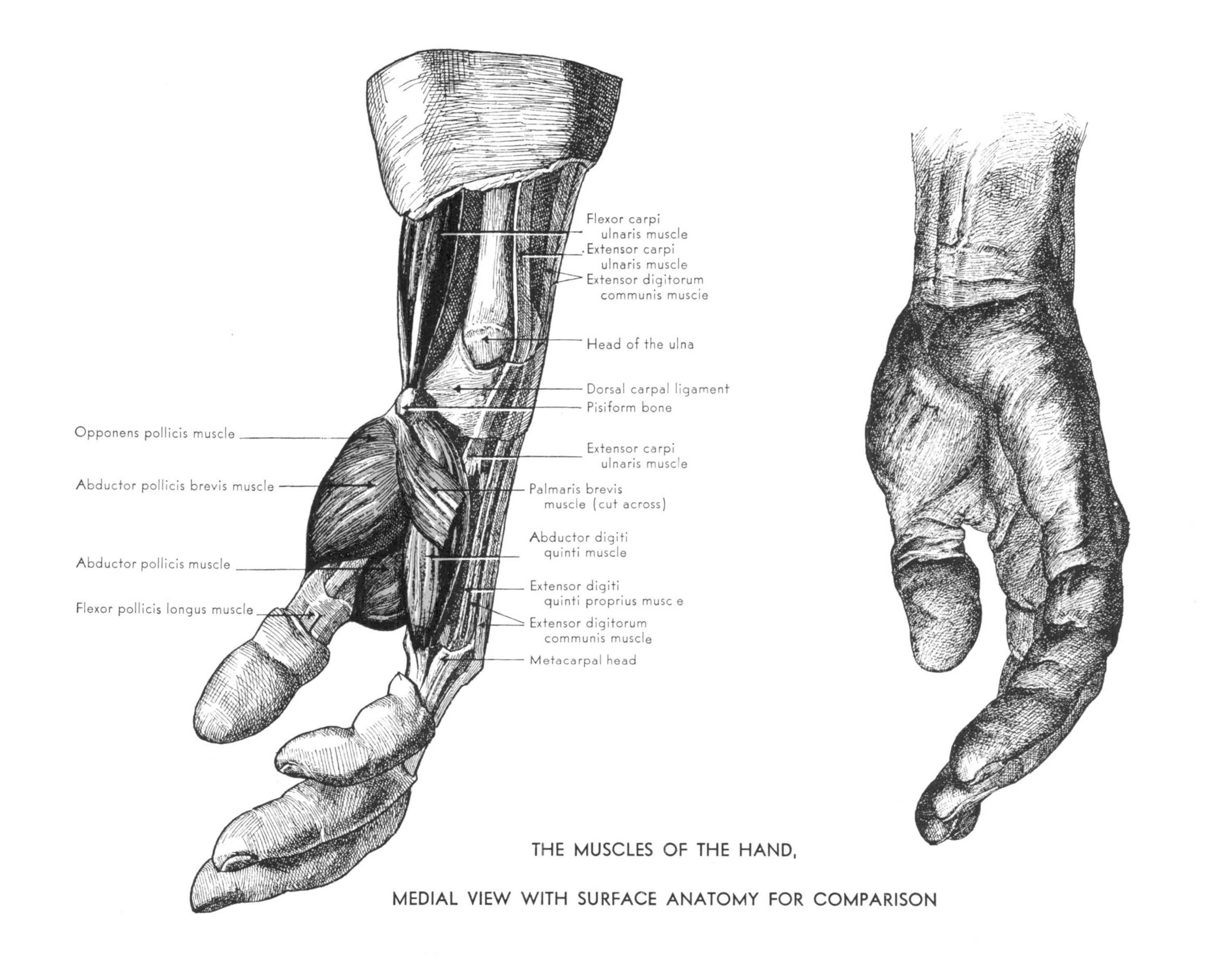

THE MUSCLES OF THE HAND,

MEDIAL VIEW WITH SURFACE ANATOMY FOR COMPARISON

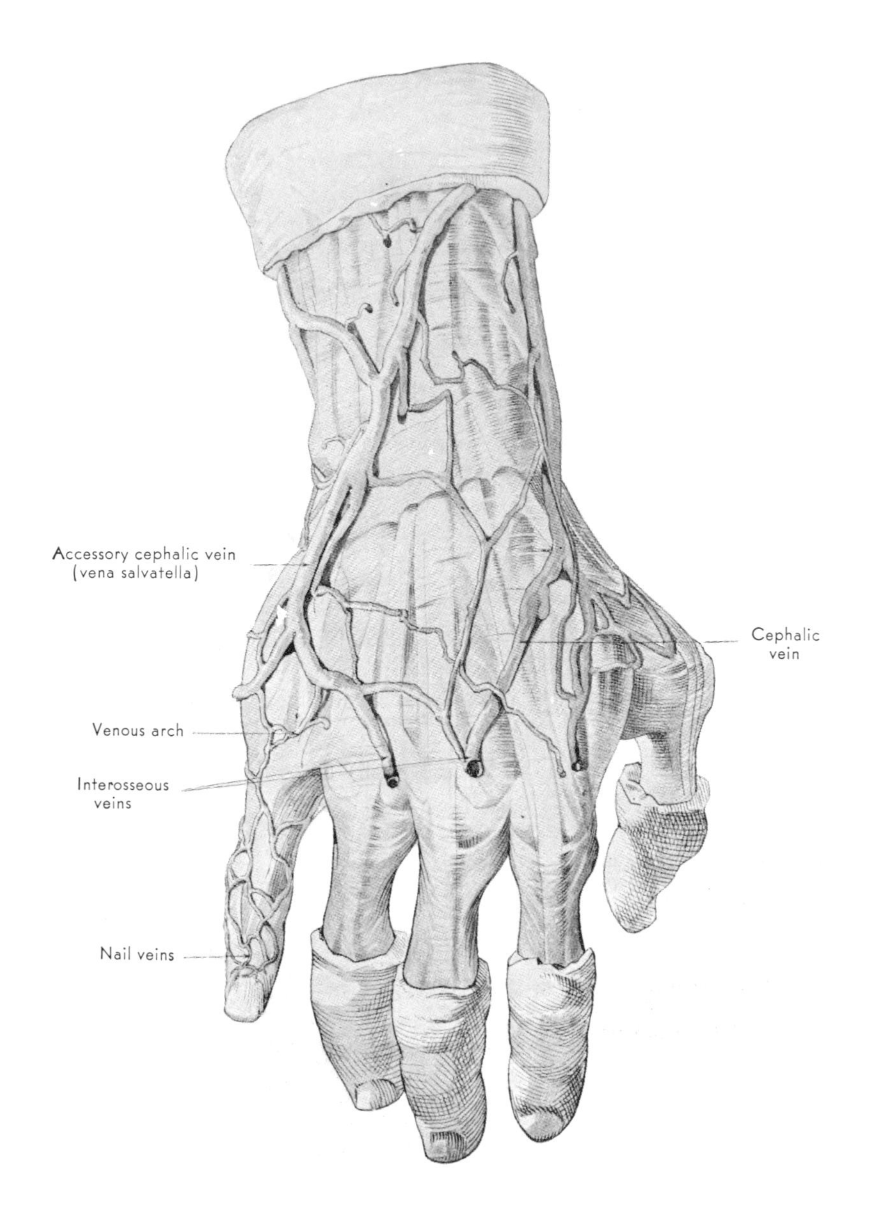

THE SUPERFICIAL VEINS OF THE HAND

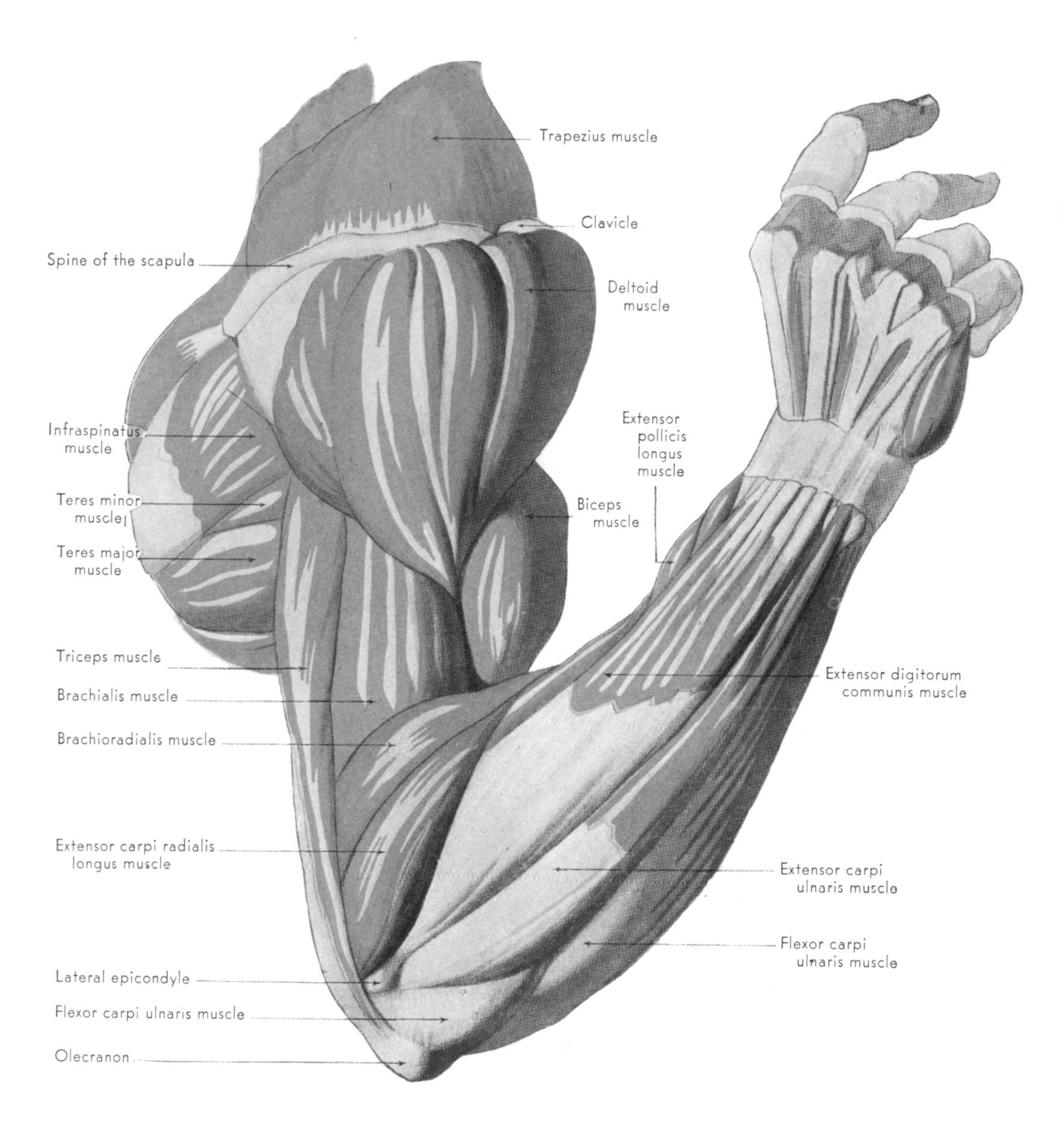

THE MUSCLES OF THE FLEXED UPPER EXTREMITY,

LATERAL VIEW WITH FOREARM PARTIALLY PRONATED

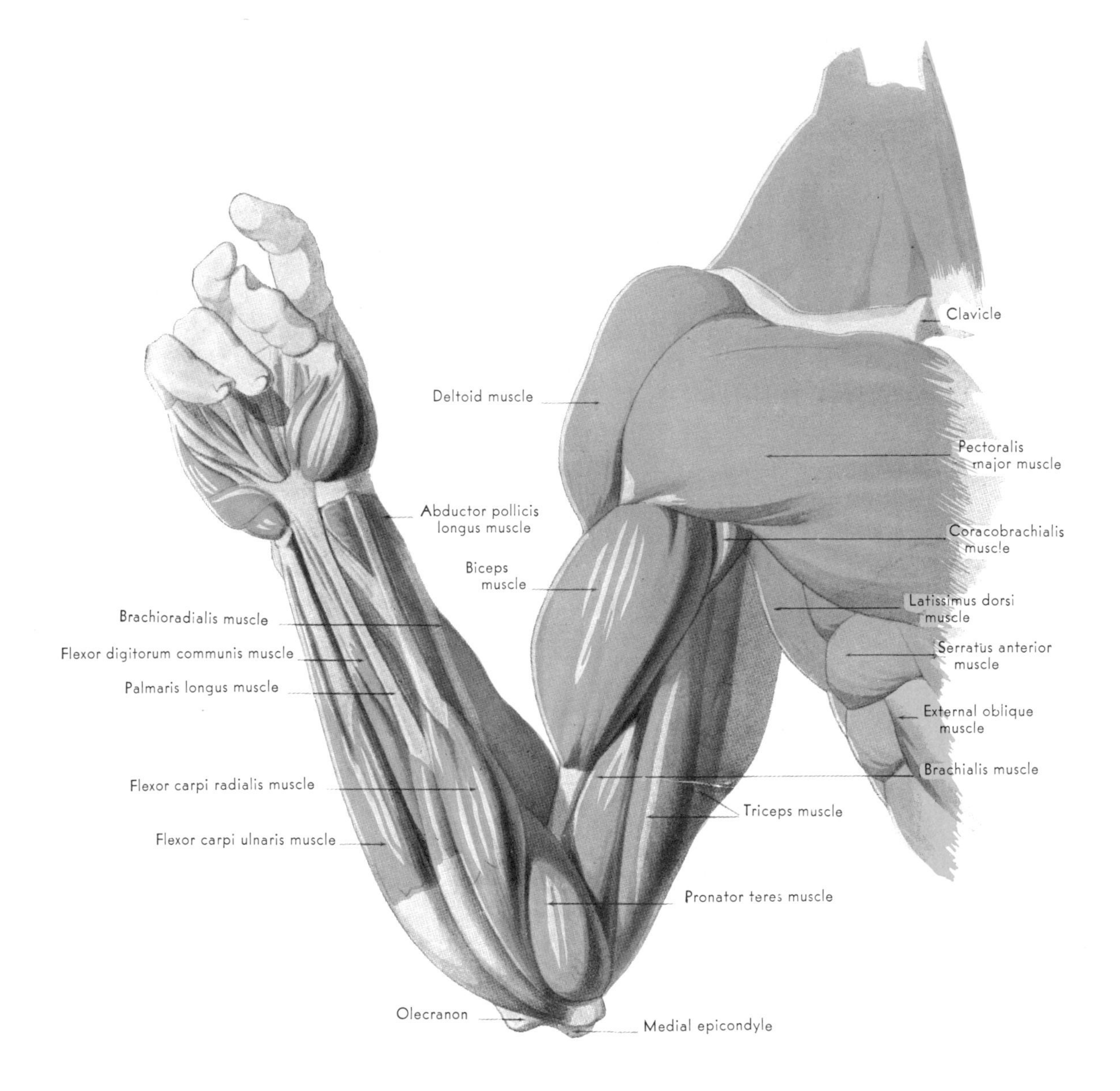

THE MUSCLES OF THE FLEXED UPPER EXTREMITY,
MEDIAL VIEW, FOREARM PARTIALLY PRONATED

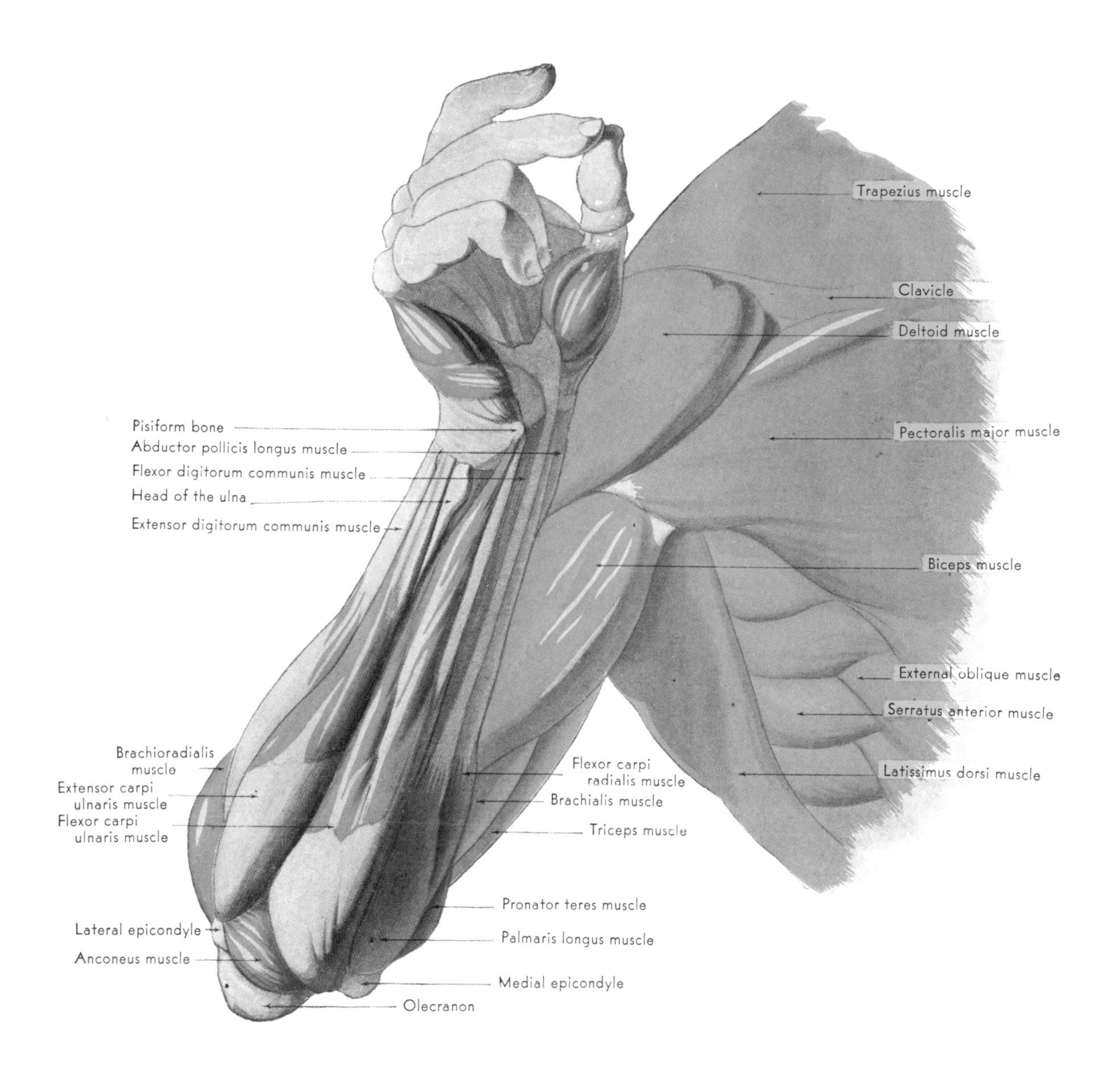

THE MUSCLES OF THE FLEXED UPPER EXTREMITY, ANTERIOR VIEW

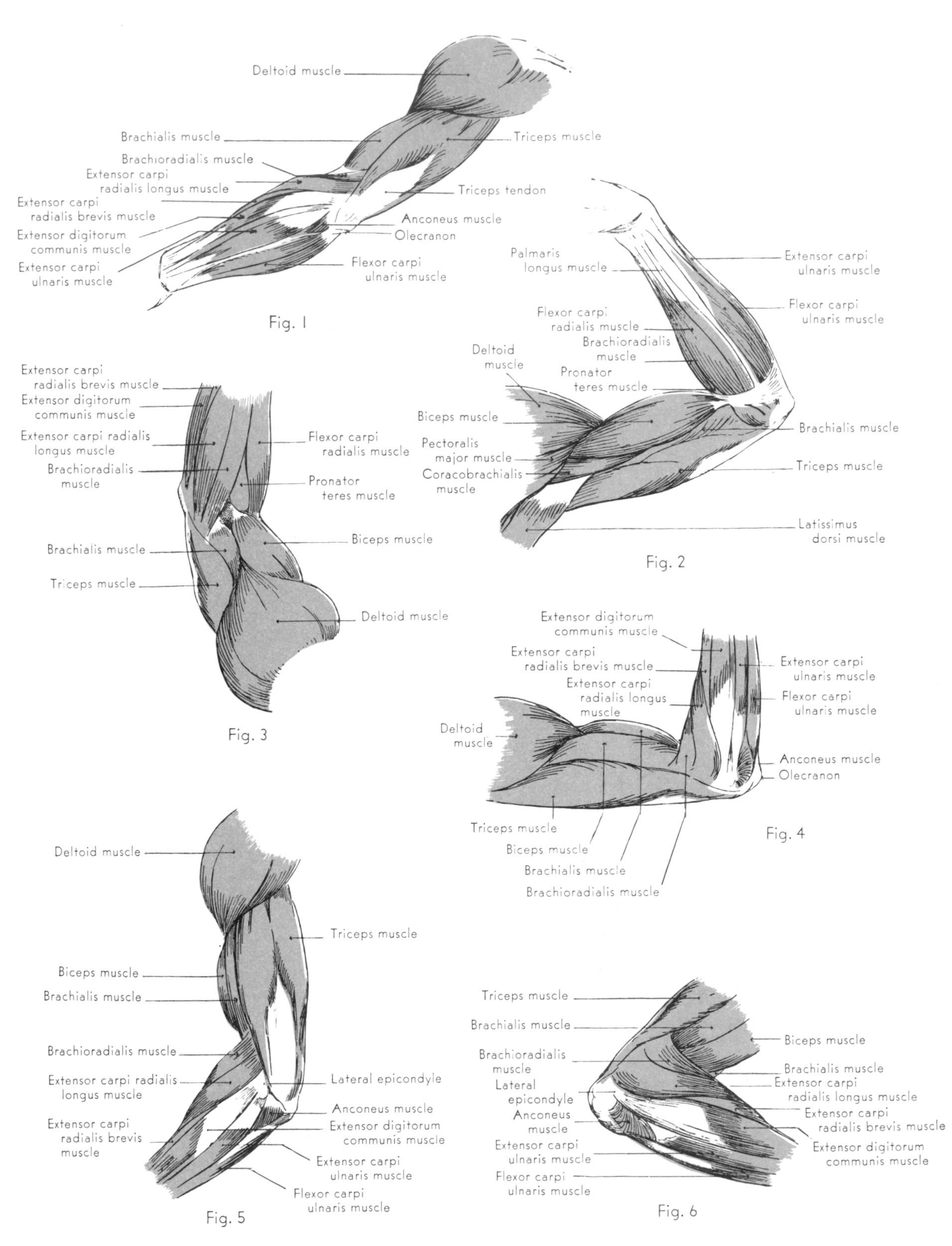

POSITIONS OF THE ARM I

Fig. 1

Fig. 2

Fig. 3

Fig. 4

Fig. 5

Fig. 6

MOTIONS OF THE SHOULDER JOINT I

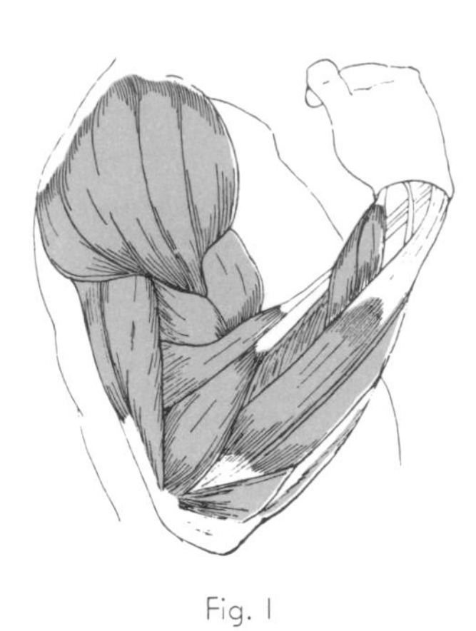

Fig. 1

Fig. 2

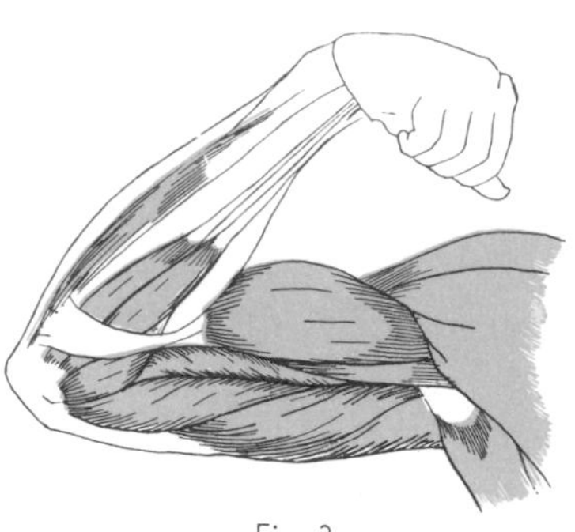

Fig. 3

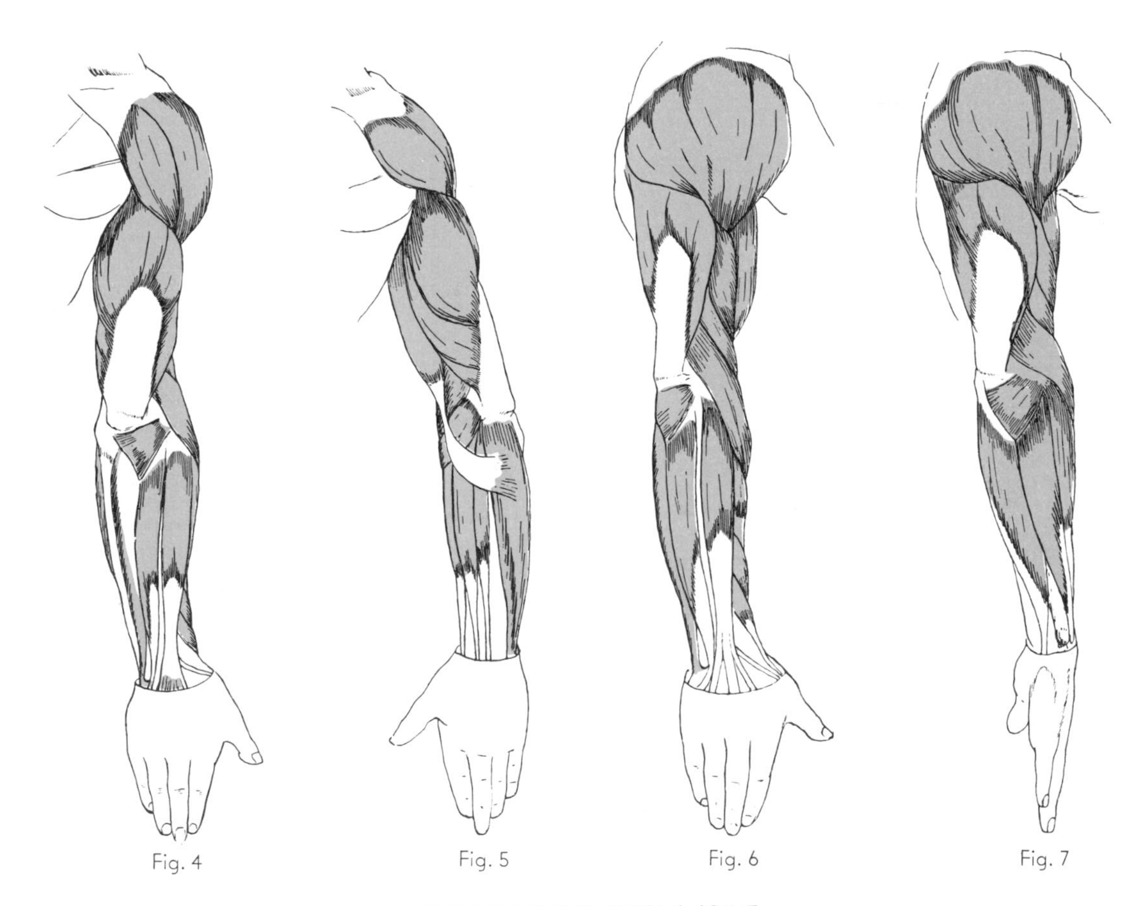

Fig. 4 Fig. 5 Fig. 6 Fig. 7

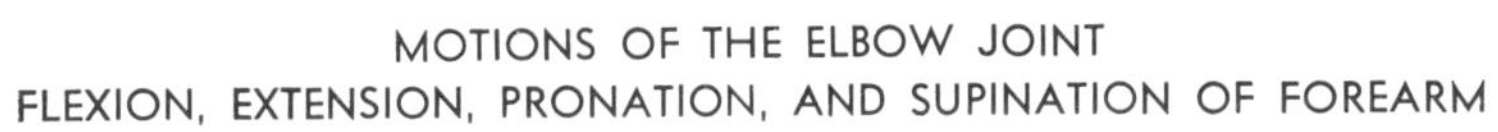

MOTIONS OF THE ELBOW JOINT

FLEXION, EXTENSION, PRONATION, AND SUPINATION OF FOREARM

PLATE 66

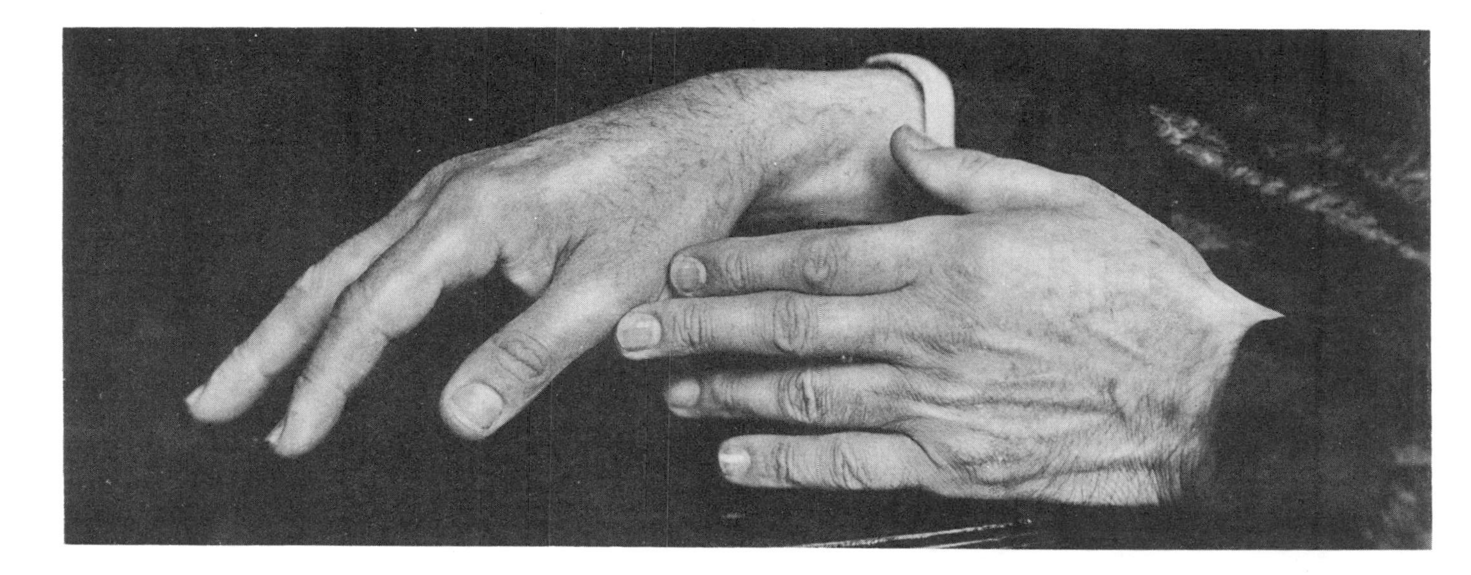

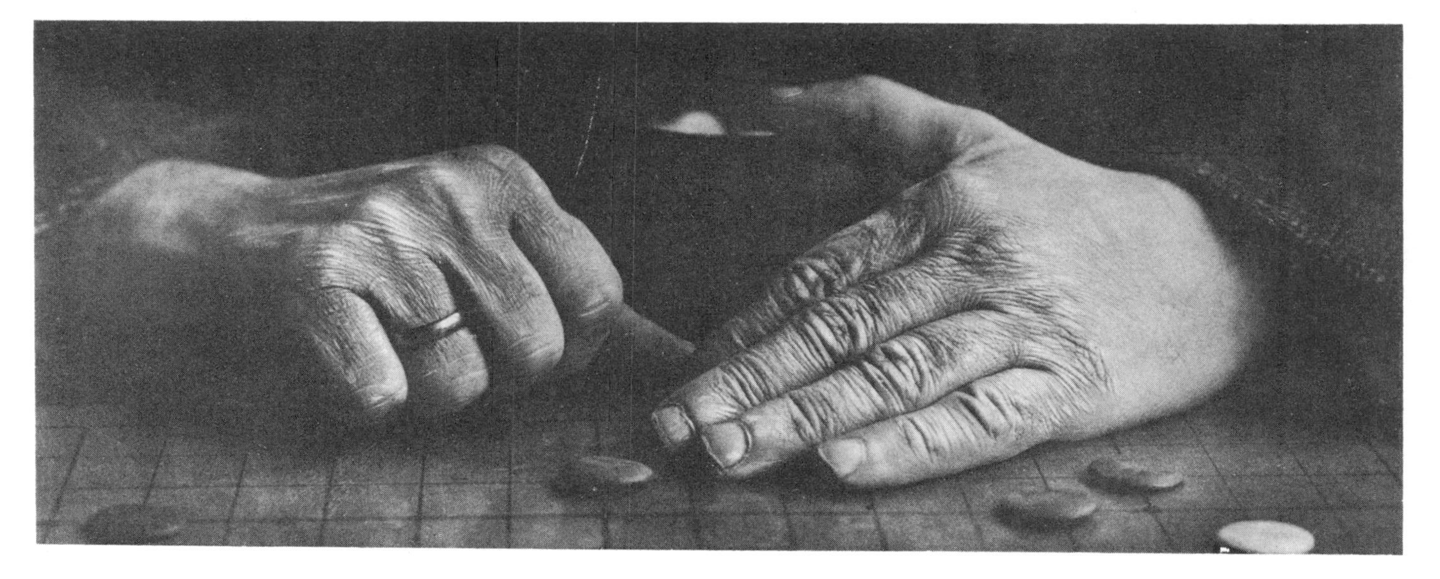

POSITIONS OF THE HAND I

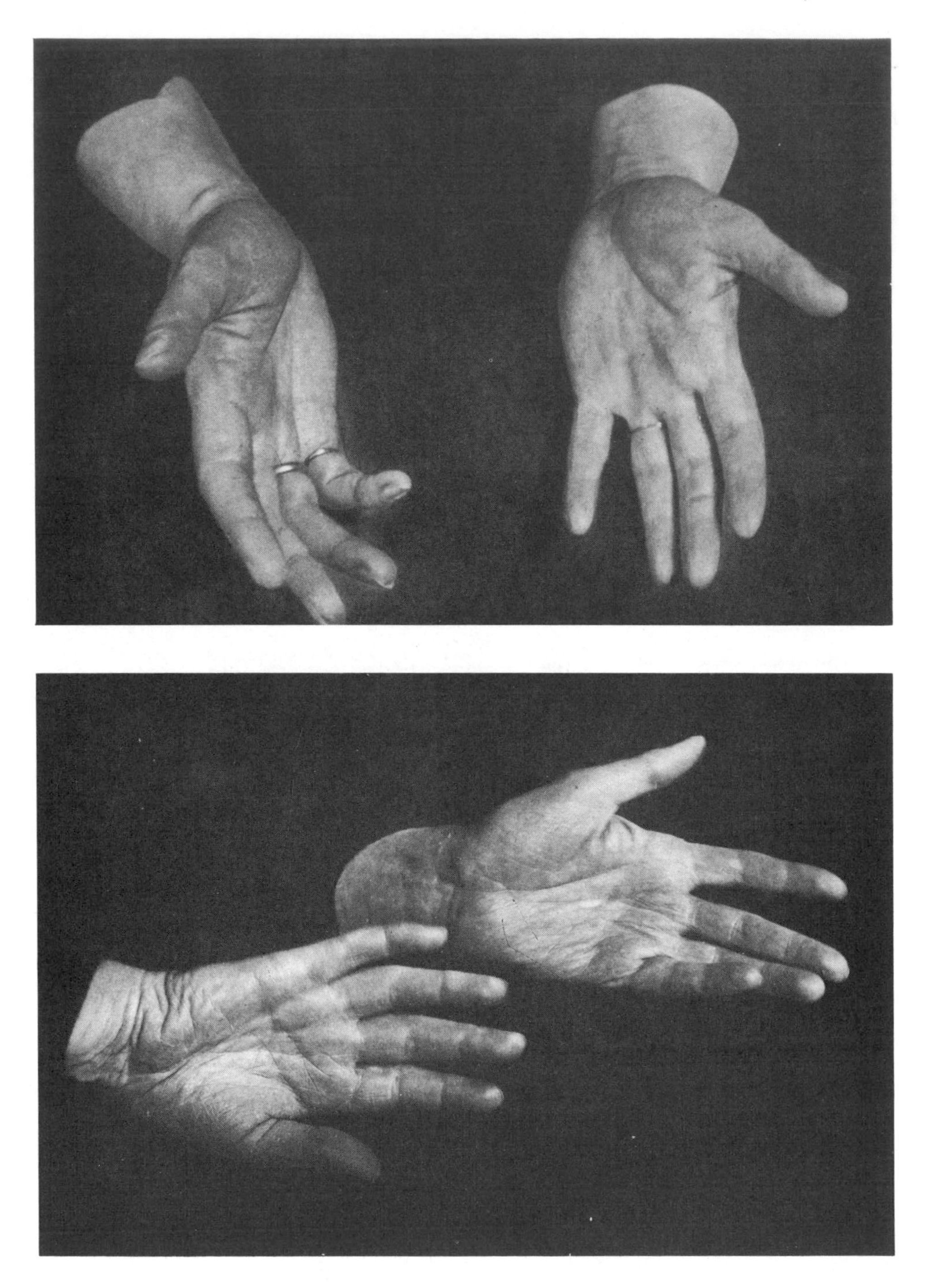

POSITIONS OF THE HAND II

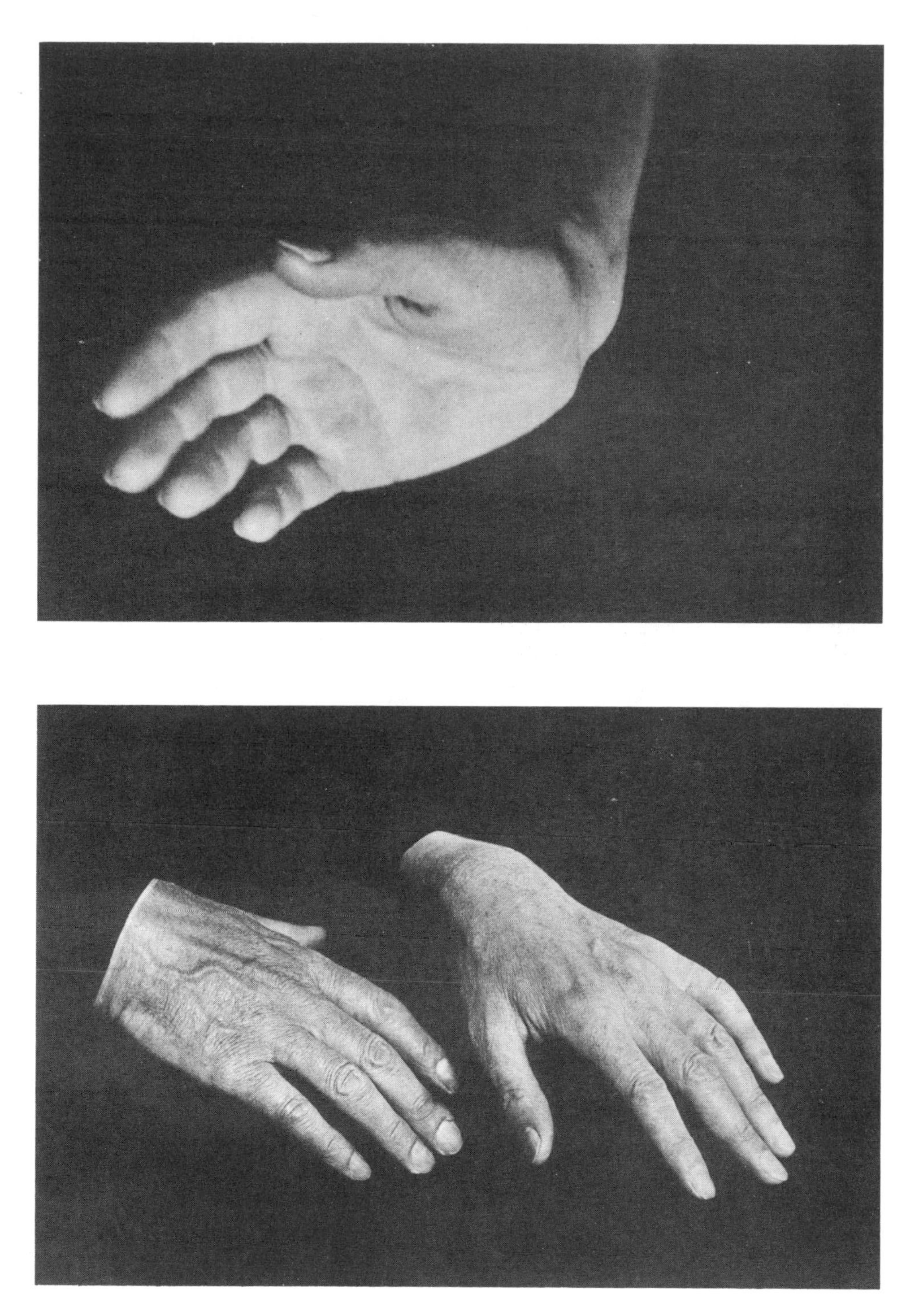

POSITIONS OF THE HAND III

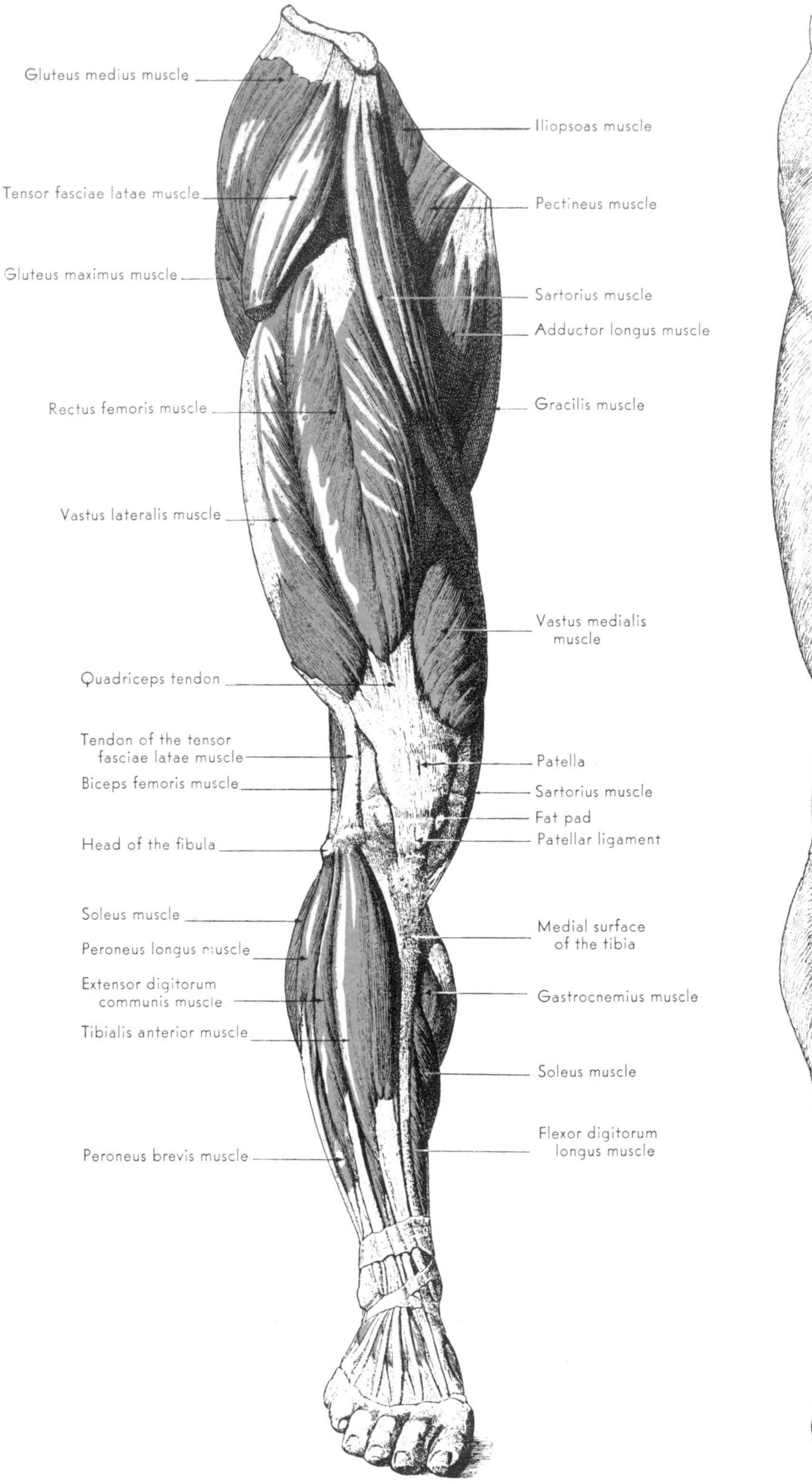

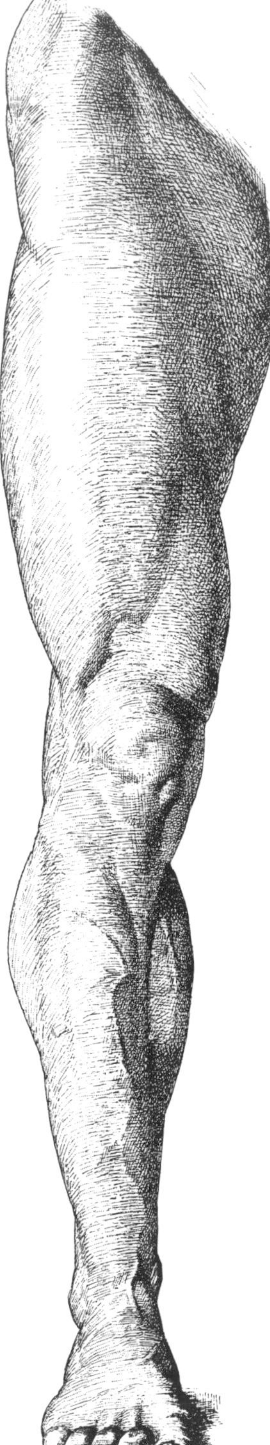

PLATE 70

THE MUSCLES OF THE LOWER EXTREMITY,
ANTERIOR VIEW WITH SURFACE ANATOMY FOR COMPARISON

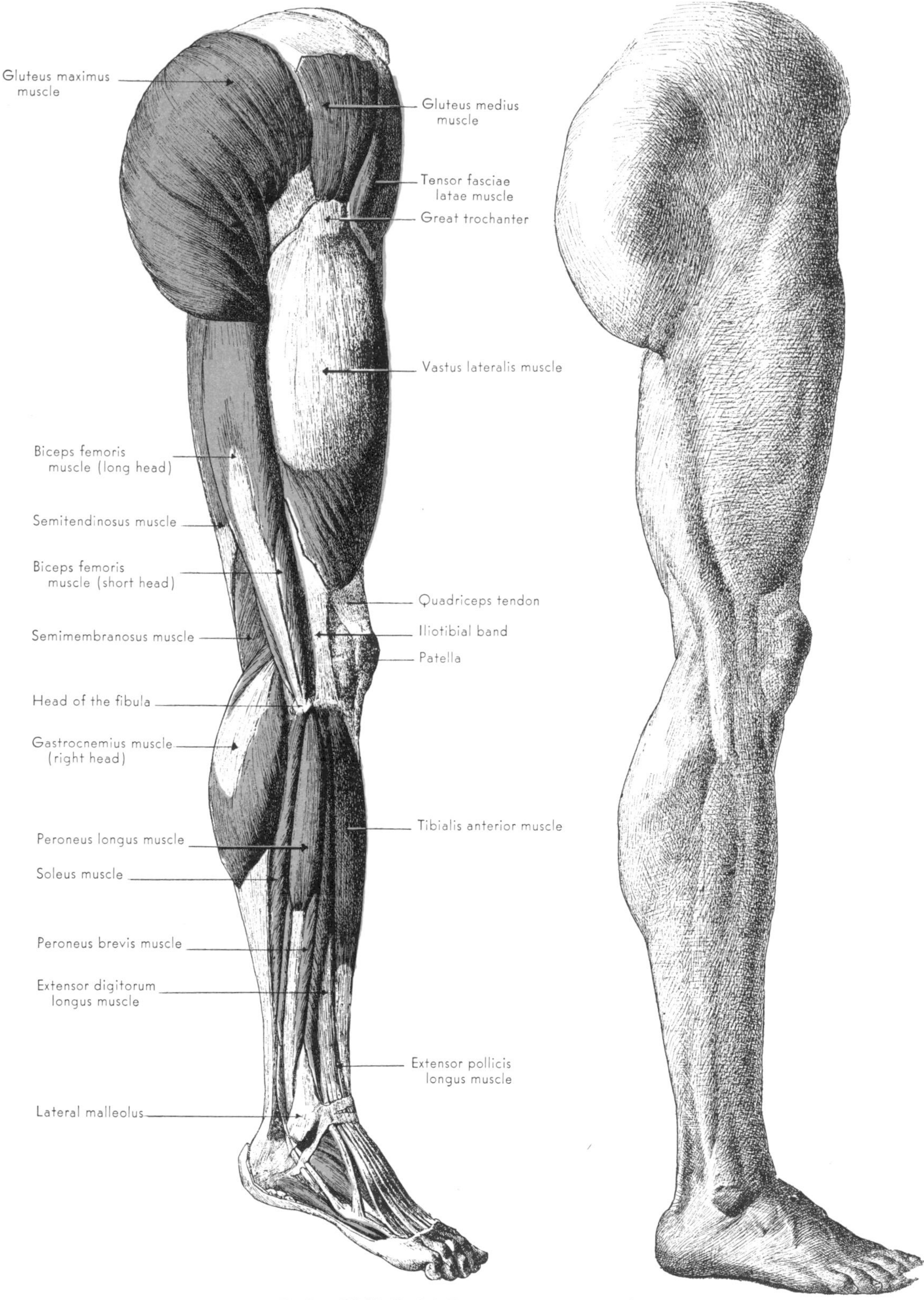

THE MUSCLES OF THE LOWER EXTREMITY,
LATERAL VIEW, WITH SURFACE ANATOMY

PLATE 71

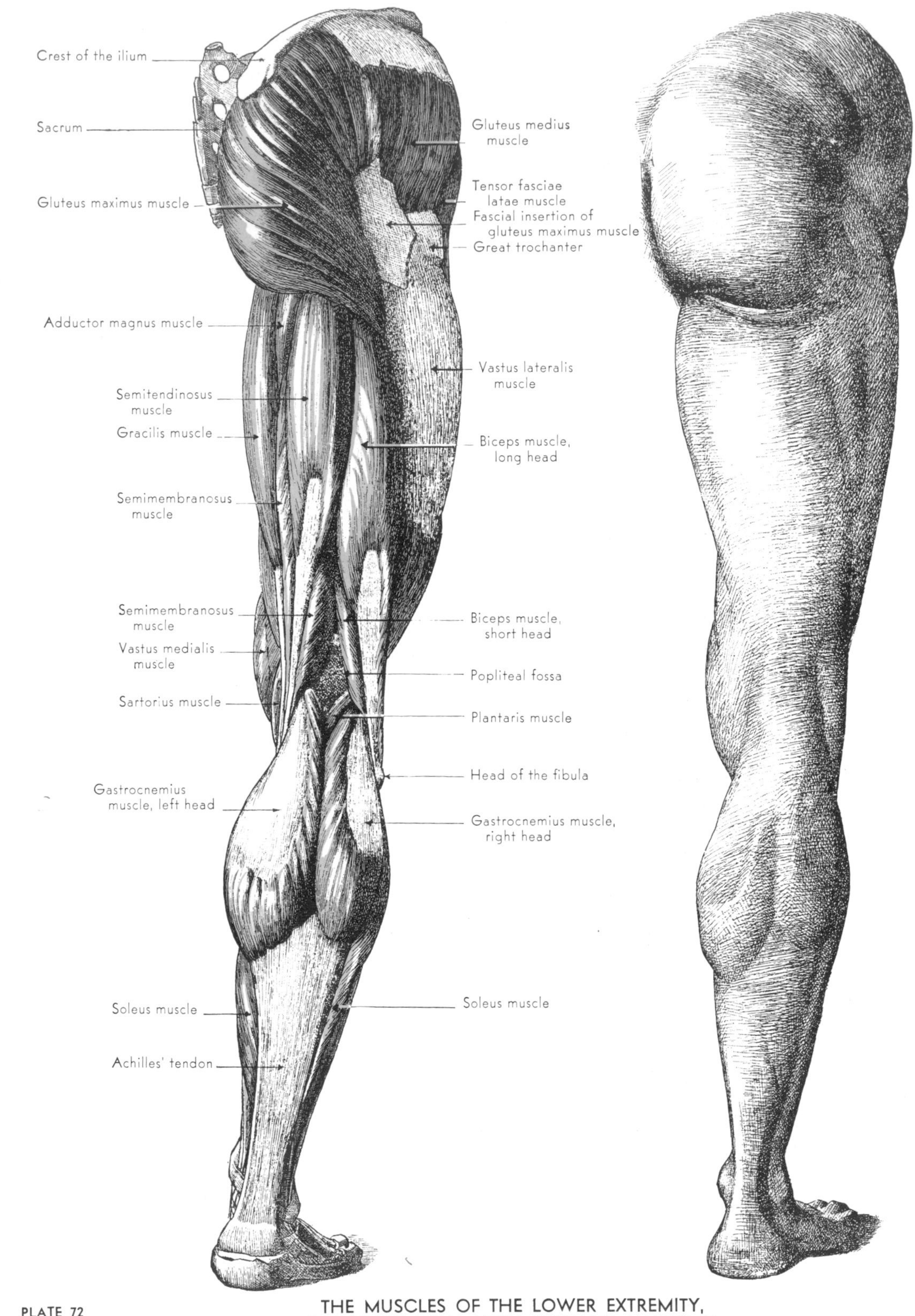

PLATE 72

THE MUSCLES OF THE LOWER EXTREMITY, POSTERIOR VIEW, WITH SURFACE ANATOMY

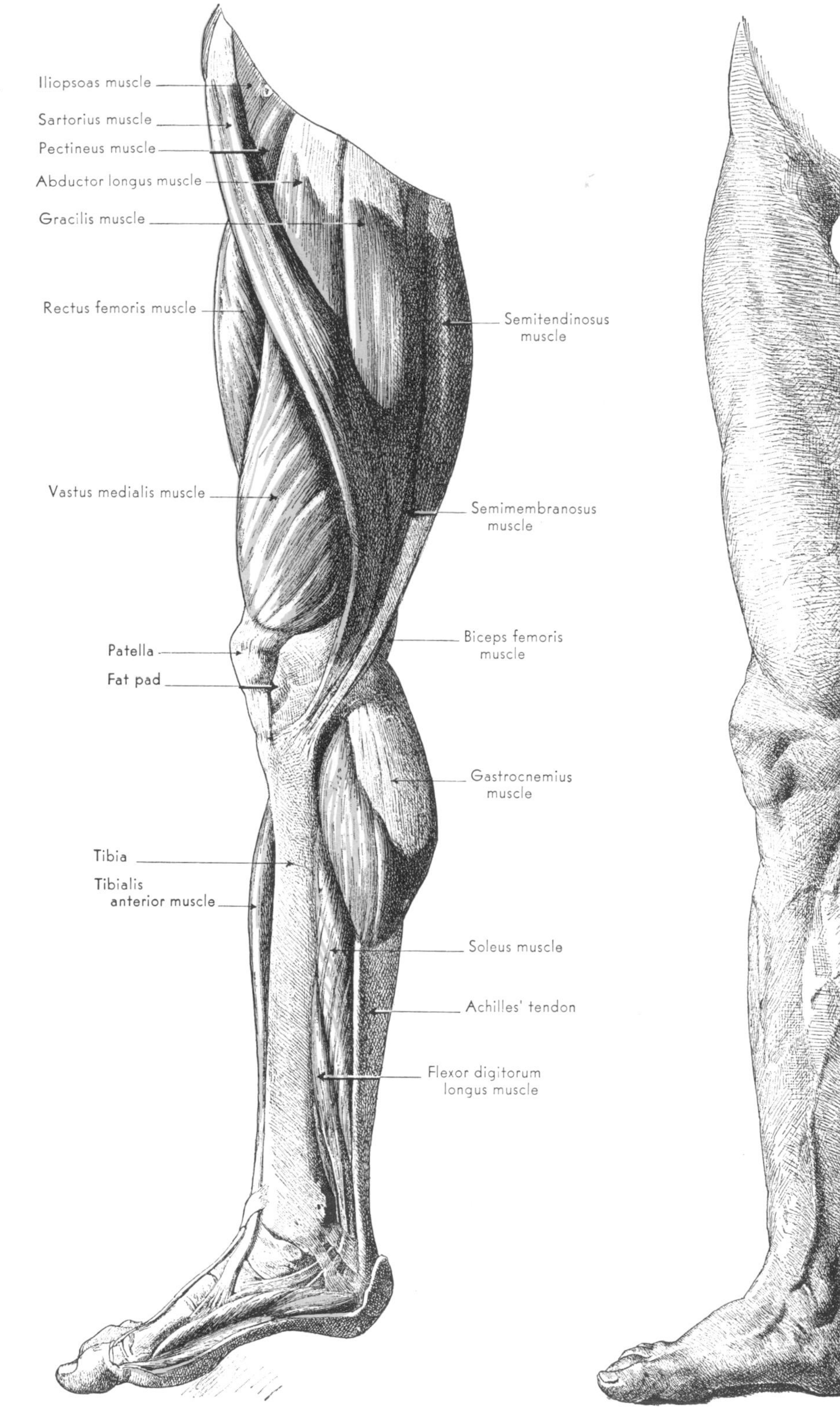

THE MUSCLES OF THE LOWER EXTREMITY,
MEDIAL VIEW, WITH SURFACE ANATOMY

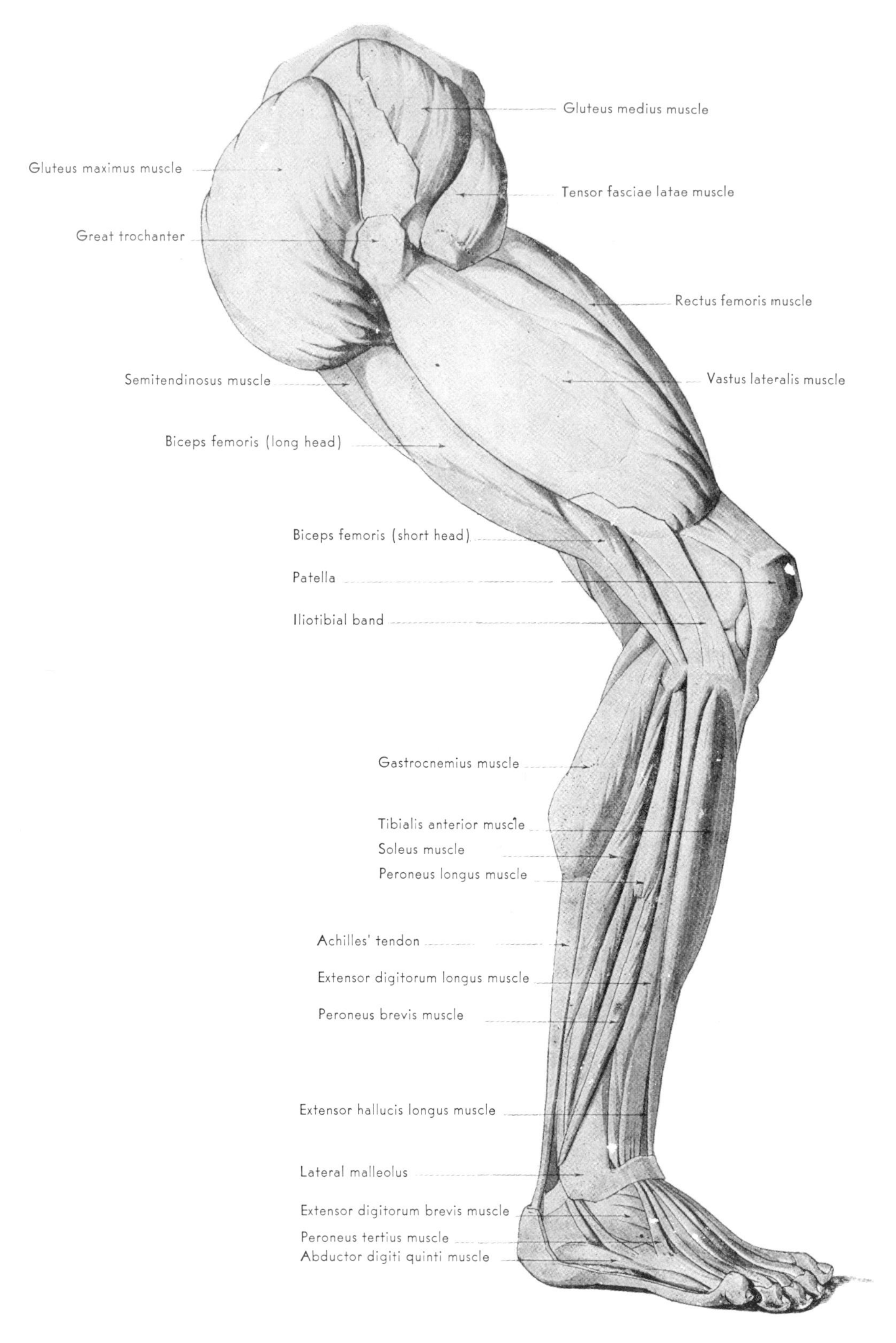

PLATE 74 THE MUSCLES OF THE FLEXED LOWER EXTREMITY, LATERAL VIEW

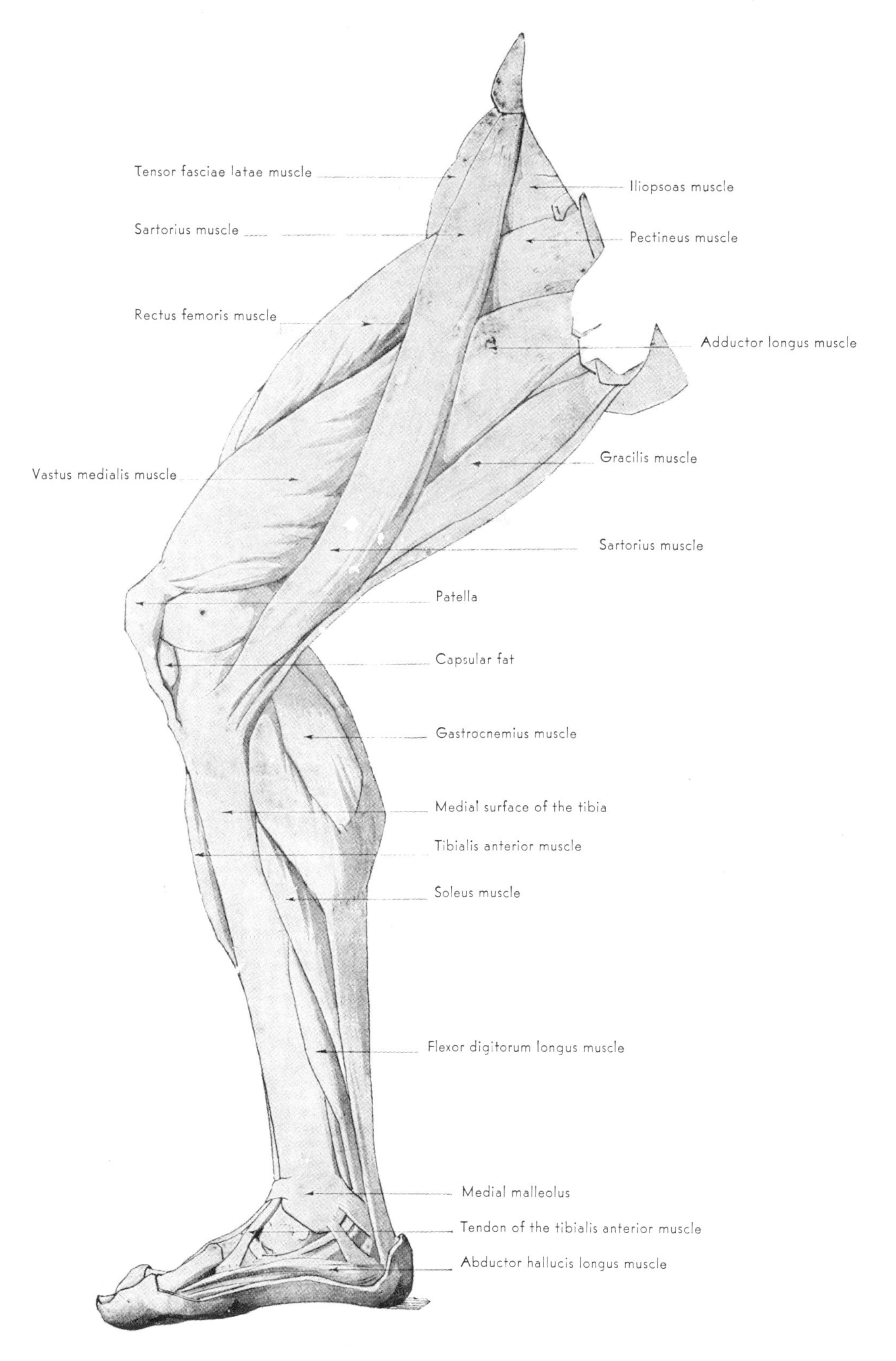

THE MUSCLES OF THE FLEXED LOWER EXTREMITY, MEDIAL VIEW

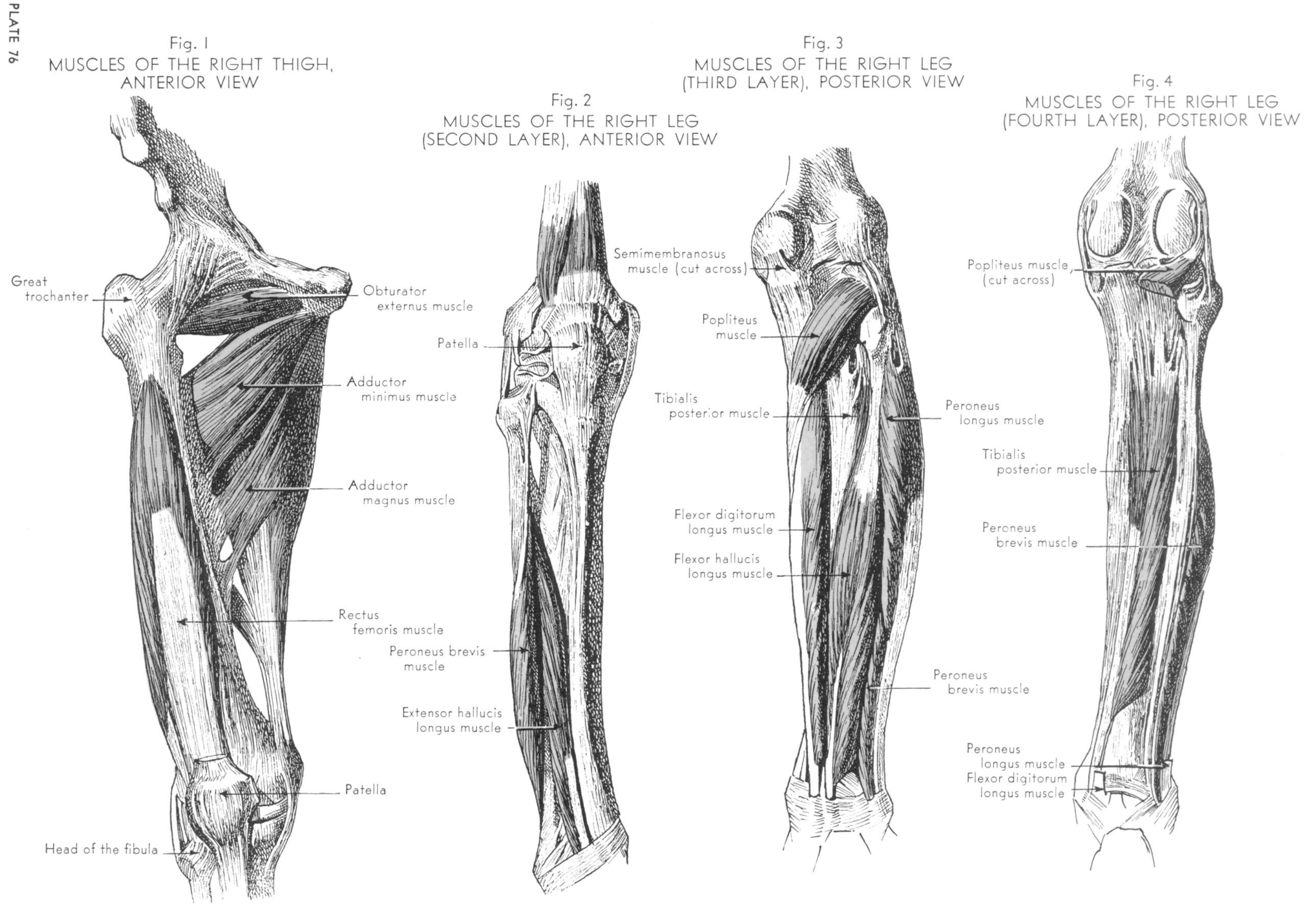

THE MUSCLE LAYERS OF THE THIGH AND LEG

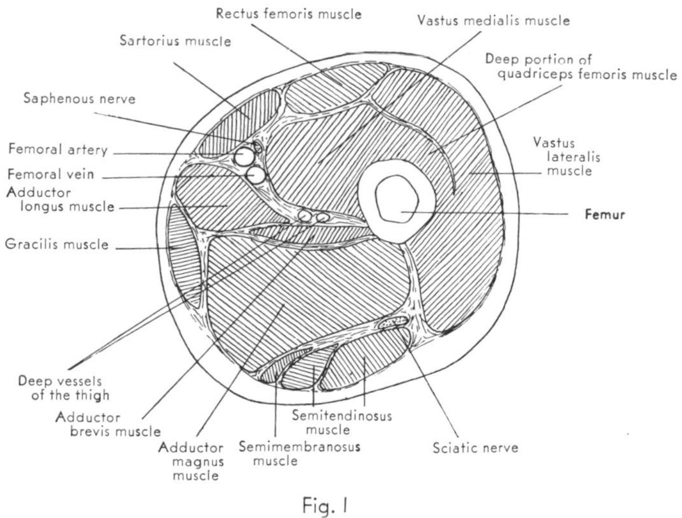

Fig. 1
CROSS-SECTION THROUGH THE THIGH

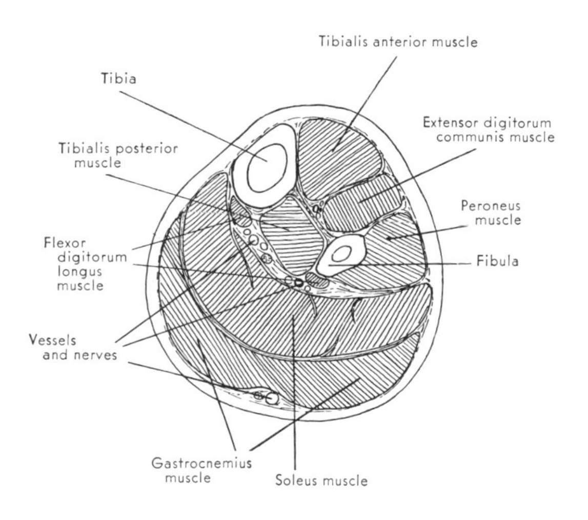

Fig. 2
CROSS-SECTION THROUGH THE LEG

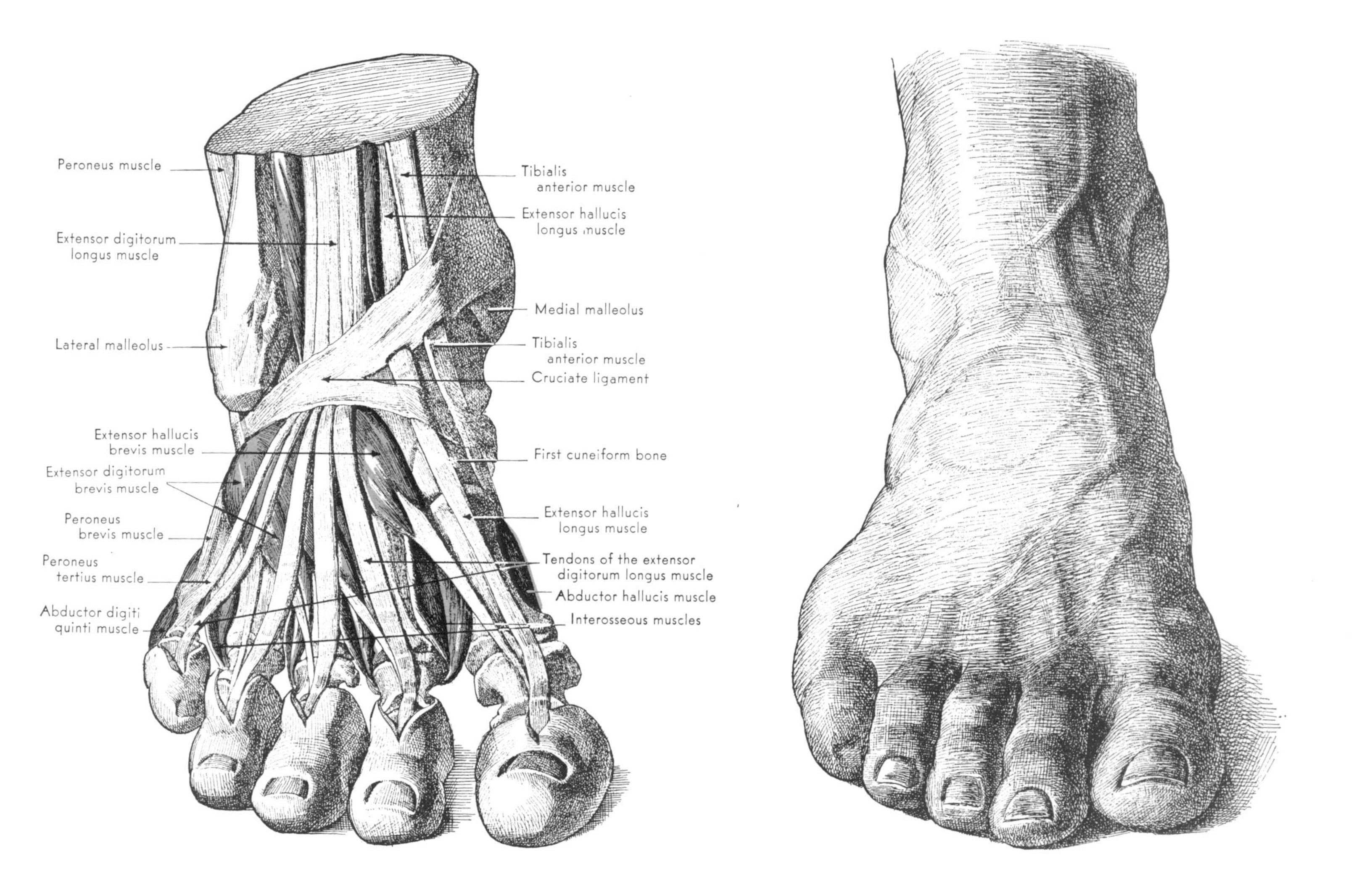

THE MUSCLES OF THE FOOT, ANTERIOR VIEW, WITH SURFACE ANATOMY

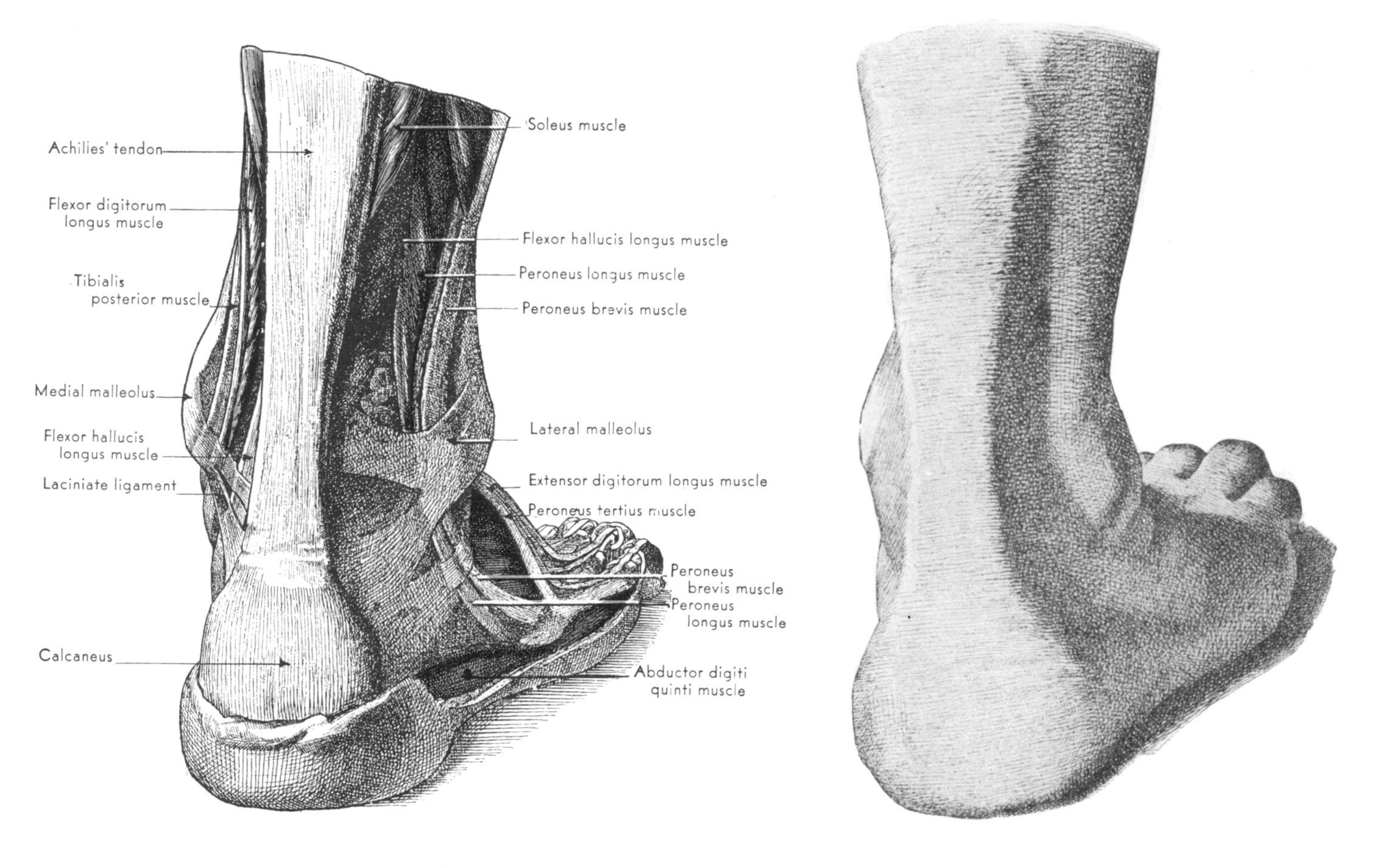

THE MUSCLES OF THE FOOT, POSTERIOR VIEW, WITH SURFACE ANATOMY

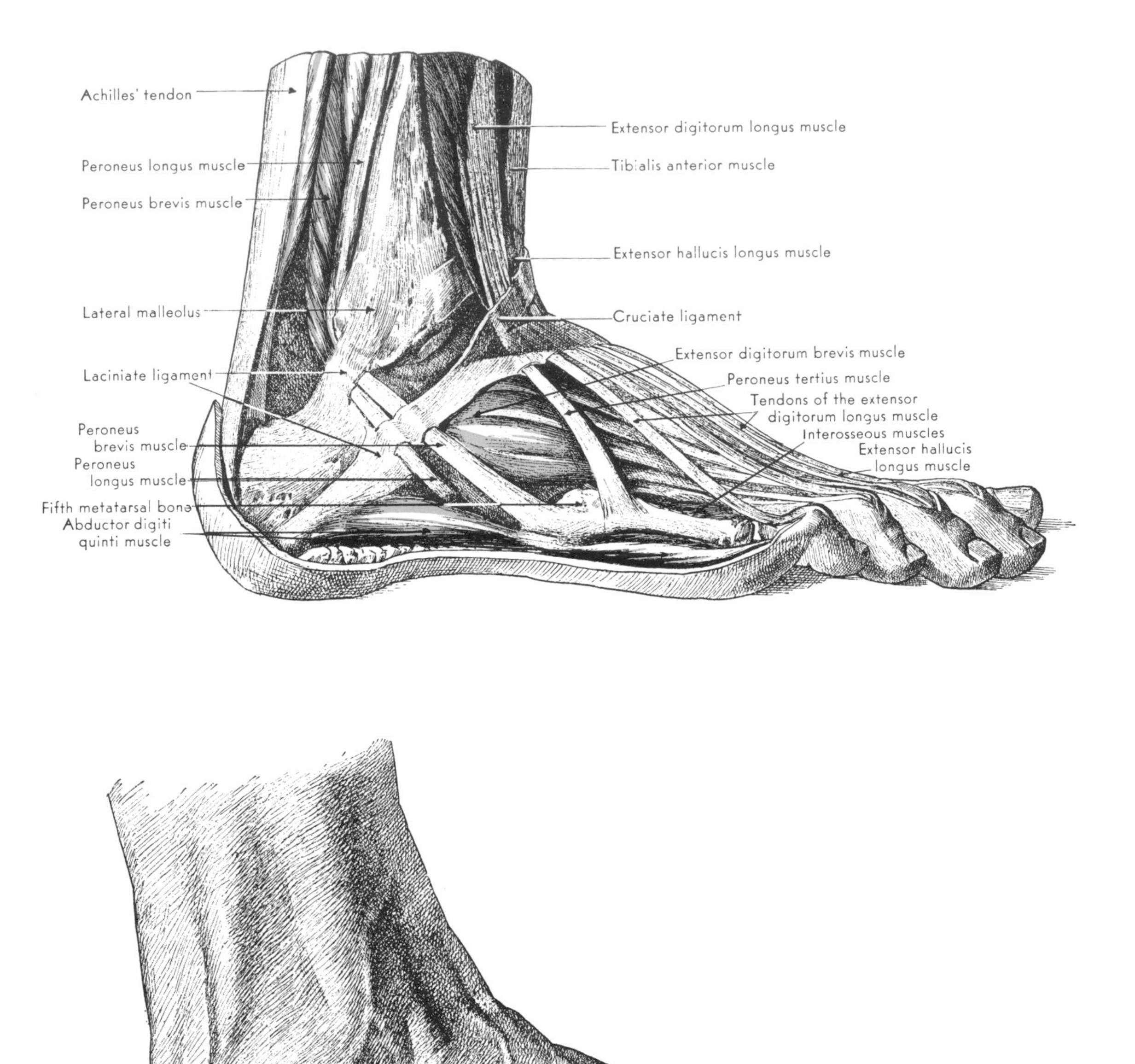

THE MUSCLES OF THE FOOT, LATERAL VIEW, WITH SURFACE ANATOMY

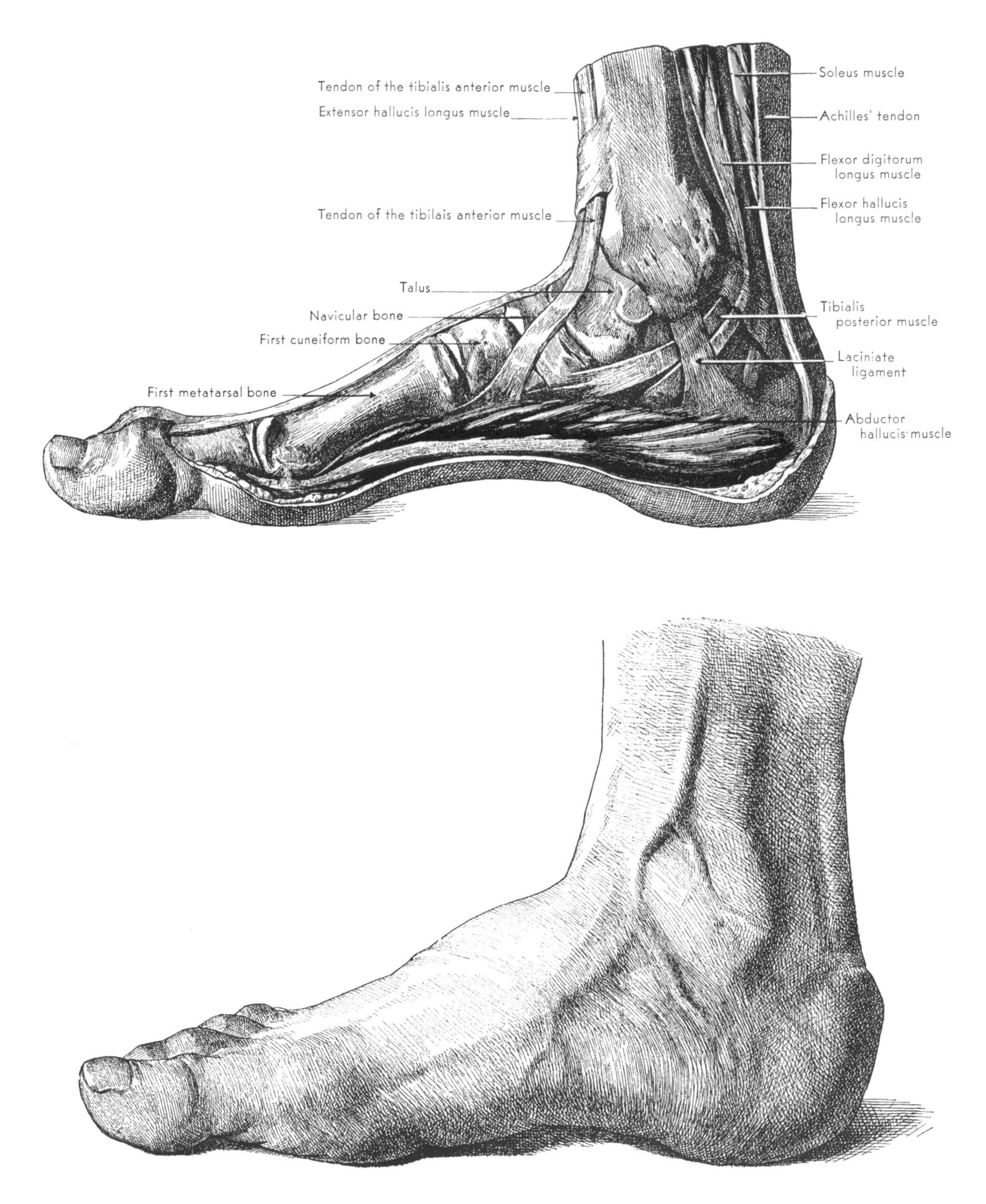

THE MUSCLES OF THE FOOT, MEDIAL VIEW, WITH SURFACE ANATOMY

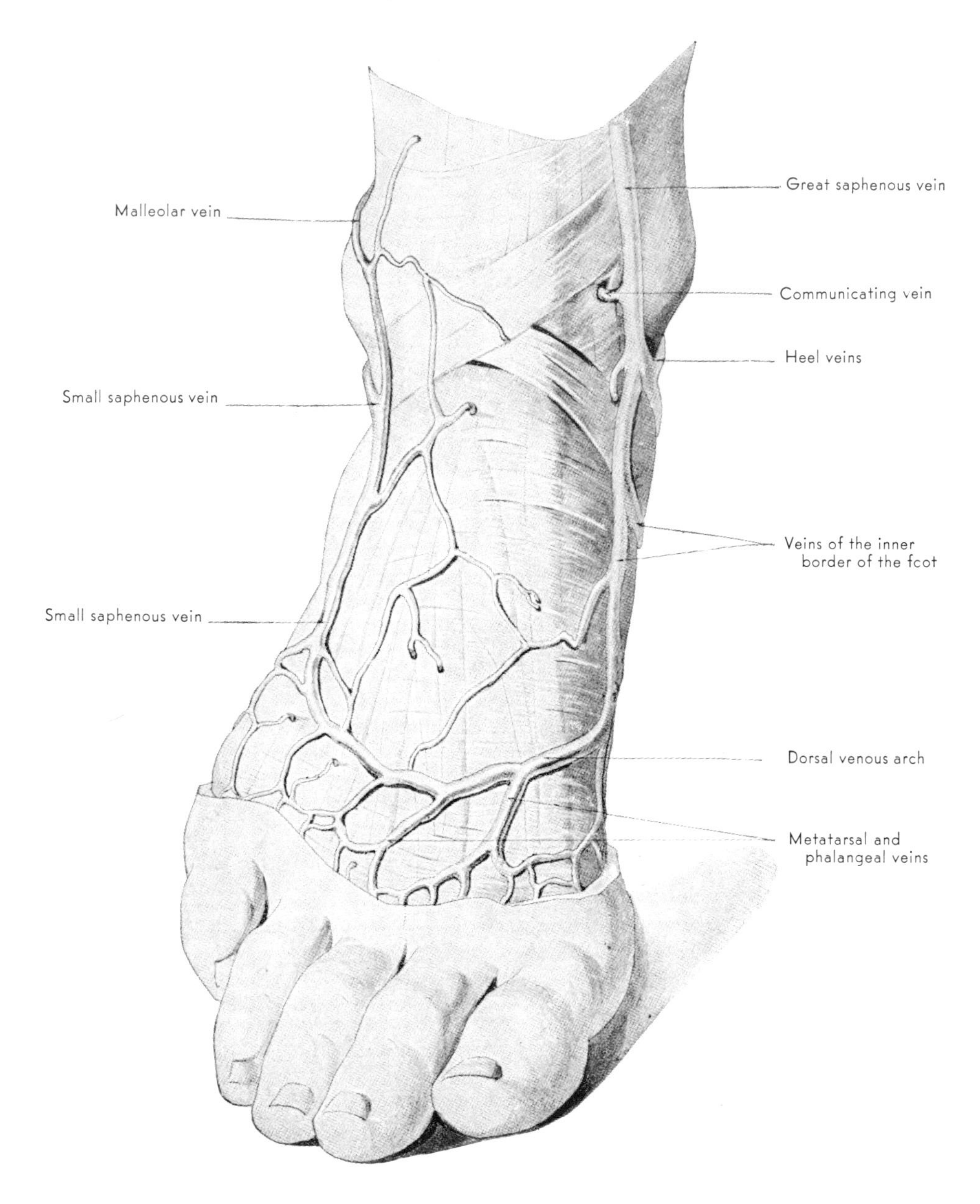

SUPERFICIAL VEINS OF THE FOOT

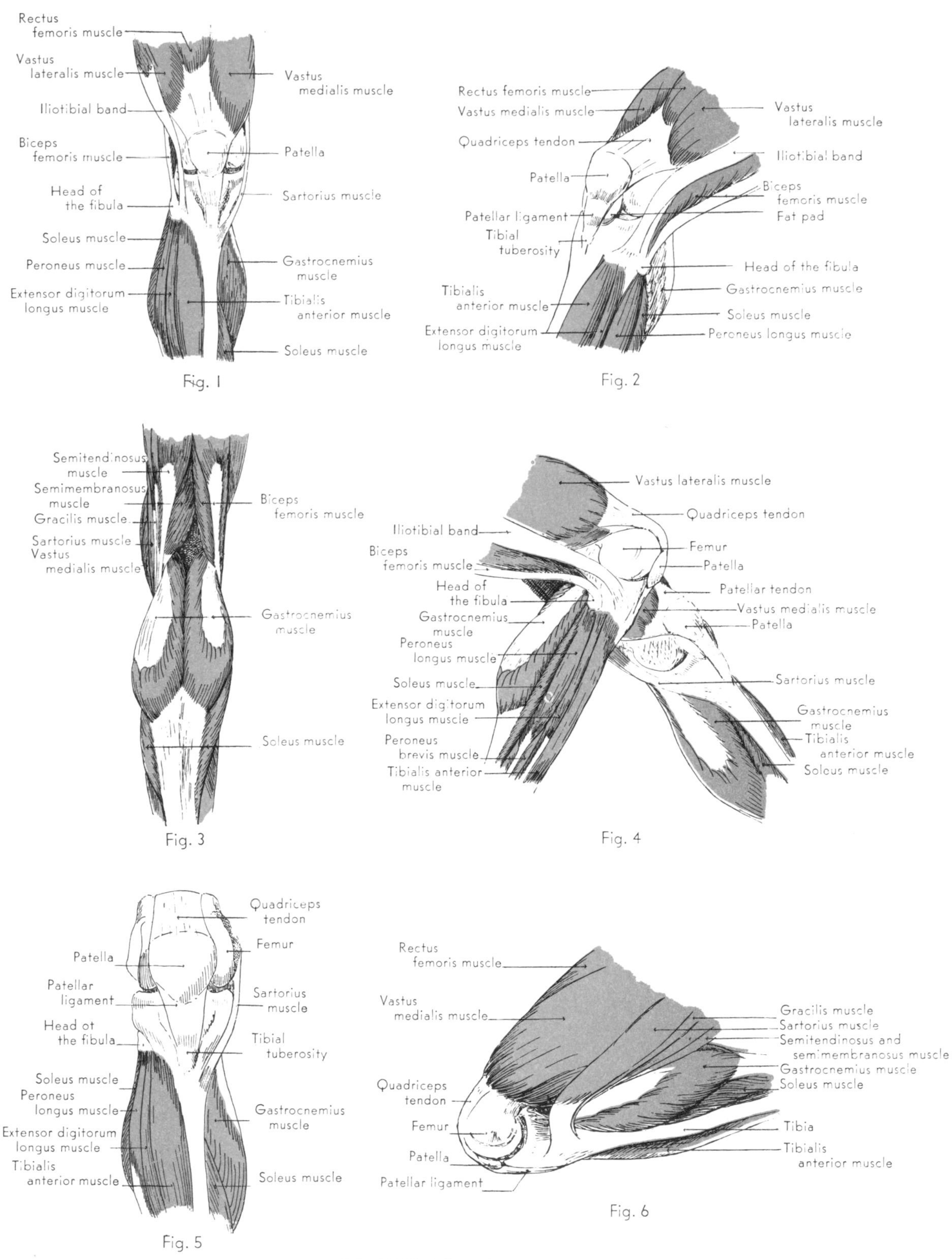

POSITIONS OF THE KNEE JOINT

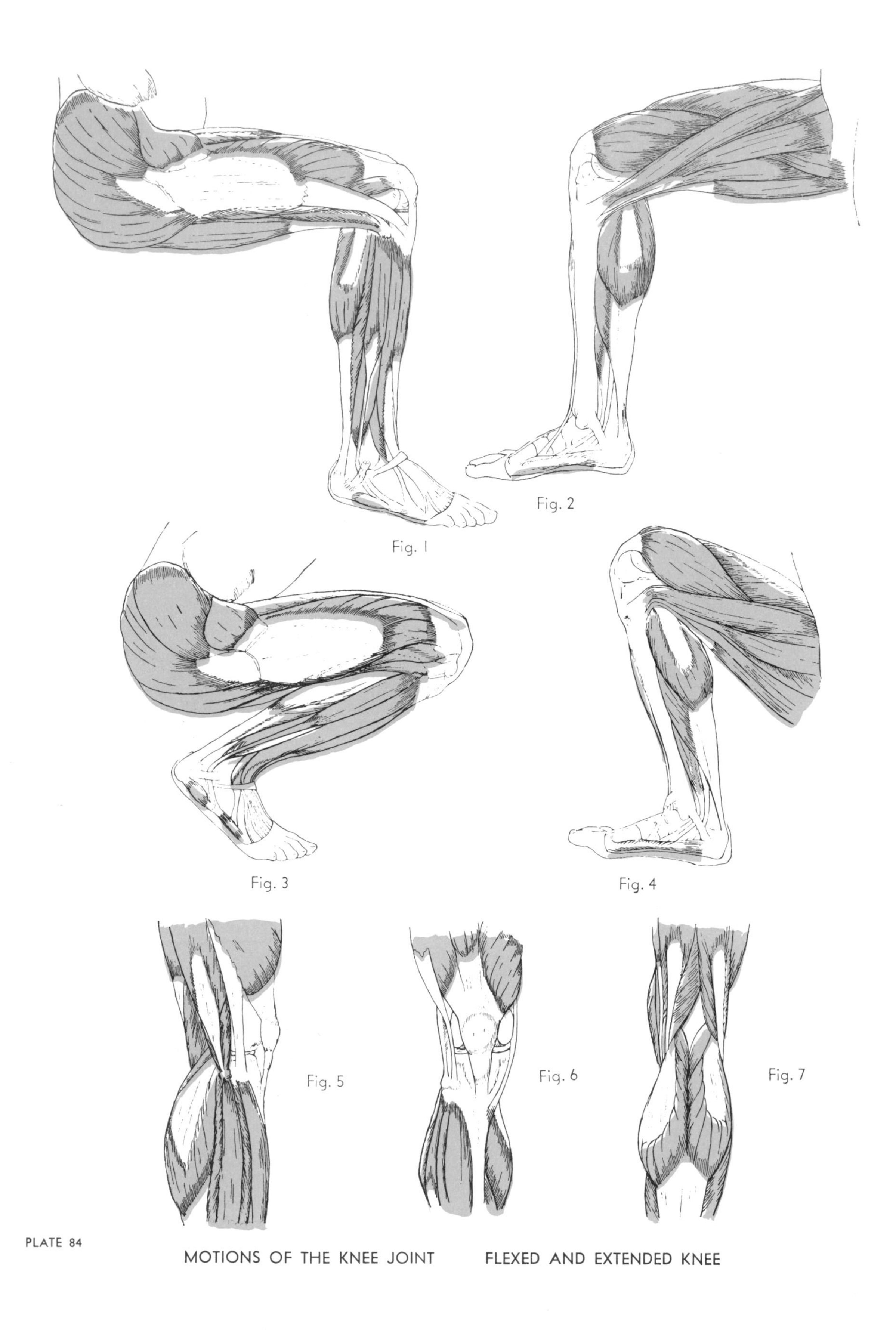

PLATE 84

MOTIONS OF THE KNEE JOINT FLEXED AND EXTENDED KNEE

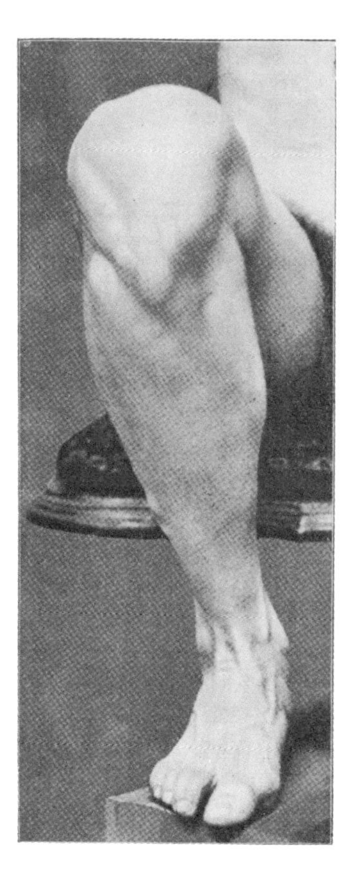

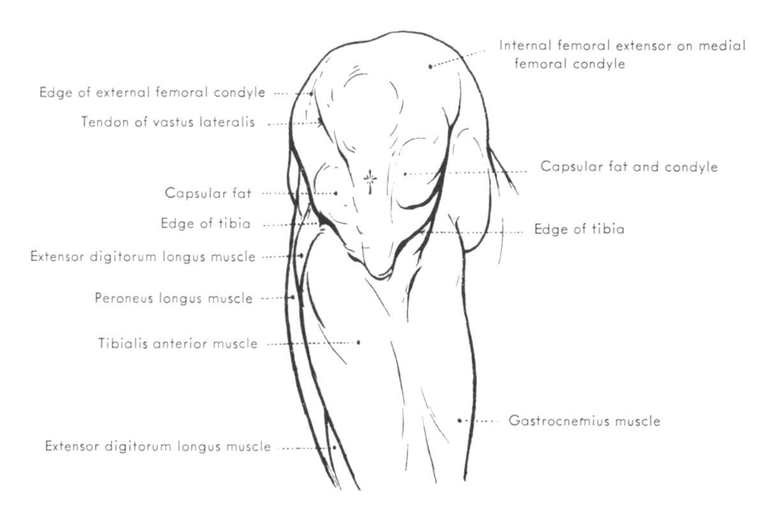

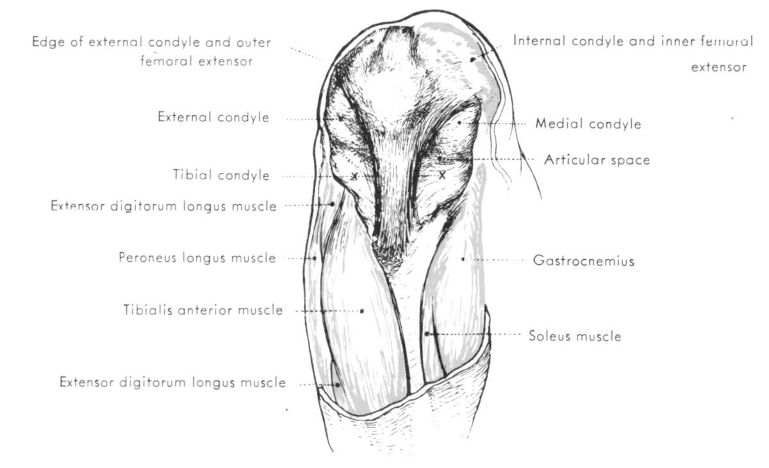

KOLLMANN, ANATOMY OF THE KNEE, EXTREME ANGLE OF BEND

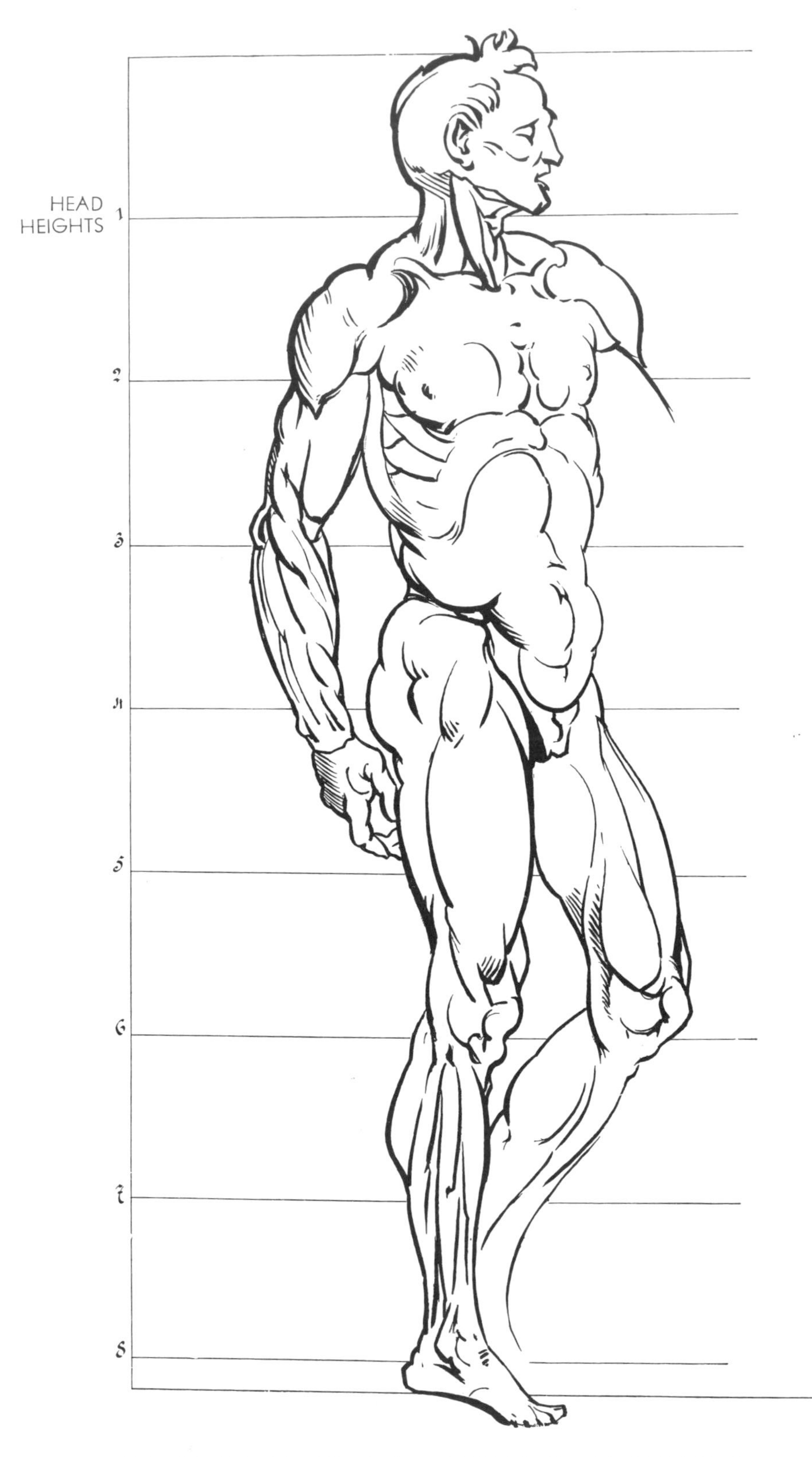

PLATE 86 RELATIVE PROPORTIONS, MALE ADULT (AFTER MICHELANGELO)

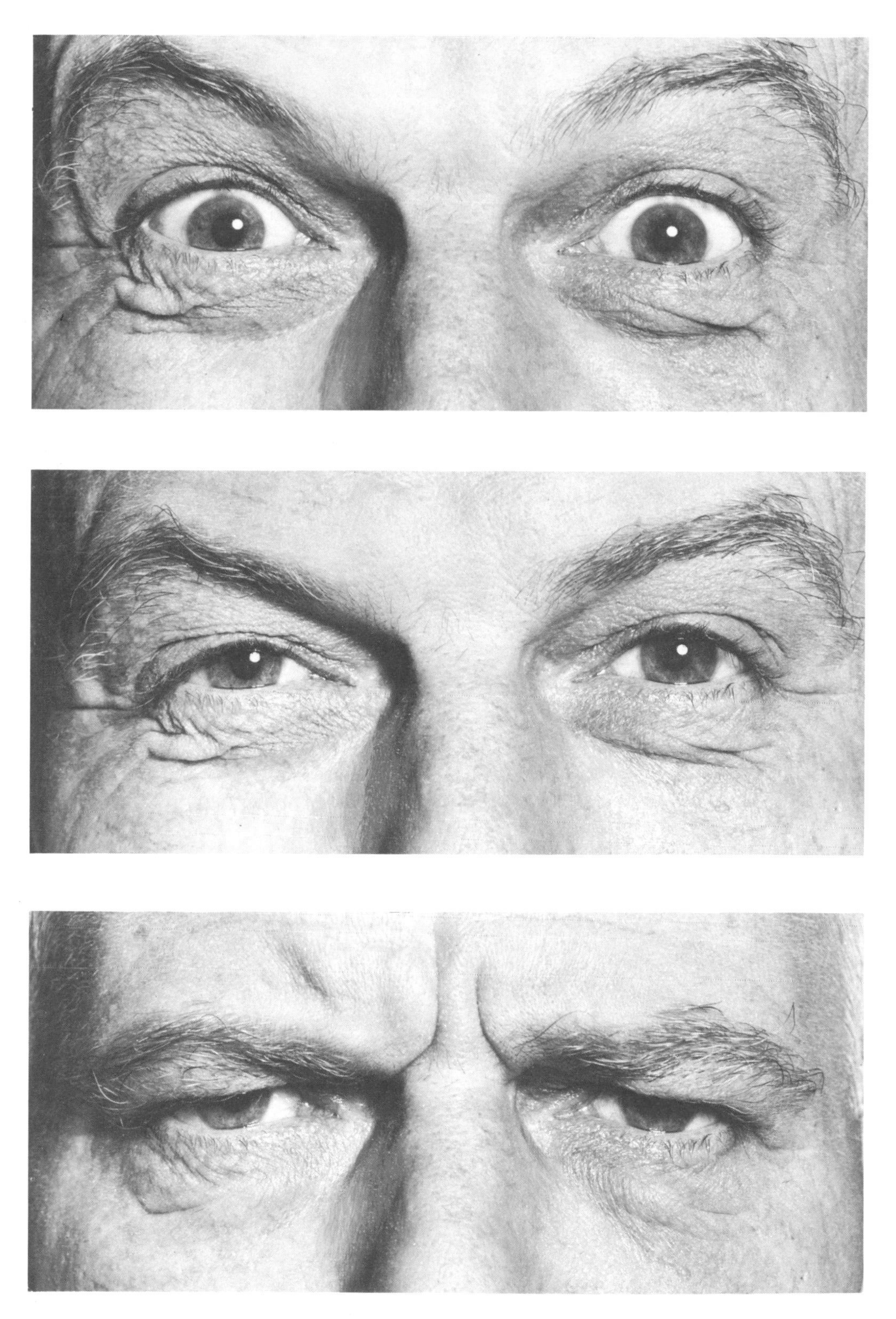

Reproduced through the courtesy of Air Express

PHOTOGRAPHS SHOWING FOLDS AROUND THE EYES

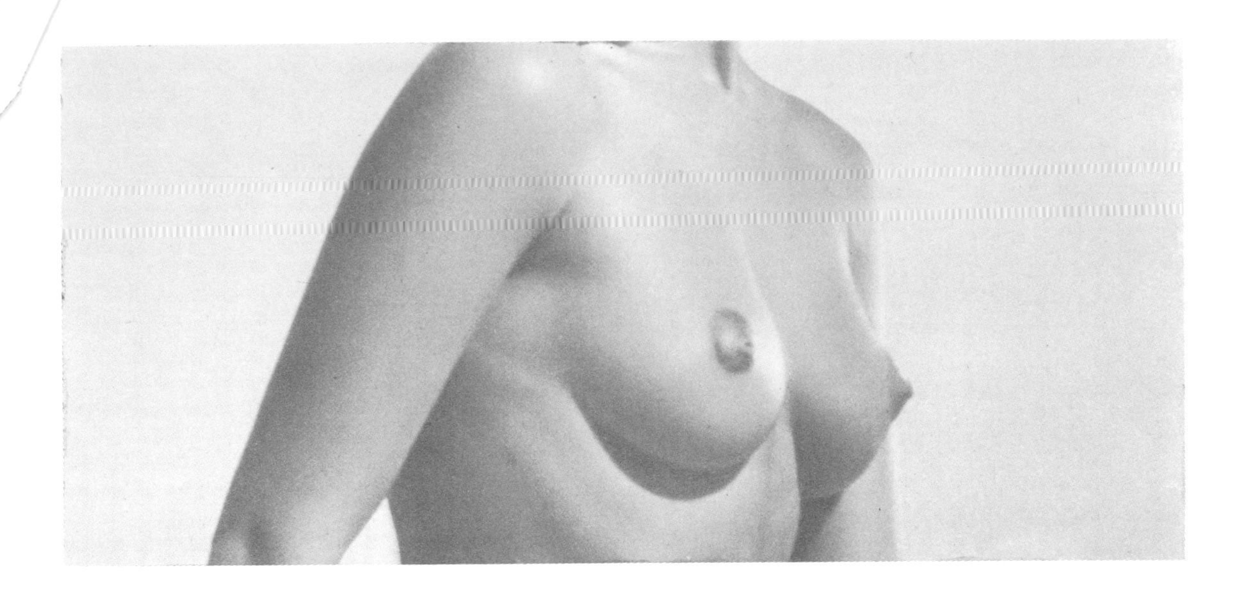

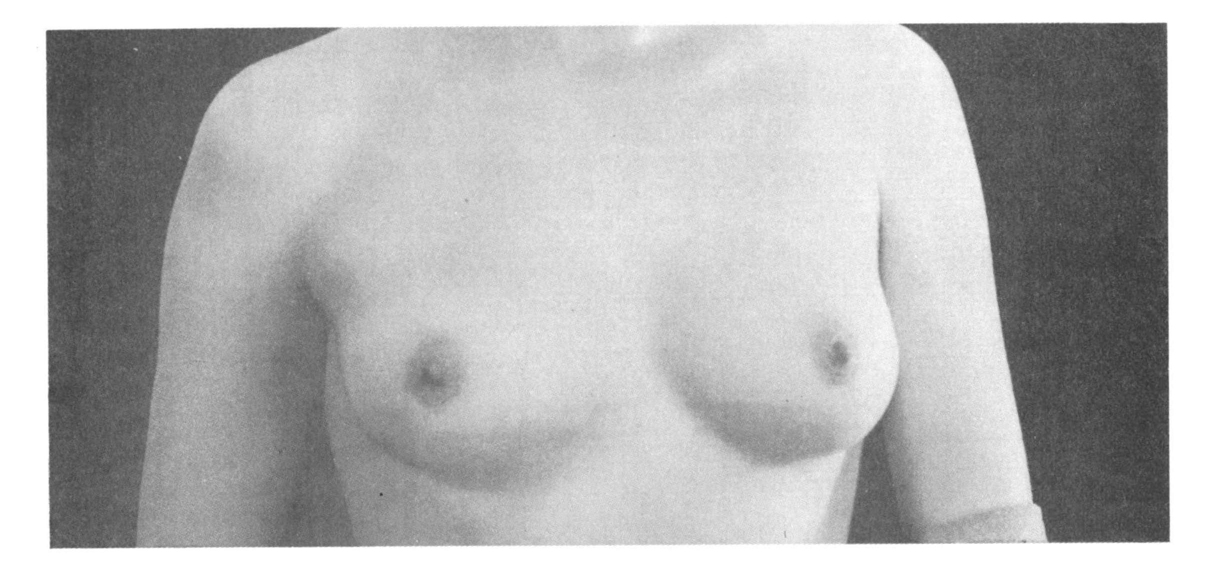

BREASTS I

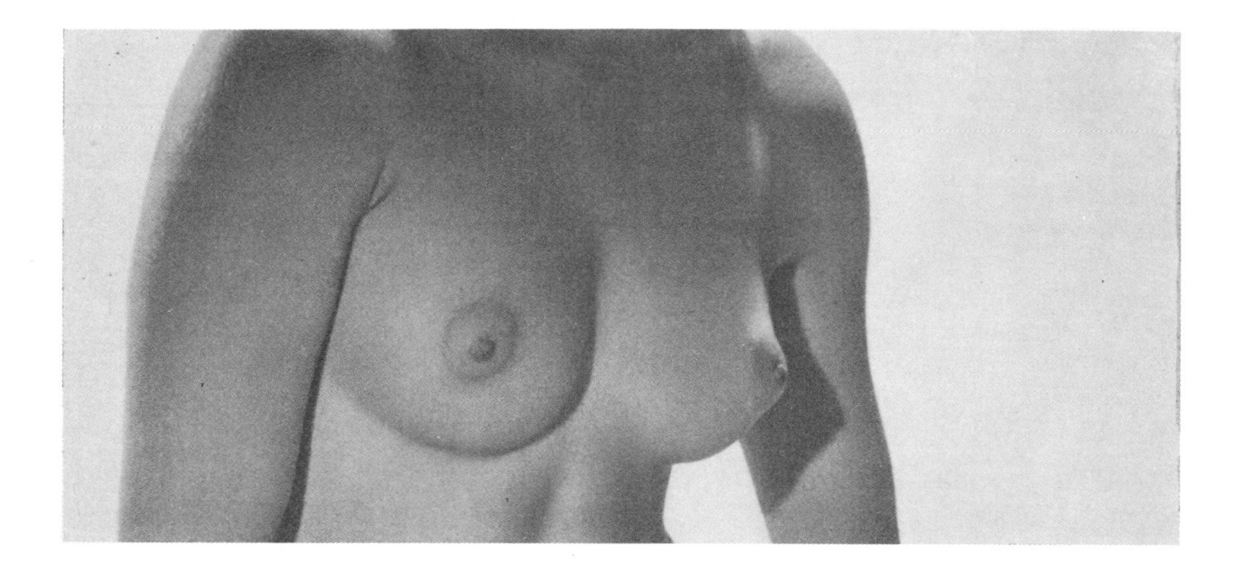

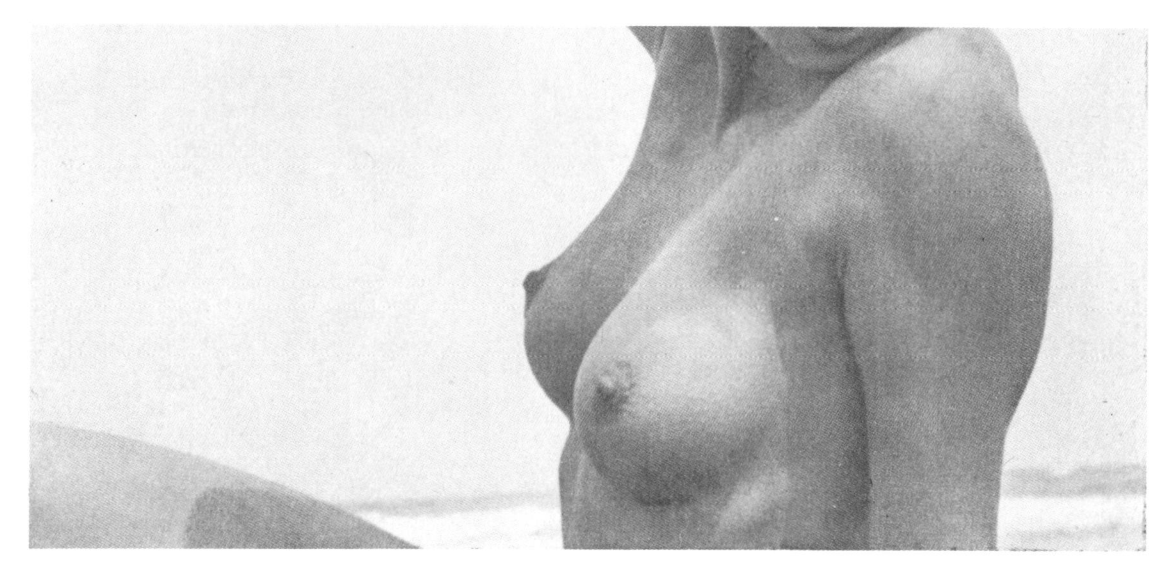

BREASTS II

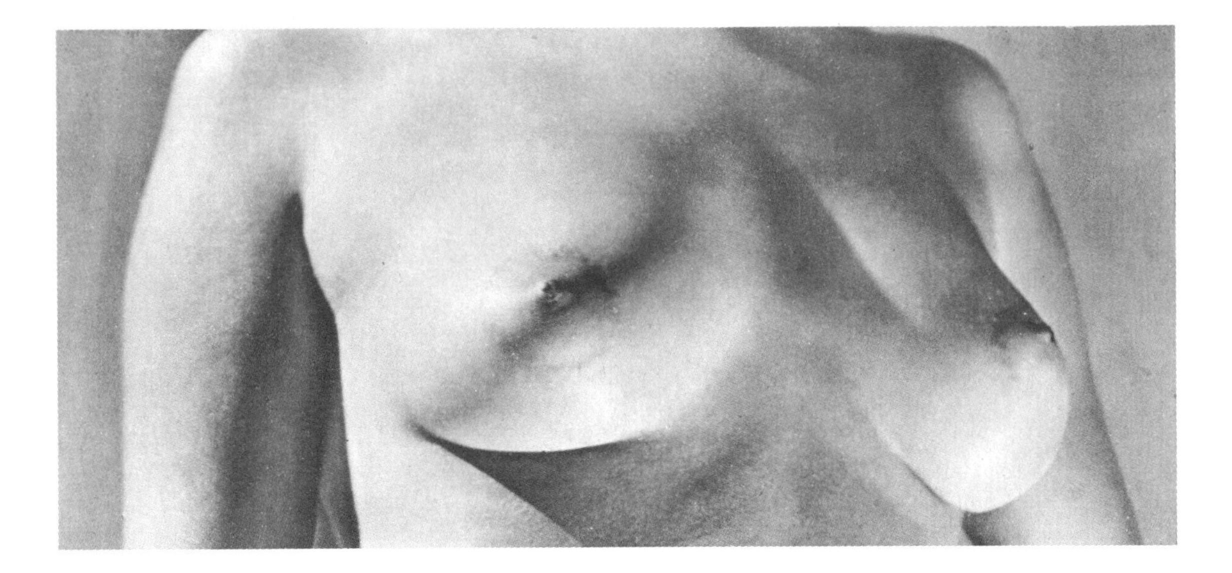

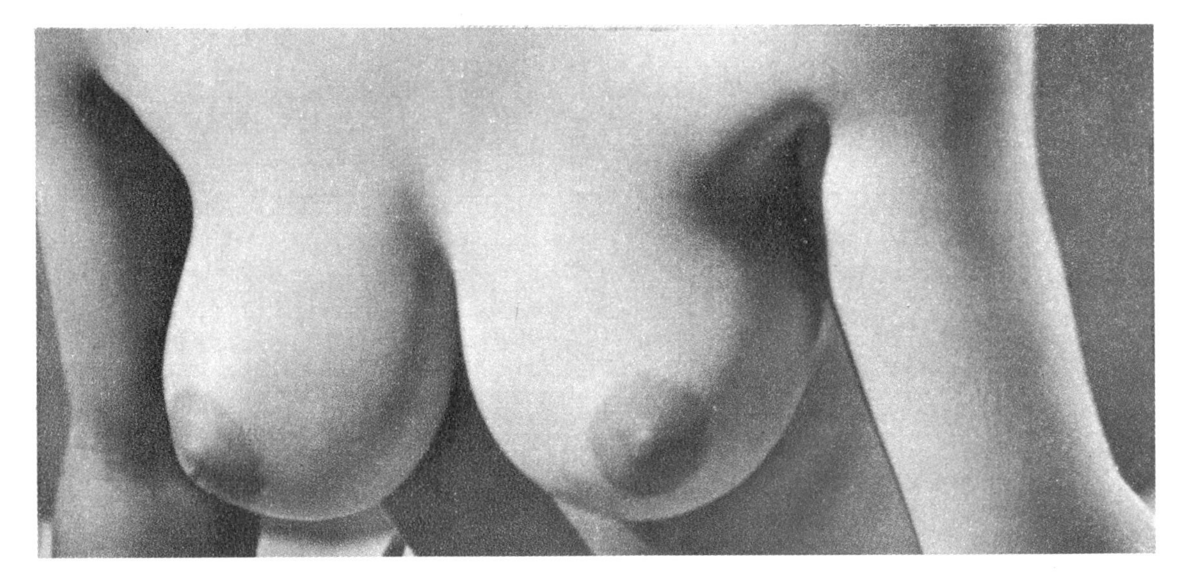

BREASTS III

A Selection of Hands
By
Heidi Lenssen

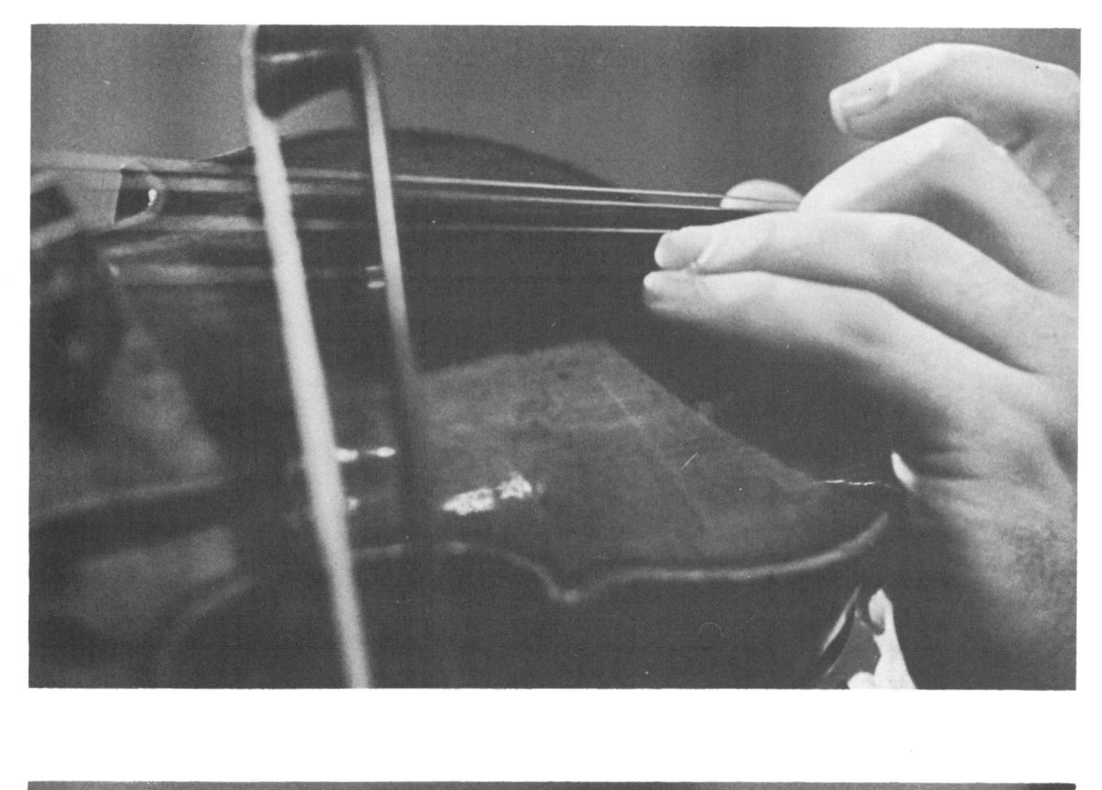

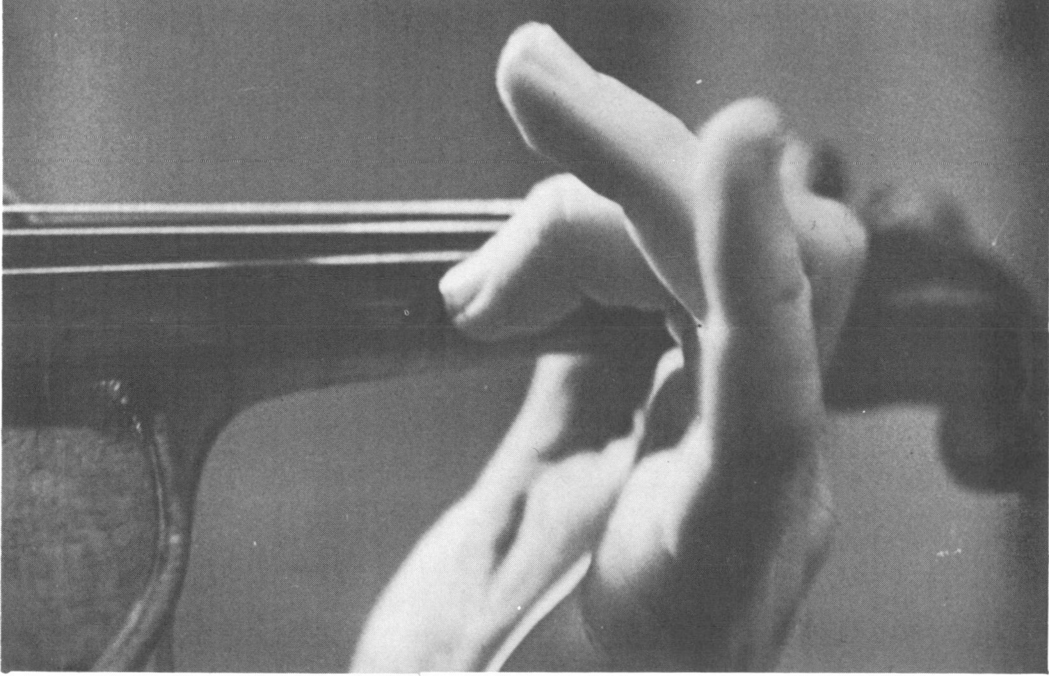

Courtesy of S. Hurok

HANDS OF THE VIOLINIST ISAAC STERN

Courtesy of S. Hurok

HANDS OF THE VIOLINIST ISAAC STERN

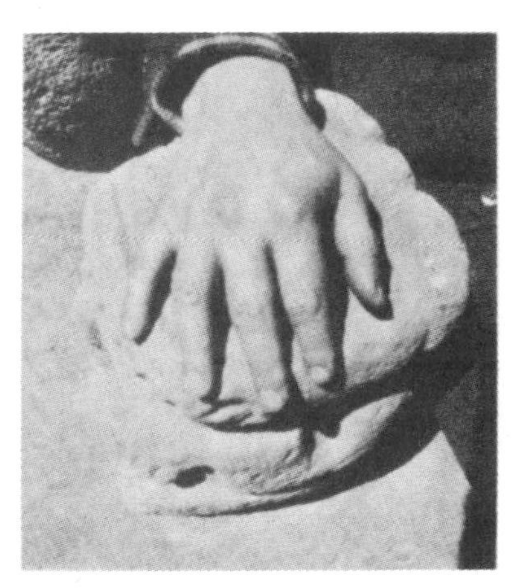

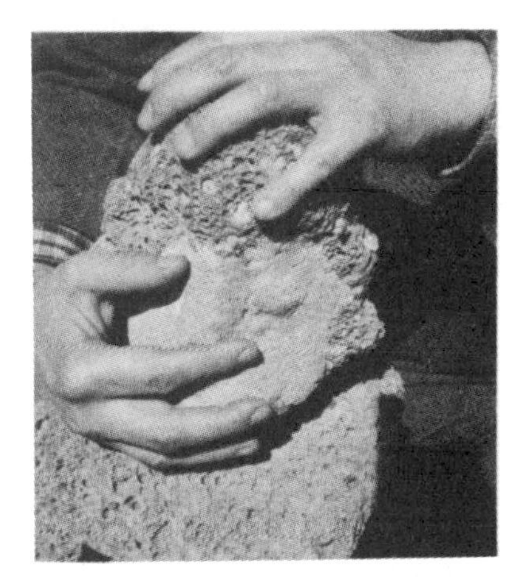

Photo by Fritz Henle

HANDS OF DIEGO RIVERA

Photo by Fritz Henle

HANDS OF AN OLD LABORER (ONE OF THE LAST SURVIVING SLAVES)

Photo by Fritz Henle

HANDS OF A SILVERSMITH

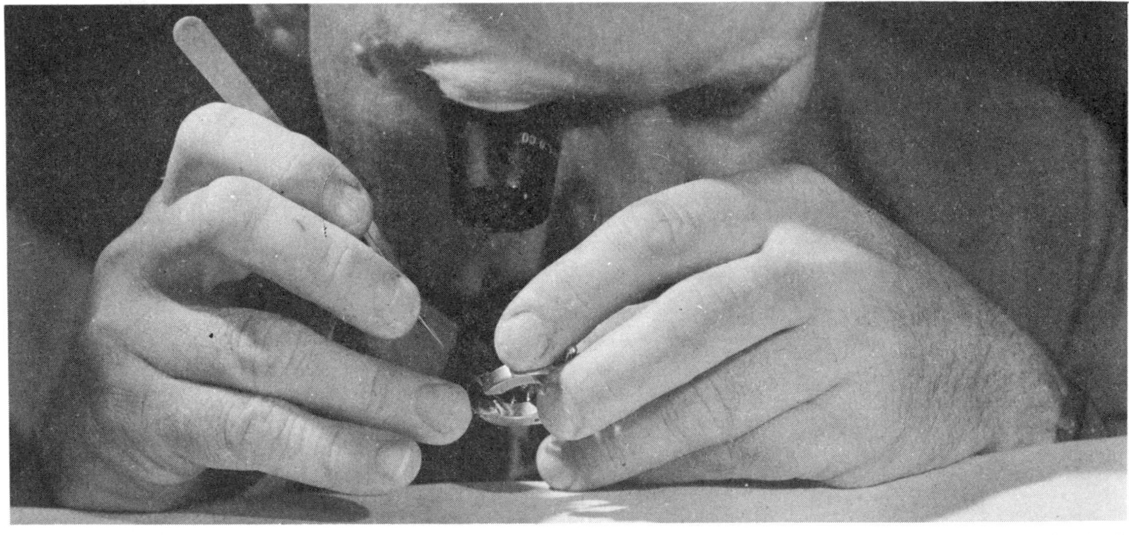

Photo by Fritz Henle

FINE MECHANIC AT WORK

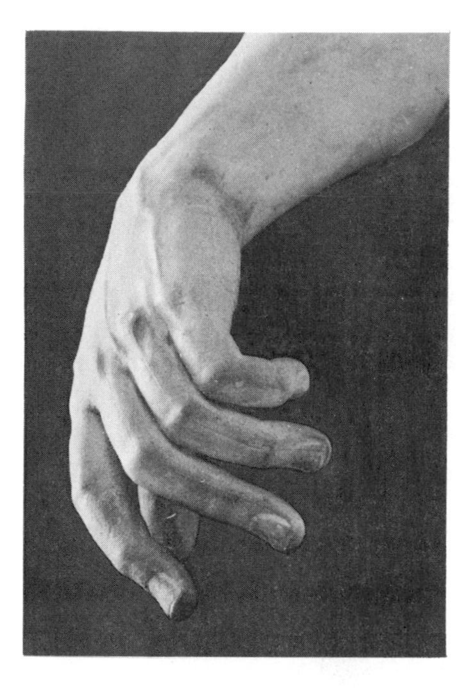

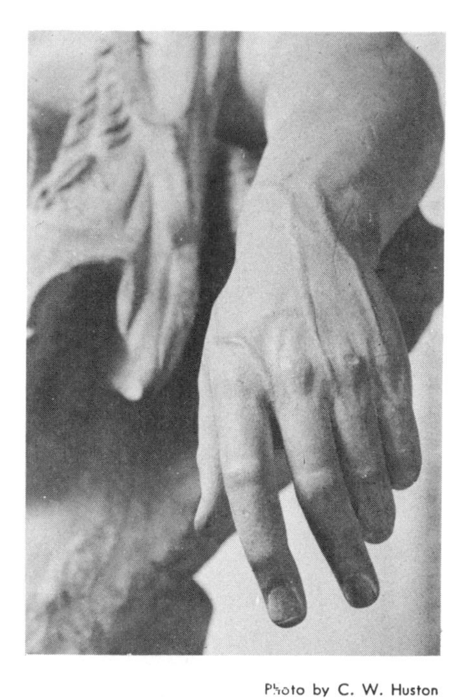

Photo by C. W. Huston

HANDS OF A YOUNG MAN, GREEK (*c.* 350 B.C.)

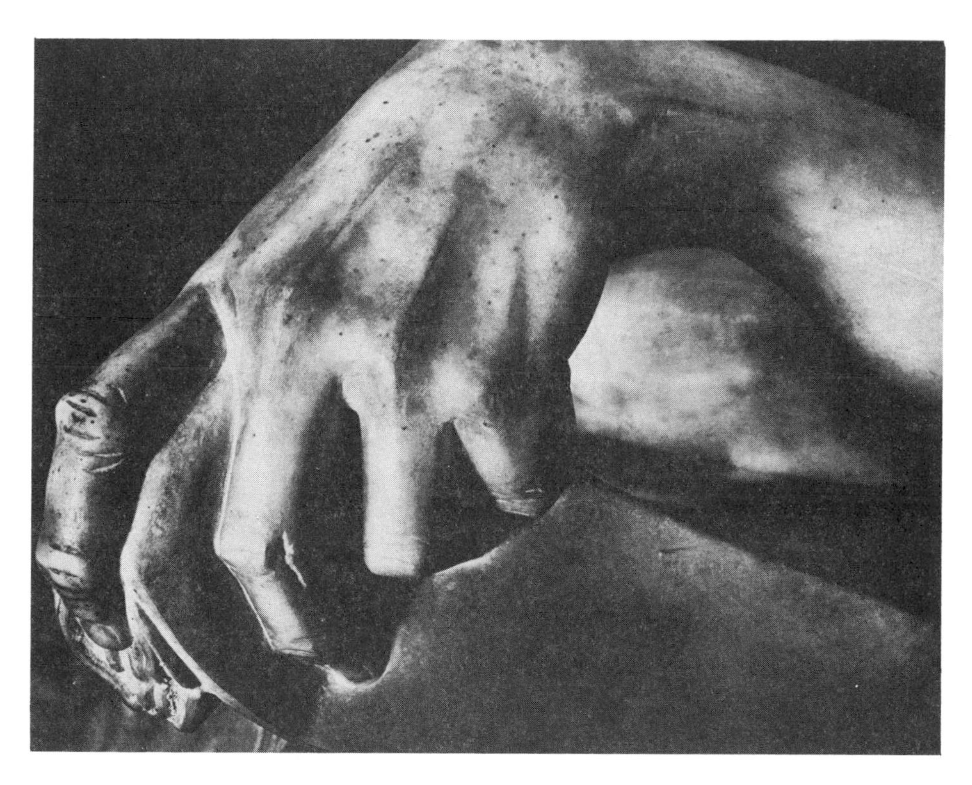

Photo by C. W. Huston

DONATELLO, "ST. JOHN" (DETAIL)

VERROCCHIO, "MARIA AND CHILD" (DETAIL)

VERROCCHIO, "ST. JOHN AND AN ANGEL" (DETAIL)

VAN DER GOES, "ADORATION OF JESUS" (DETAIL)

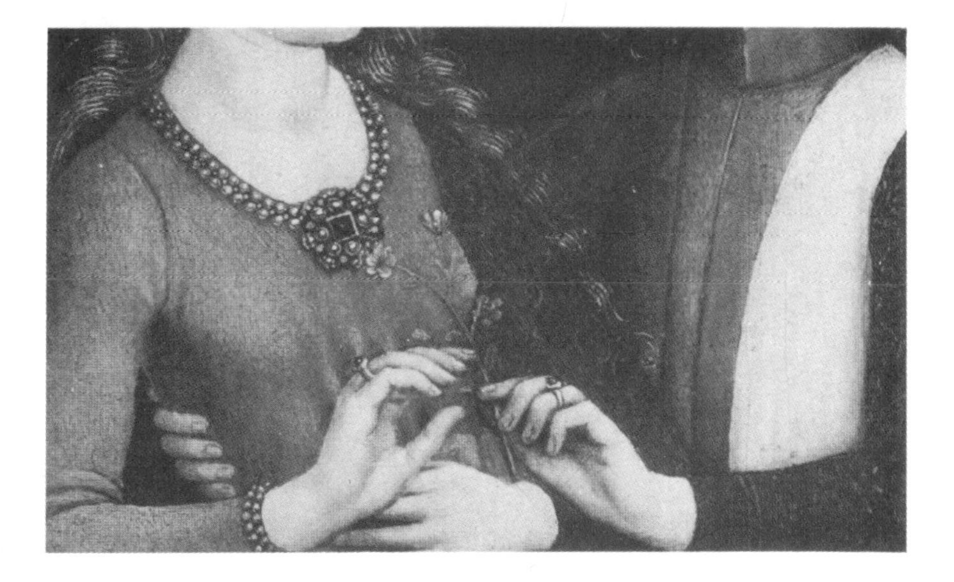

SOUTH GERMAN (*c.* 1470) (DETAIL)

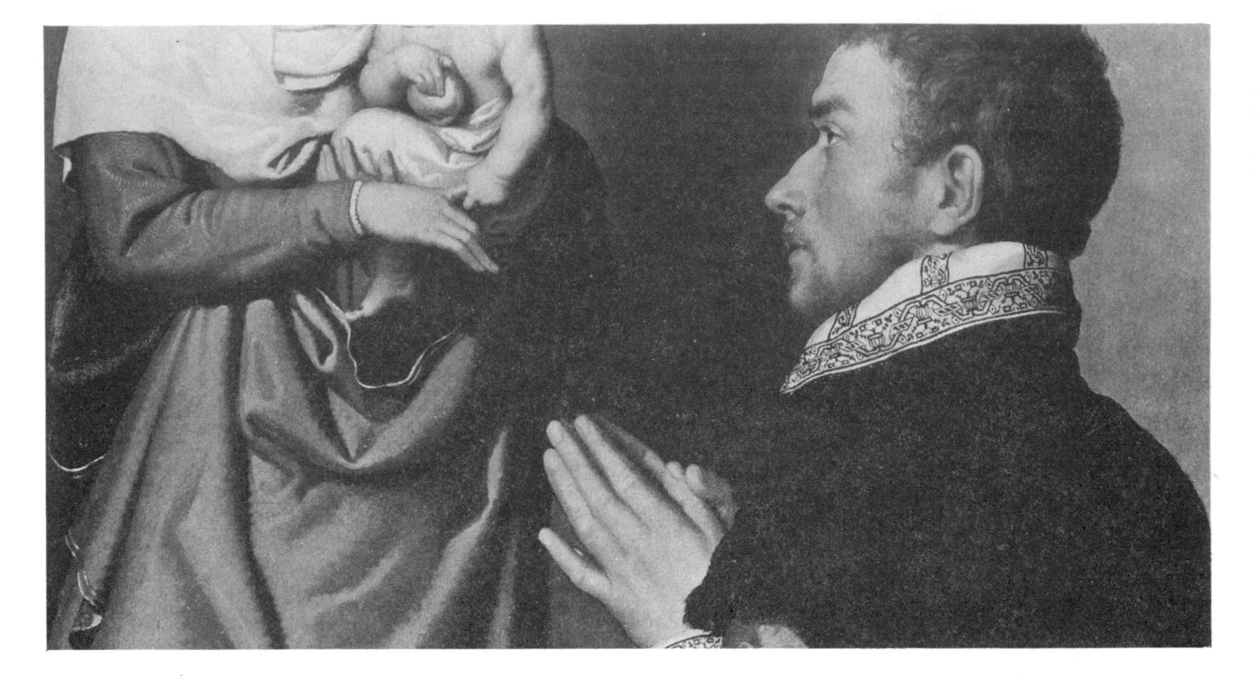

MORONI, "GENTLEMAN IN ADORATION" (DETAIL)

BASSANO (DETAIL)

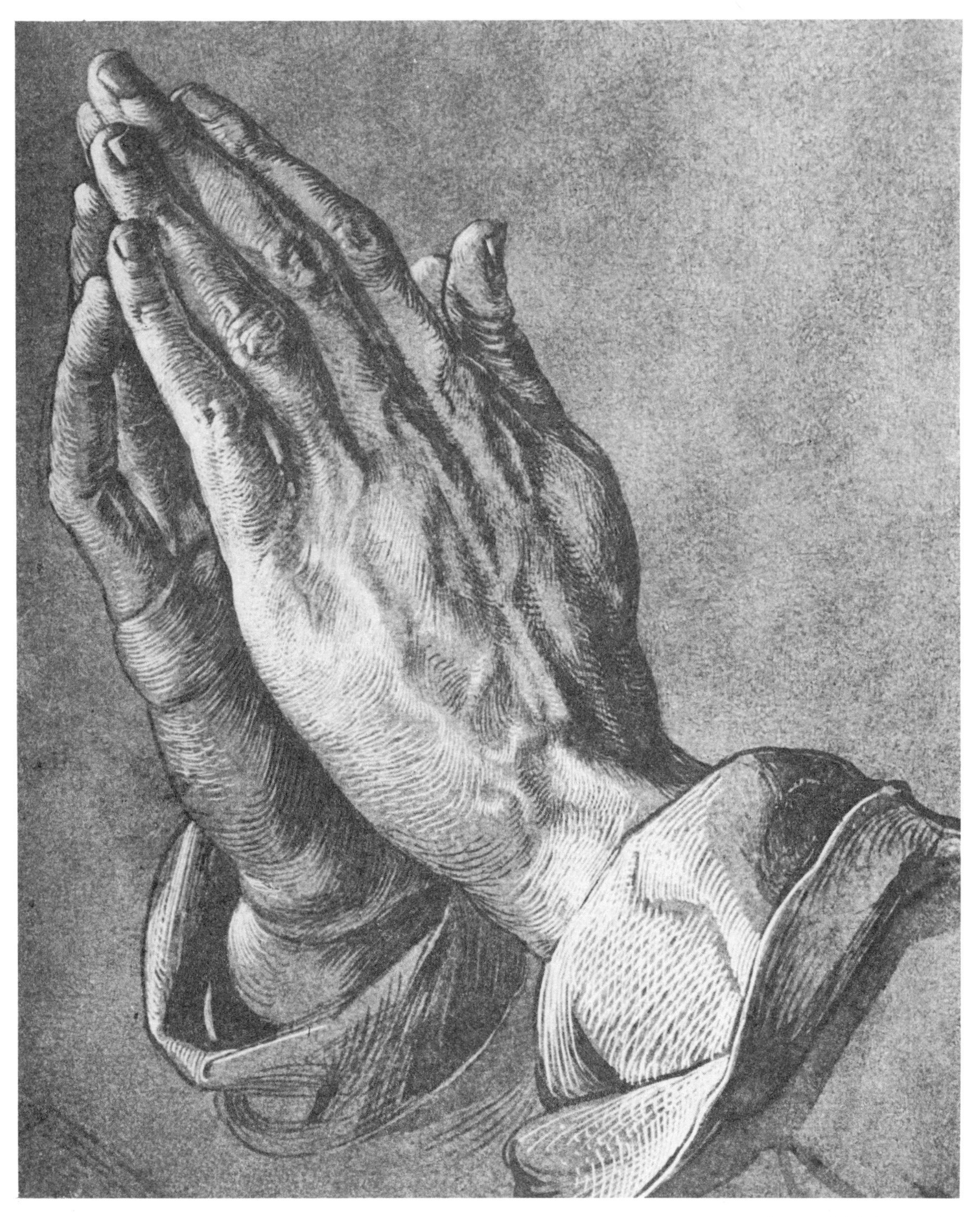

Courtesy of E. S. Herrmann, Inc.

DURER, "PRAYING HANDS"

EL GRECO, "INCOGNITO" (DETAIL)

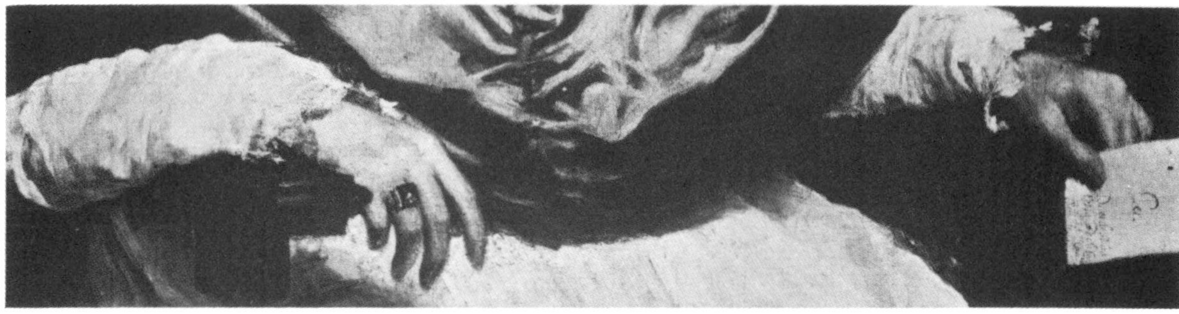

VELASQUEZ, "PORTRAIT OF INNOCENTIUS POPE" (DETAIL)

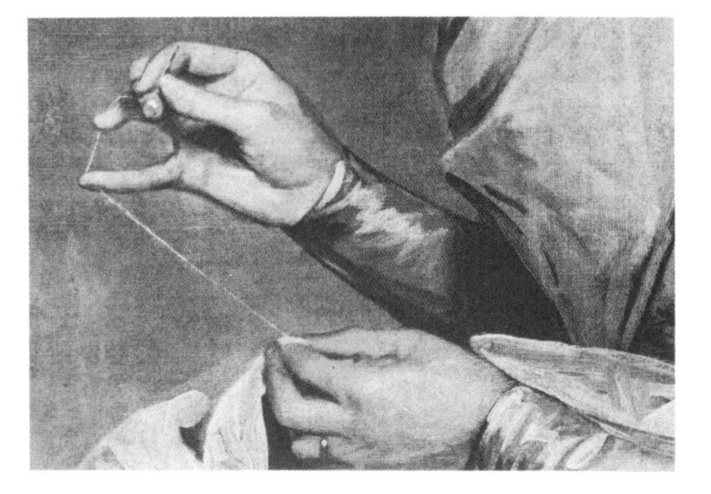

STUART, PORTRAIT (DETAIL)

A Selection of Illustrations from the Old Masters
And Historical Sources

Courtesy of the Metropolitan Museum of Art

POLLAIUOLO, "BATTLE OF THE NAKED MEN"

Courtesy of the Metropolitan Museum of Art

BARBARI, "APOLLO AND DIANA"

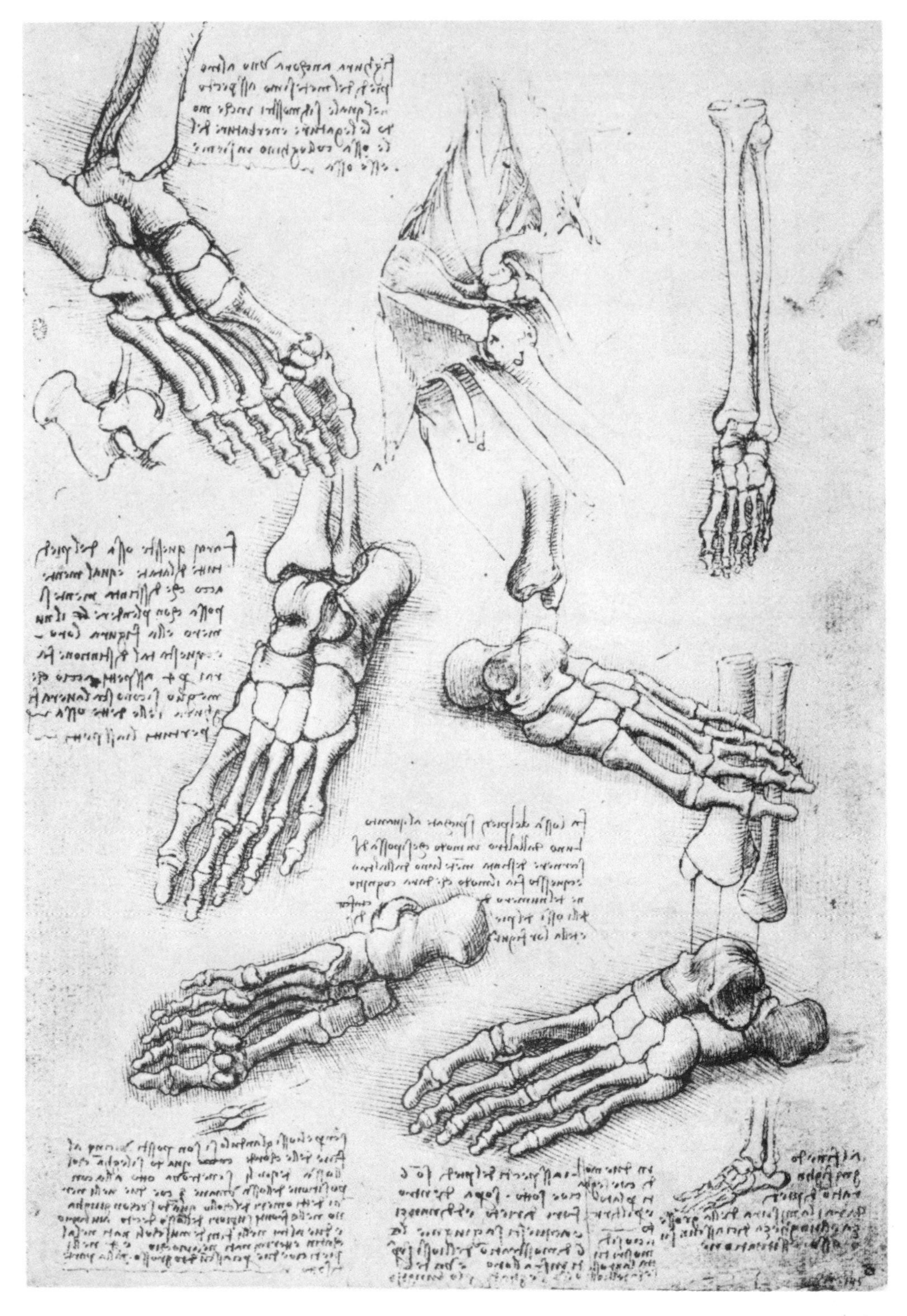

Courtesy of the Burndy Library, Norwalk, Conn.

LEONARDO DA VINCI, BONES OF THE FOOT AND DISSECTION OF THE SHOULDER JOINT.

Courtesy of the Burndy Library, Norwalk, Conn.

LEONARDO DA VINCI, MUSCLES OF THE TRUNK

Courtesy of the Burndy Library, Norwalk, Conn.

LEONARDO DA VINCI, MUSCLES OF THE TRUNK AND LEG

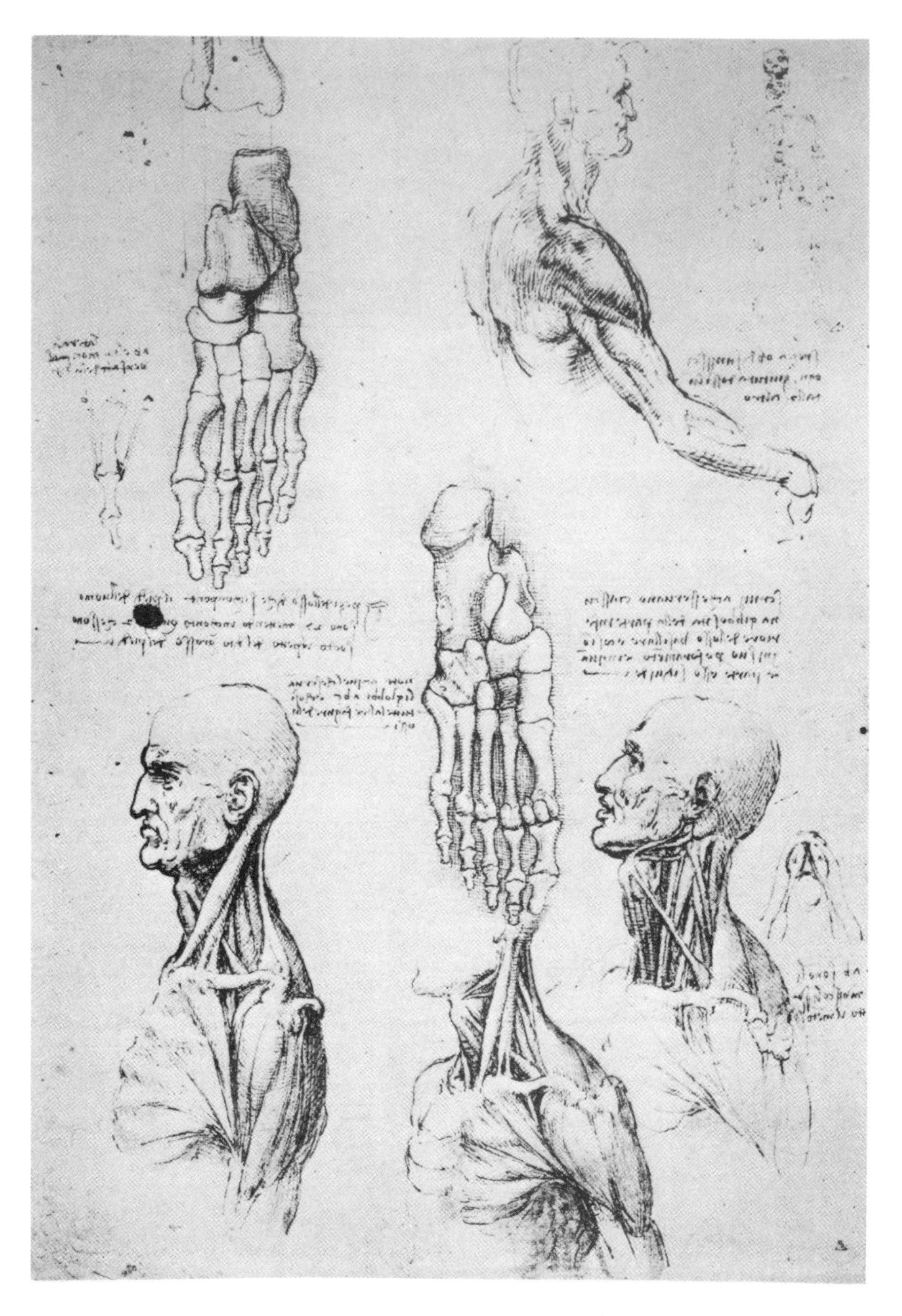

Courtesy of the Burndy Library, Norwalk, Conn.

PLATE 112 LEONARDO DA VINCI, BONES OF THE FOOT AND MUSCLES OF THE HEAD, NECK, AND SHOULDER

Courtesy of the Burndy Library, Norwalk, Conn.

LEONARDO DA VINCI, SURFACE ANATOMY OF THE SHOULDER

Courtesy of the Burndy Library, Norwalk, Conn.

PLATE 114 LEONARDO DA VINCI, MUSCLES OF THE SHOULDER I

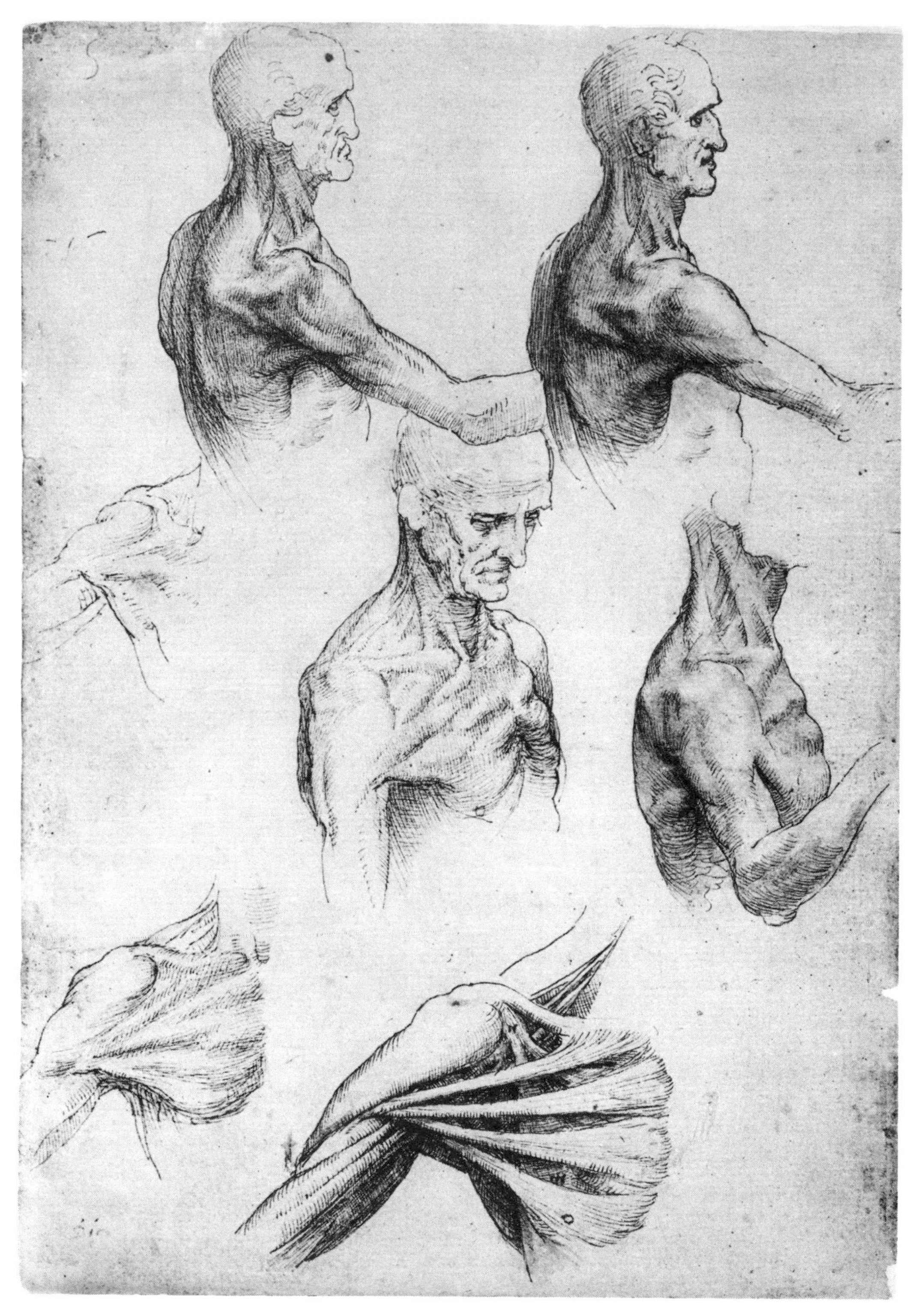

Courtesy of the Burndy Library, Norwalk, Conn.

LEONARDO DA VINCI, MUSCLES OF THE SHOULDER II

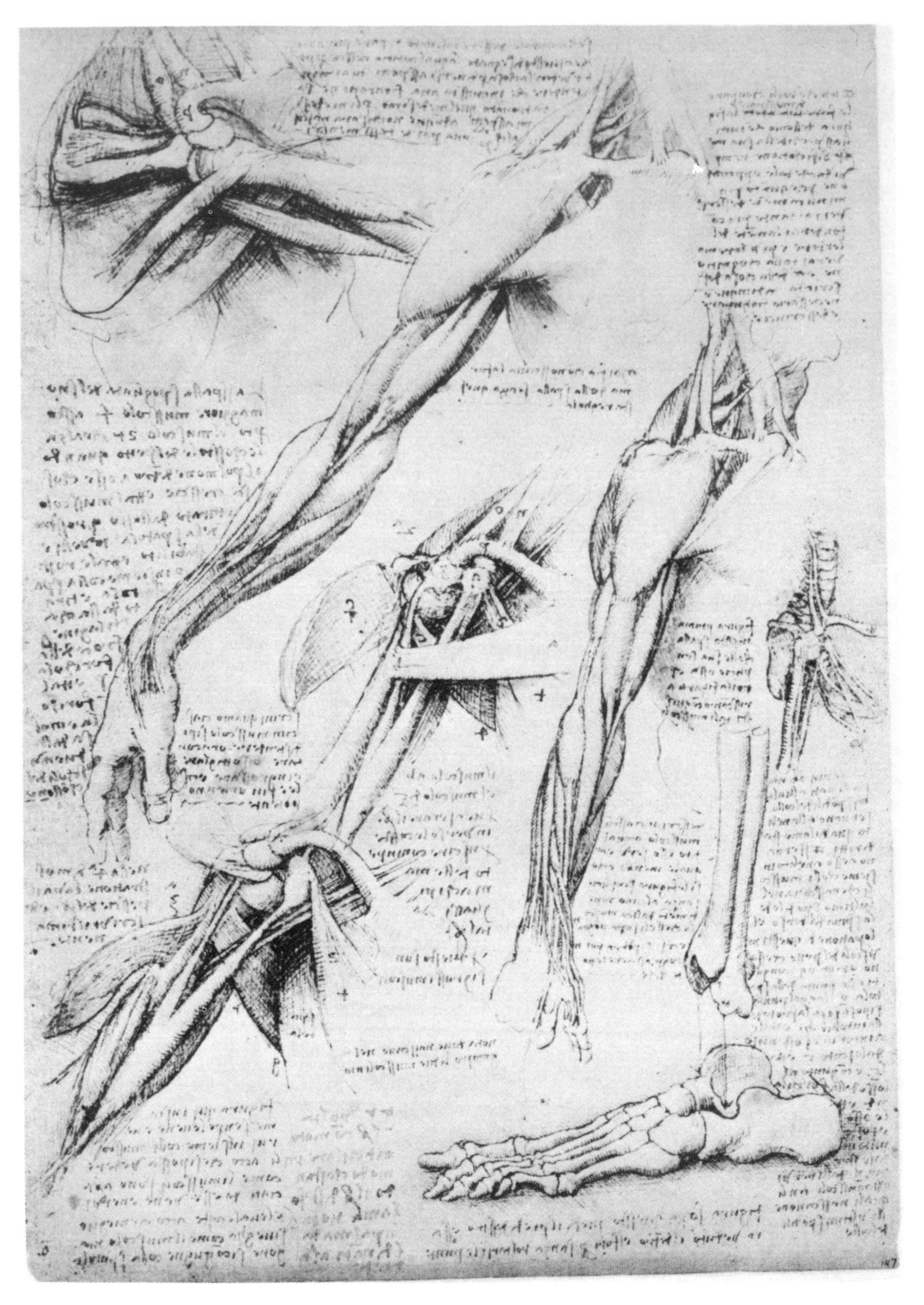

Courtesy of the Burndy Library, Norwalk, Conn.

LEONARDO DA VINCI, MUSCLES OF THE SHOULDER III

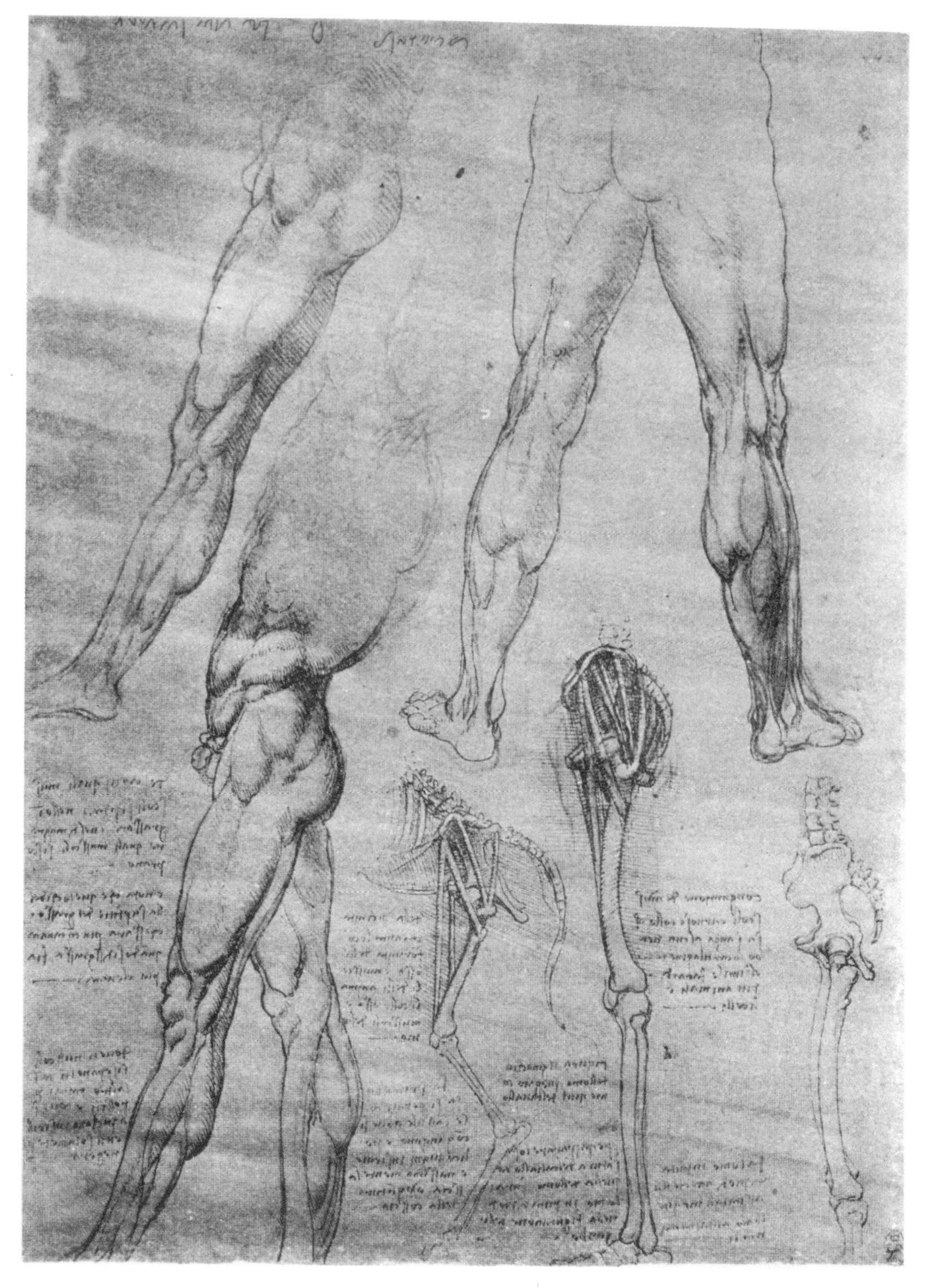

Courtesy of the Burndy Library, Norwalk, Conn.

LEONARDO DA VINCI, SURFACE ANATOMY OF THE LOWER EXTREMITY

PLATE 117

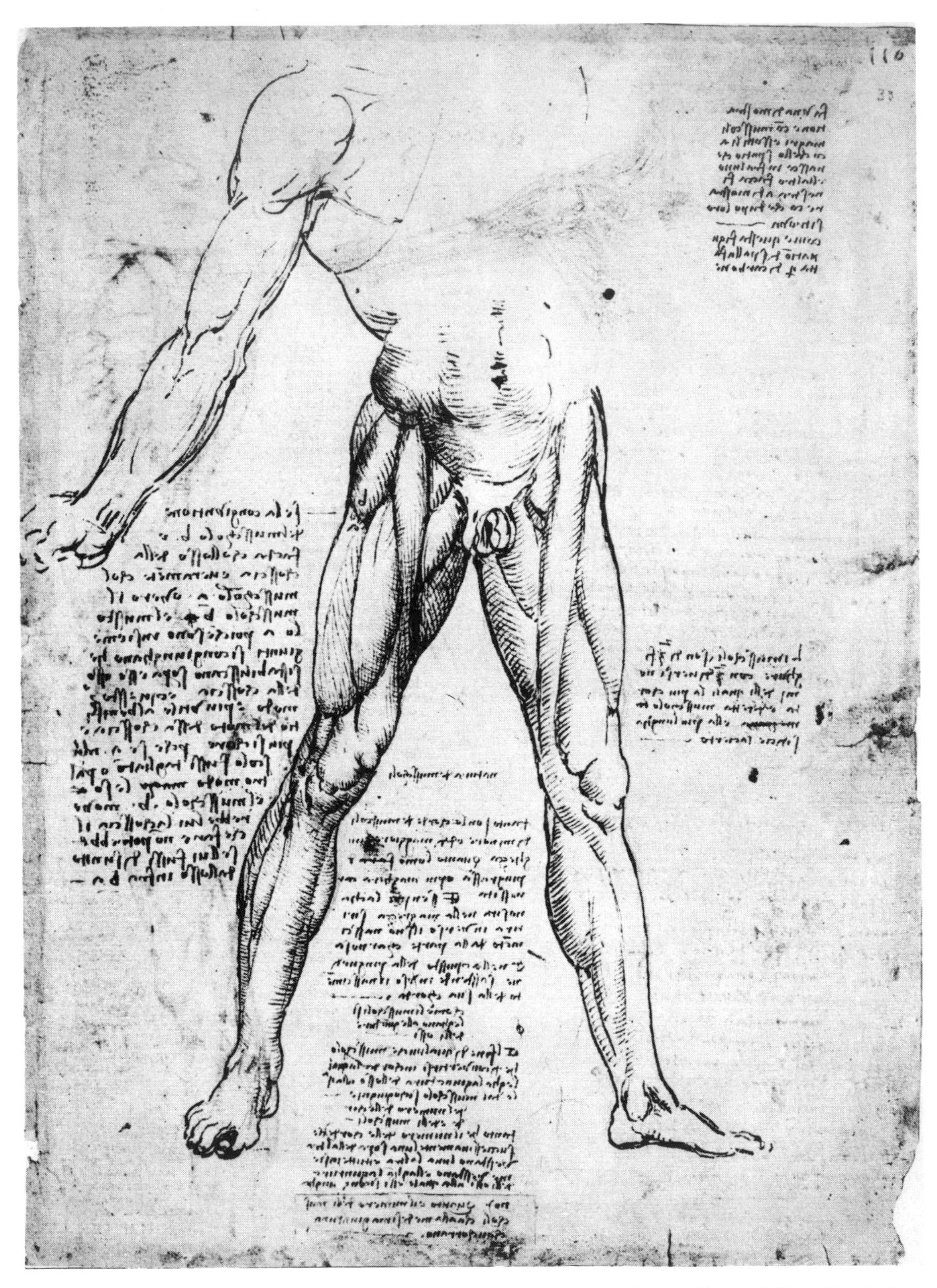

Courtesy of the Burndy Library, Norwalk, Conn.

LEONARDO DA VINCI, MUSCLES OF THE LOWER EXTREMITY I

PLATE 118

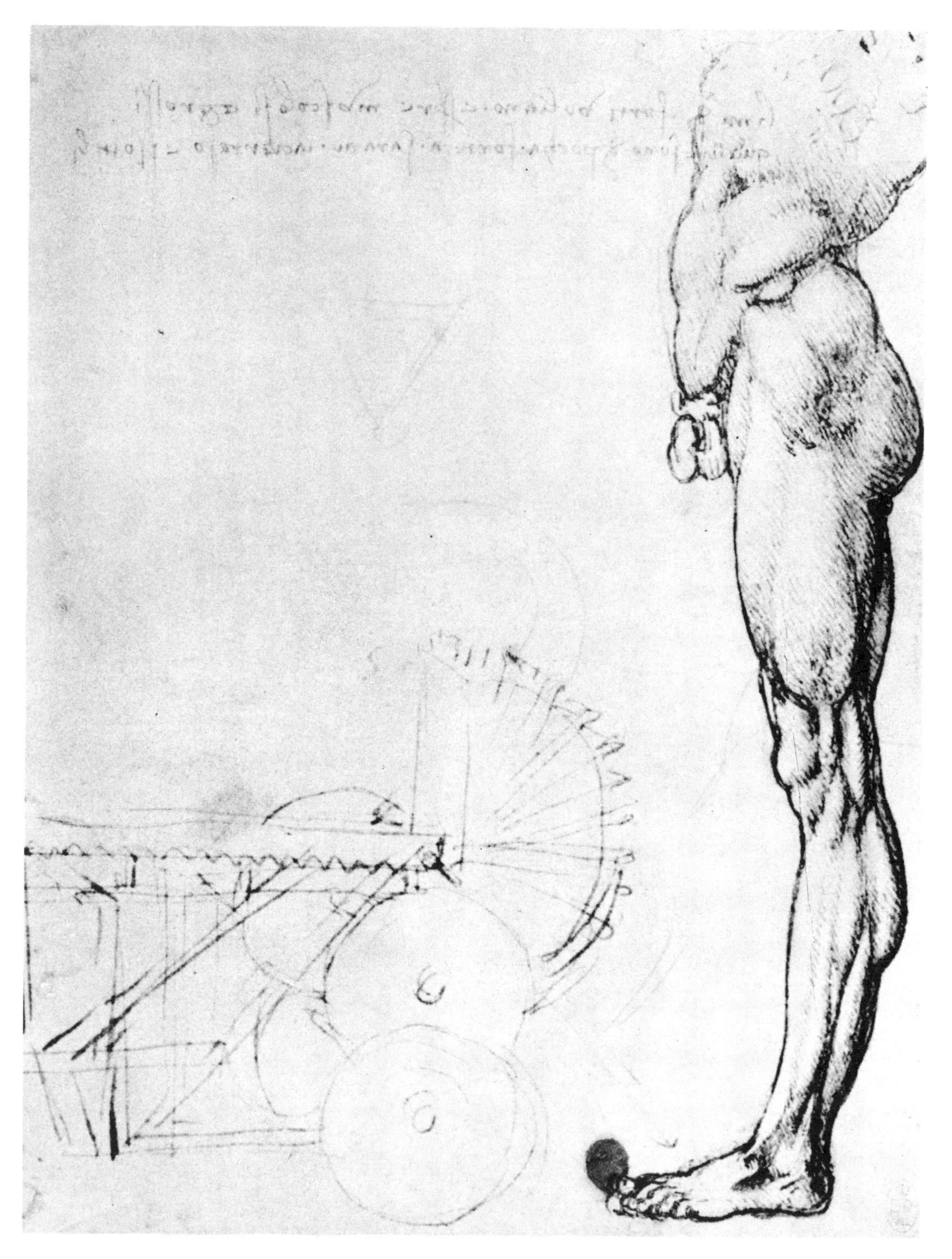

Courtesy of the Burndy Library, Norwalk, Conn.

LEONARDO DA VINCI, MUSCLES OF THE LOWER EXTREMITY II

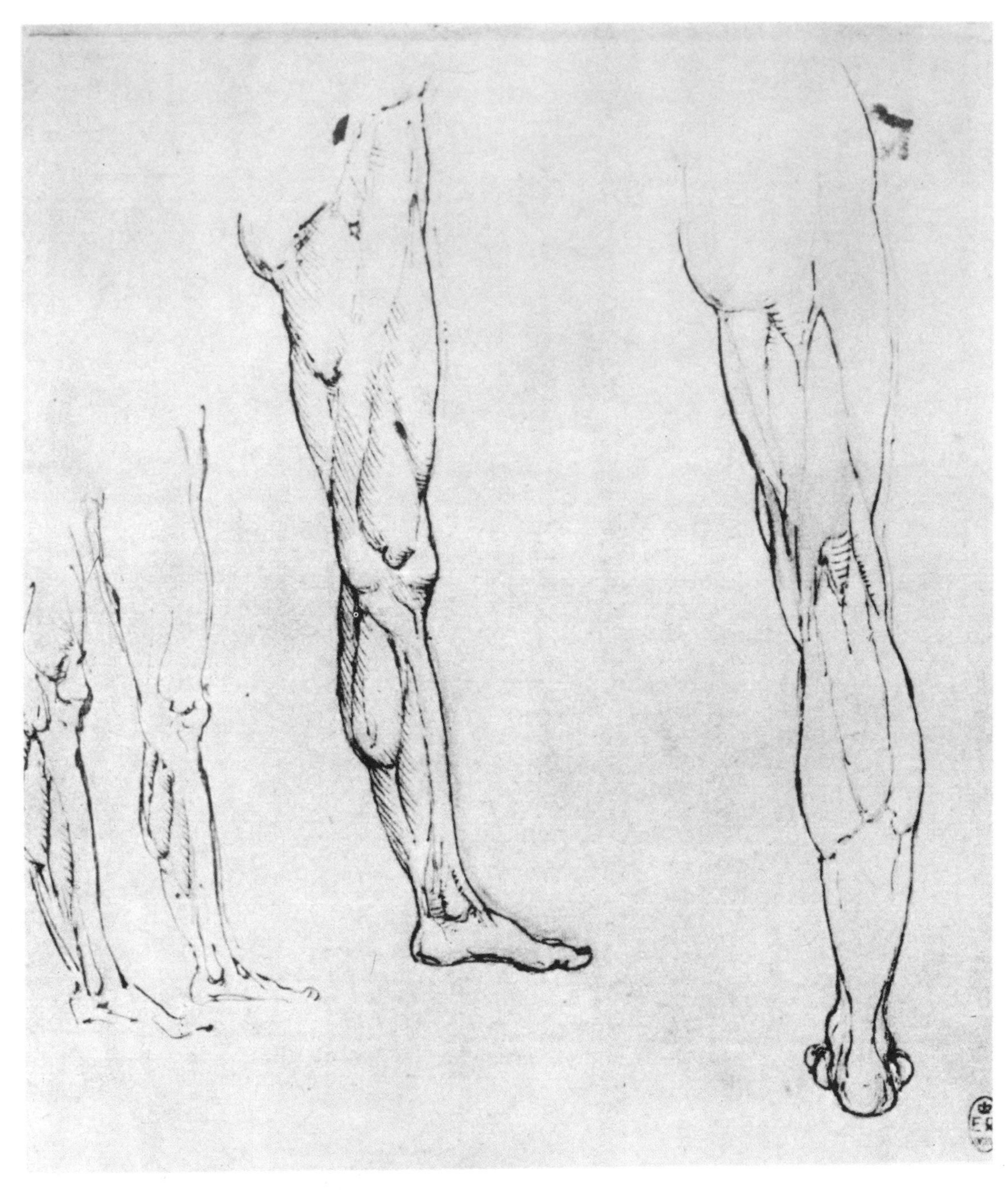

Courtesy of the Burndy Library, Norwalk, Conn.

LEONARDO DA VINCI, MUSCLES OF THE LOWER EXTREMITY III

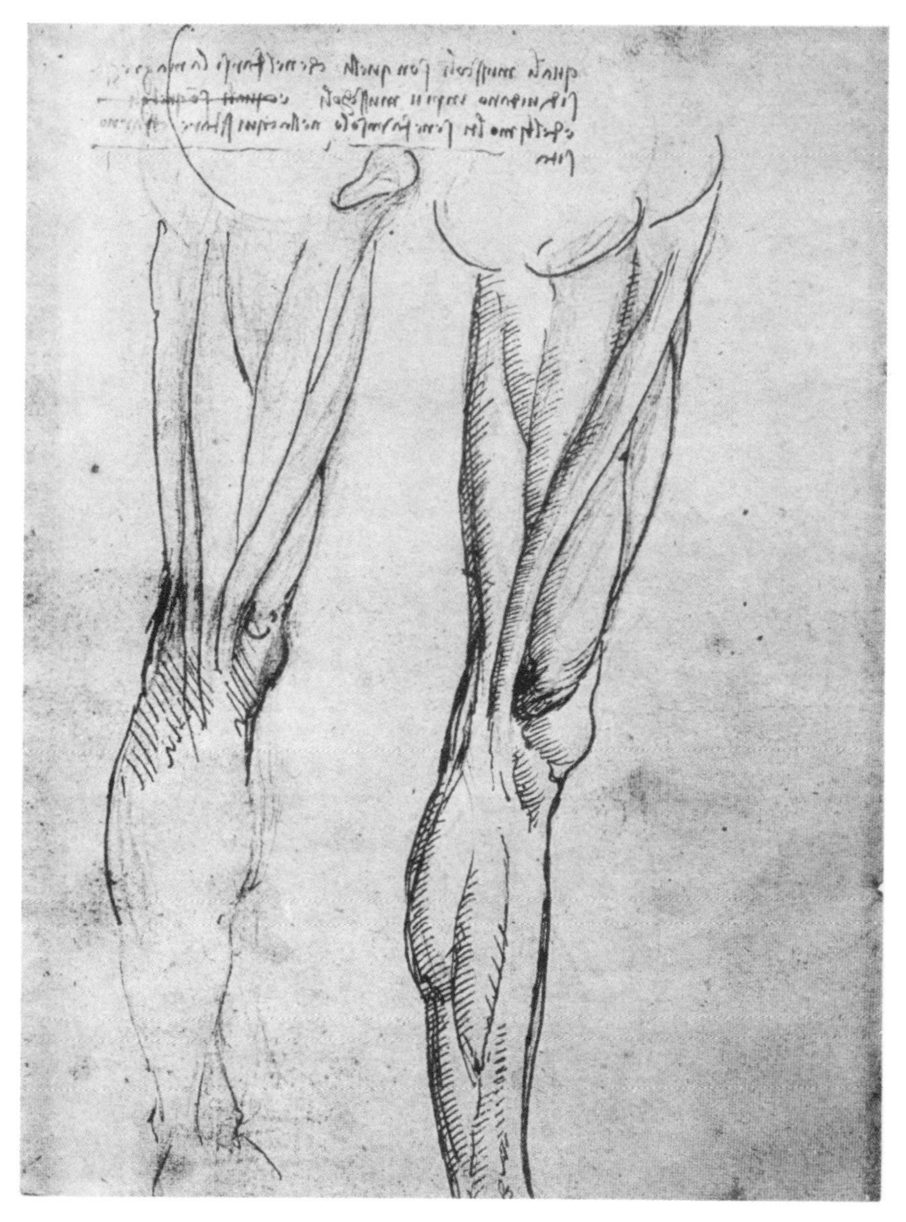

Courtesy of the Burndy Library, Norwalk, Conn.

LEONARDO DA VINCI, MUSCLES OF THE LOWER EXTREMITY IV

Courtesy of the Burndy Library, Norwalk, Conn.

LEONARDO DA VINCI, MUSCLES OF THE LOWER EXTREMITY V

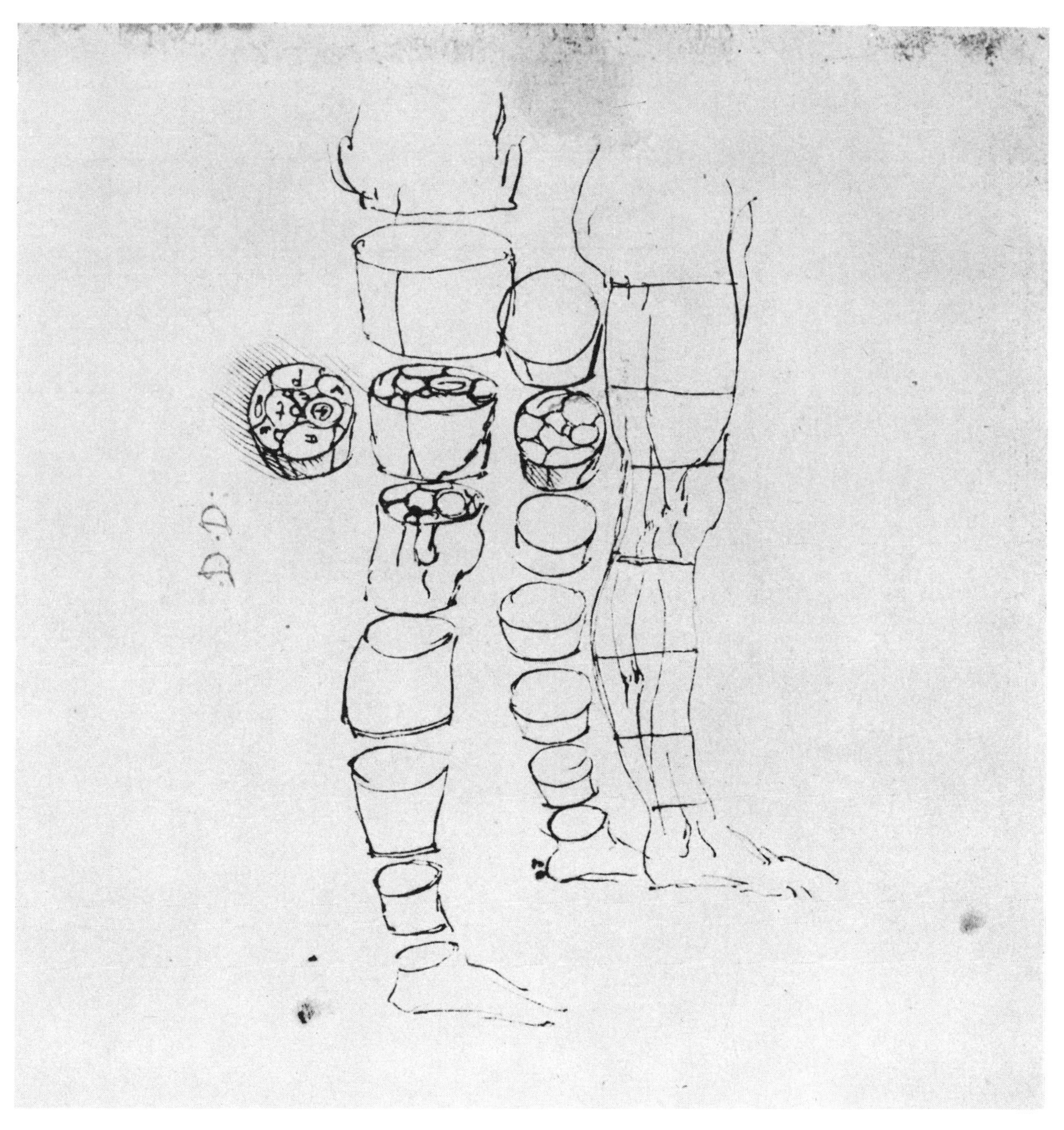

Courtesy of the Burndy Library, Norwalk, Conn.

LEONARDO DA VINCI, MUSCLES OF THE LOWER EXTREMITY VI

Courtesy of the Metropolitan Museum of Art

DURER, "ADAM AND EVE"

PLATE 124

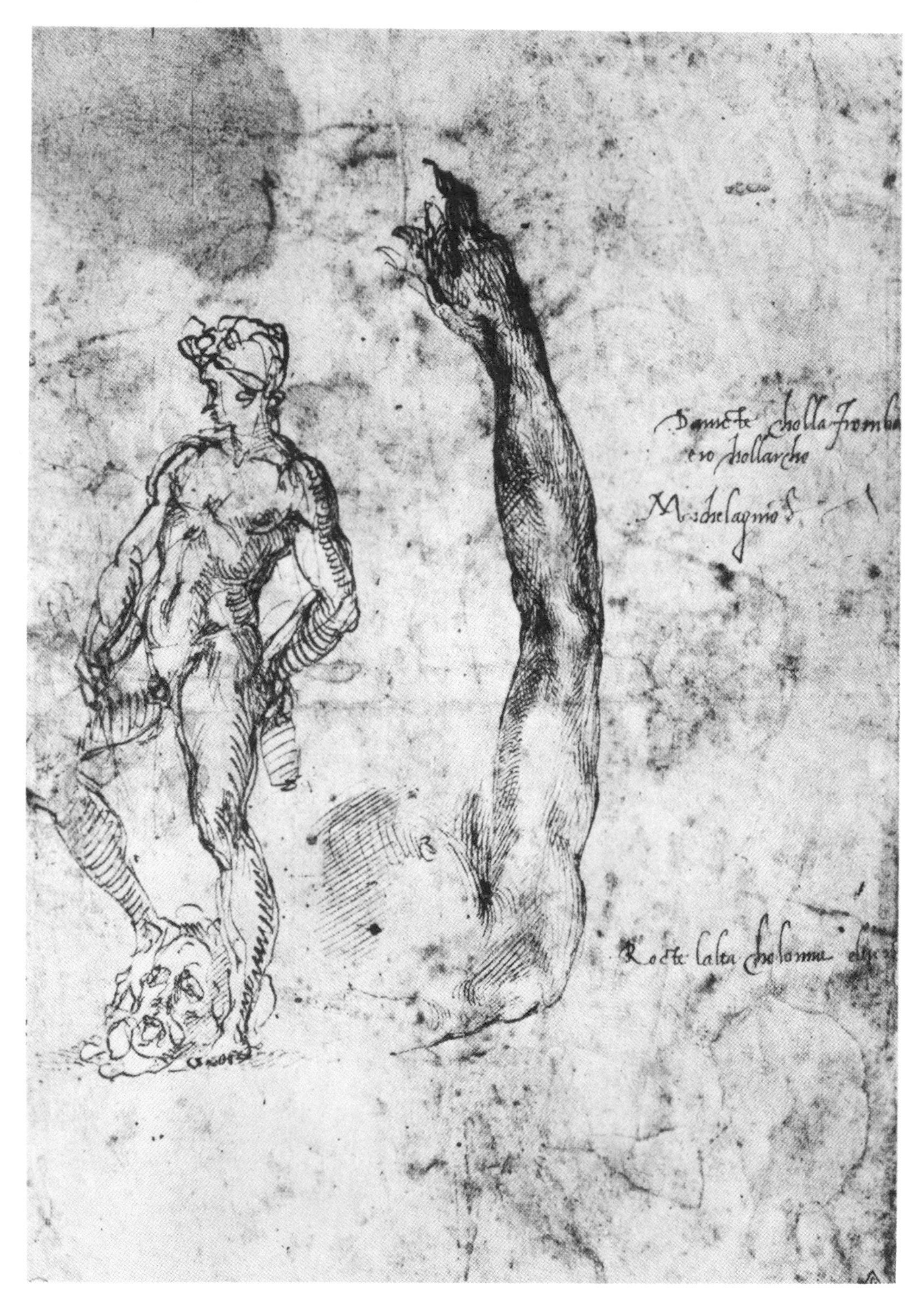

Courtesy of the Louvre, Paris

MICHELANGELO, SKETCH FOR THE BRONZE DAVID AND ARM STUDY FOR THE MARBLE DAVID

PLATE 125

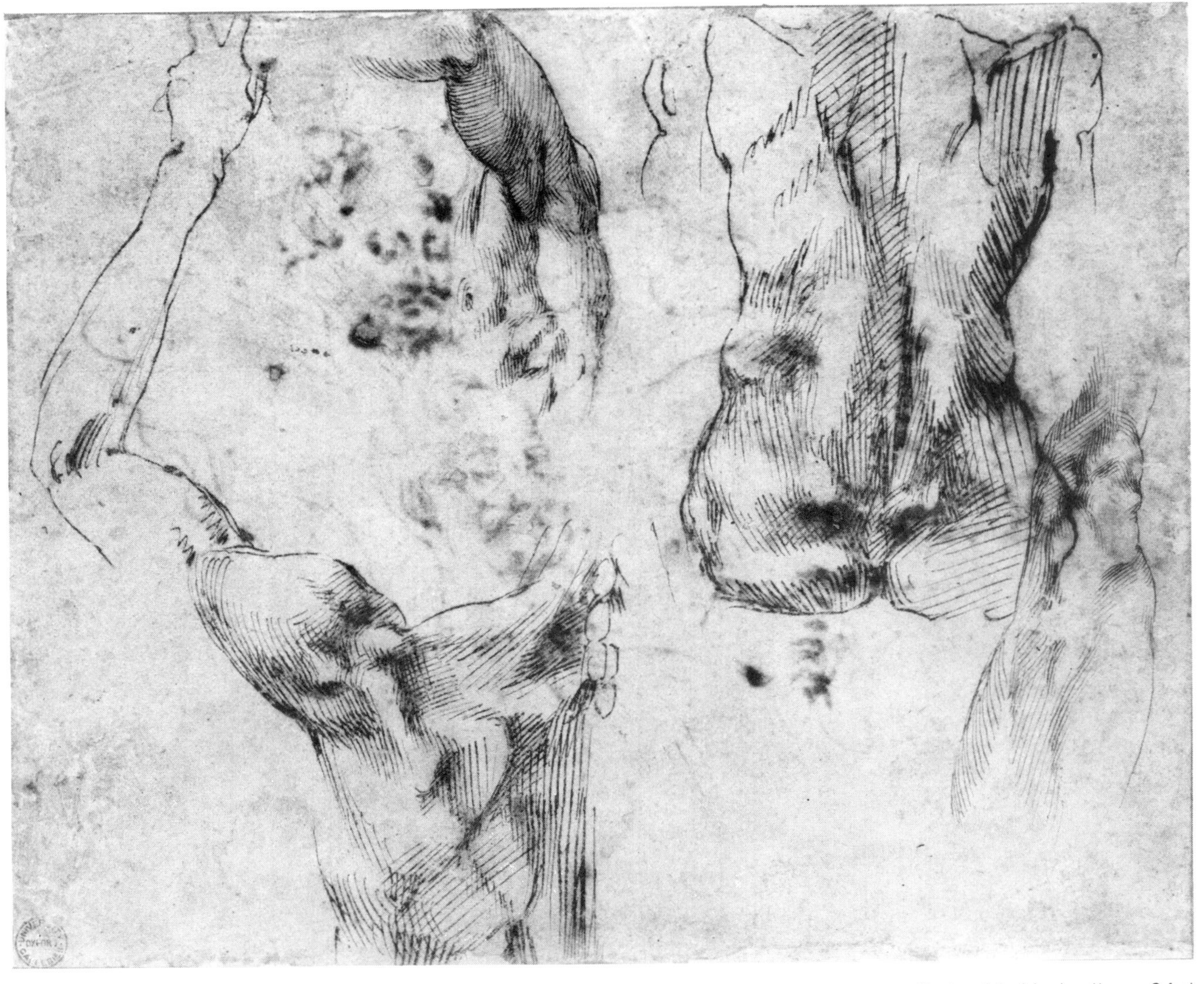

Courtesy of the Ashmolean Museum, Oxford

MICHELANGELO, STUDIES OF THE NUDE

Courtesy of Casa Buonarroti, Florence

MICHELANGELO, NUDE SEEN FROM THE BACK

PLATE 127

Courtesy of the British Museum, London

MICHELANGELO, SKETCH FOR THE BRUGES MADONNA, THREE NUDE MEN

Courtesy of the British Museum, London

MICHELANGELO, STUDIES FOR THE IGNUDI OF THE SISTINE CHAPEL CEILING

Courtesy of the Metropolitan Museum of Art

MICHELANGELO, STUDIES FOR THE LIBYAN SIBYL

PLATE 130

Courtesy of the Ashmolean Museum, Oxford

MICHELANGELO, SKETCHES FOR THE SISTINE CHAPEL CEILING AND THE TOMB OF JULIUS

PLATE 131

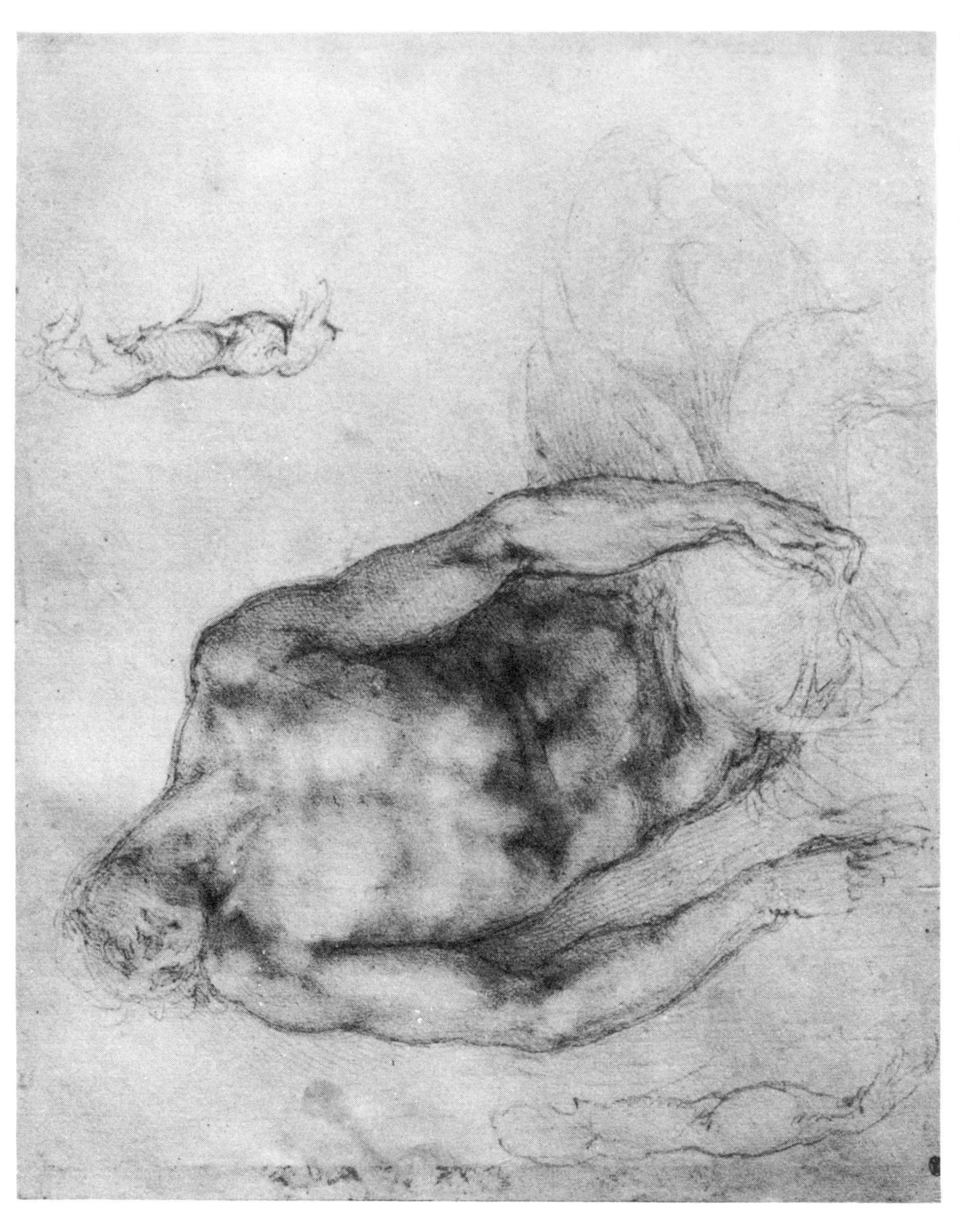

Courtesy of the Louvre, Paris

MICHELANGELO, STUDY FOR A PIETA

Courtesy of Casa Buonarroti, Florence

MICHELANGELO, STUDY FOR A RECUMBENT FIGURE IN THE MEDICI CHAPEL

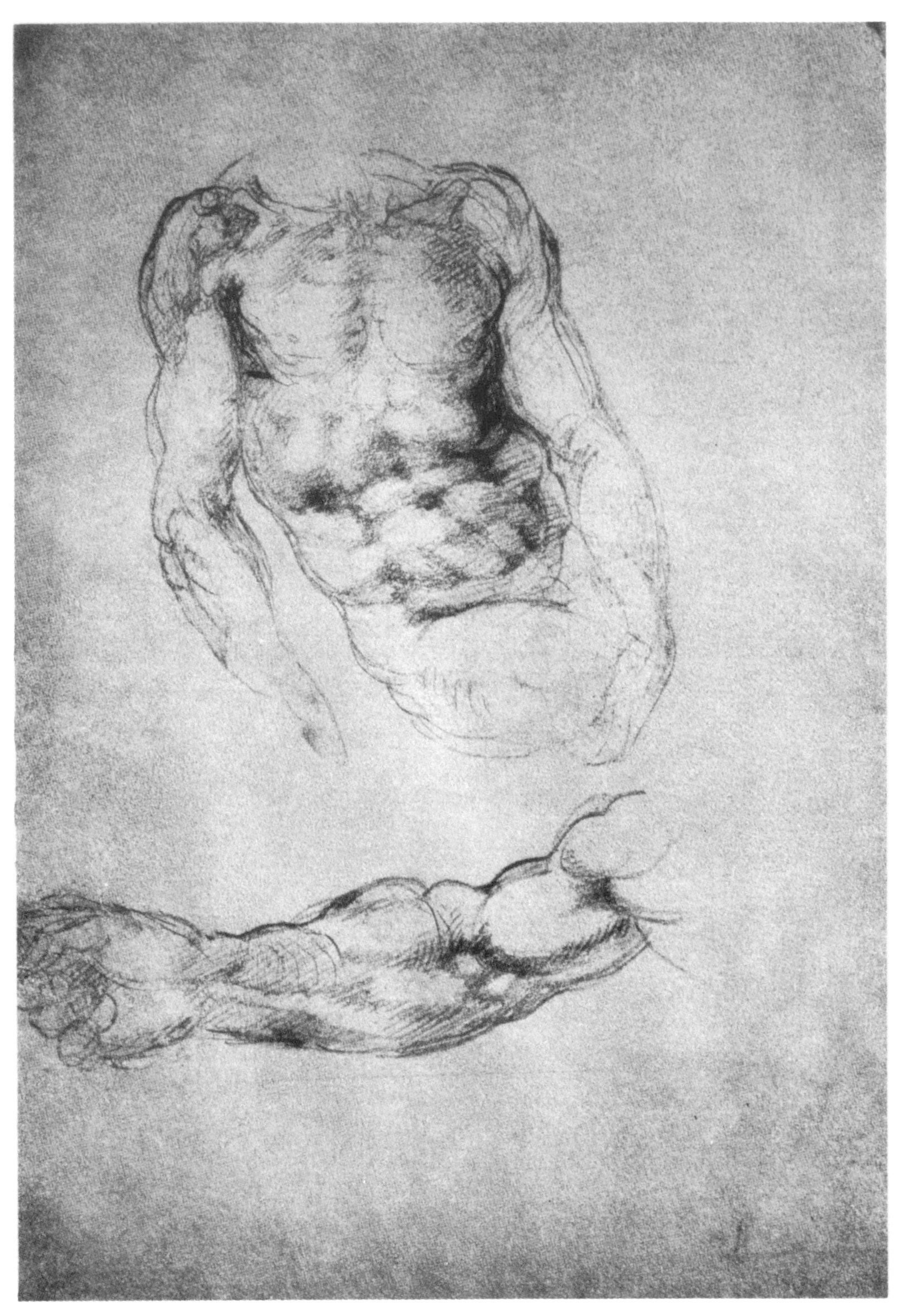

Courtesy of Casa Buonarroti, Florence

MICHELANGELO, ARM AND TORSO STUDY FOR A PIETA

Courtesy of Casa Buonarroti, Florence

MICHELANGELO, STUDY OF HEADS FOR THE LEDA

Courtesy of Casa Buonarroti, Florence

MICHELANGELO, STUDY OF A BACKGROUND FIGURE FOR THE "RISEN CHRIST"

Courtesy of the British Museum, London

MICHELANGELO, STUDY FOR THE "LAST JUDGMENT."

Courtesy of the British Museum, London

MICHELANGELO, CRUCIFIXION FOR VITTORIA COLONNA

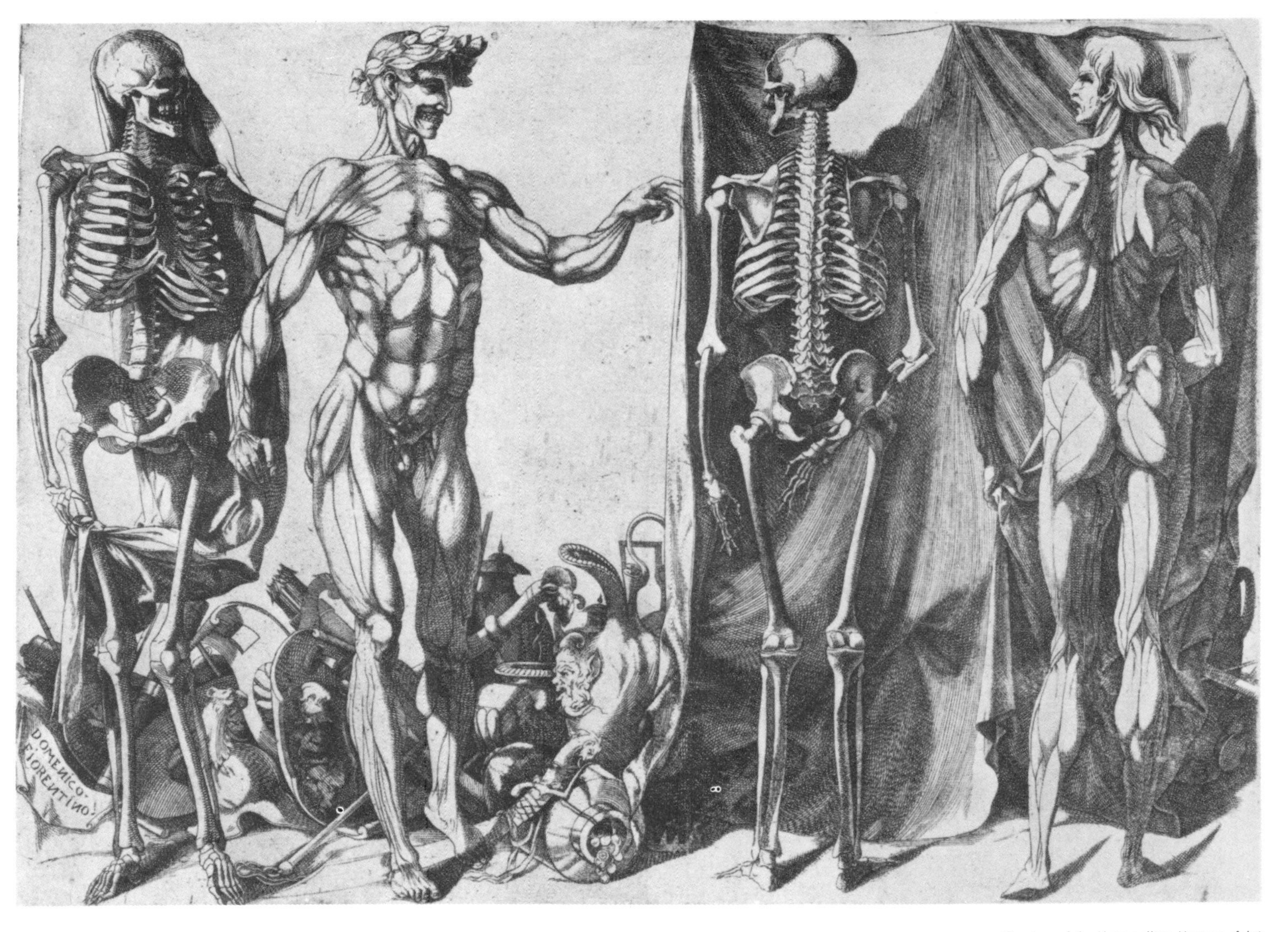

Courtesy of the Metropolitan Museum of Art

BARBIERI, DESIGN FROM AN ANATOMY BOOK

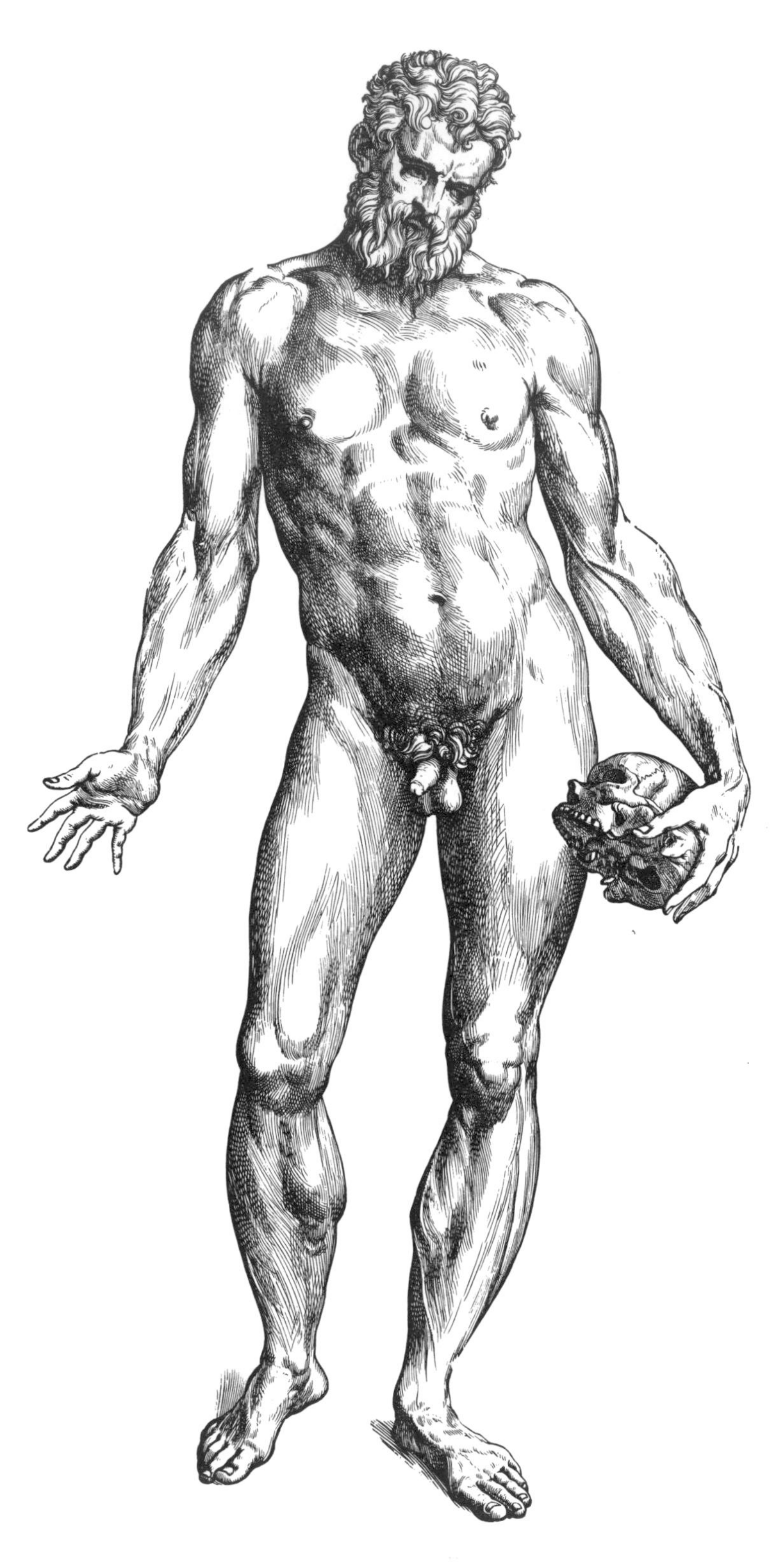

VESALIUS, MALE NUDE

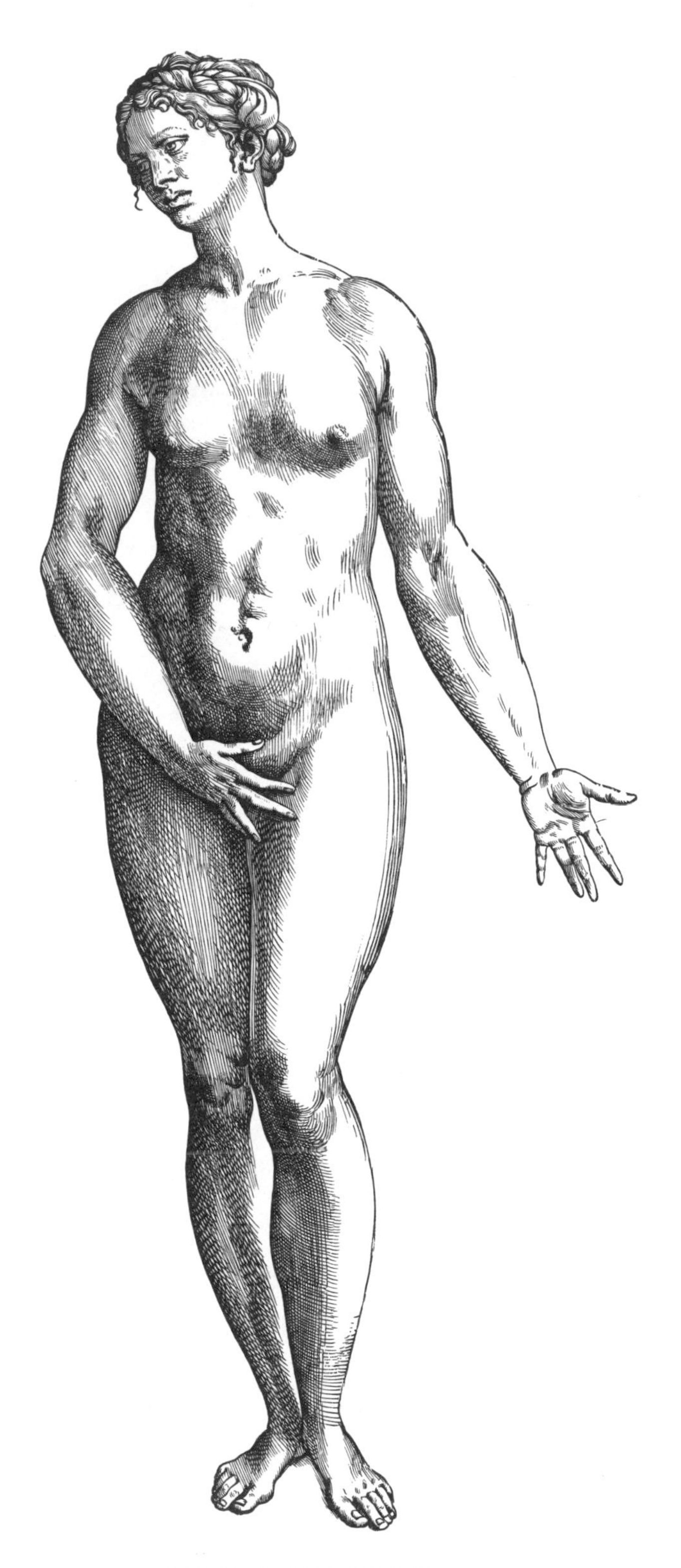

VESALIUS, FEMALE NUDE

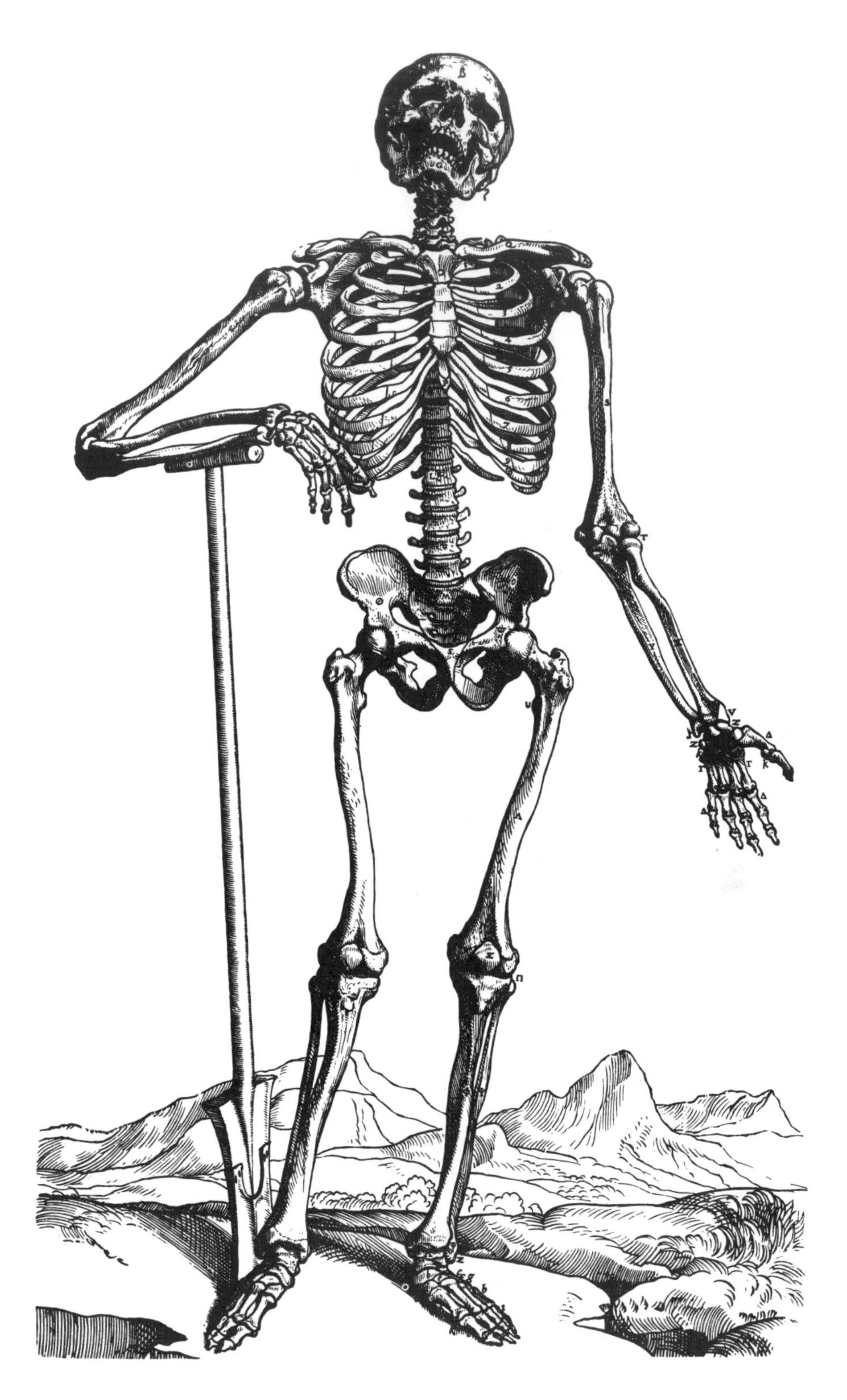

VESALIUS, SKELETON I

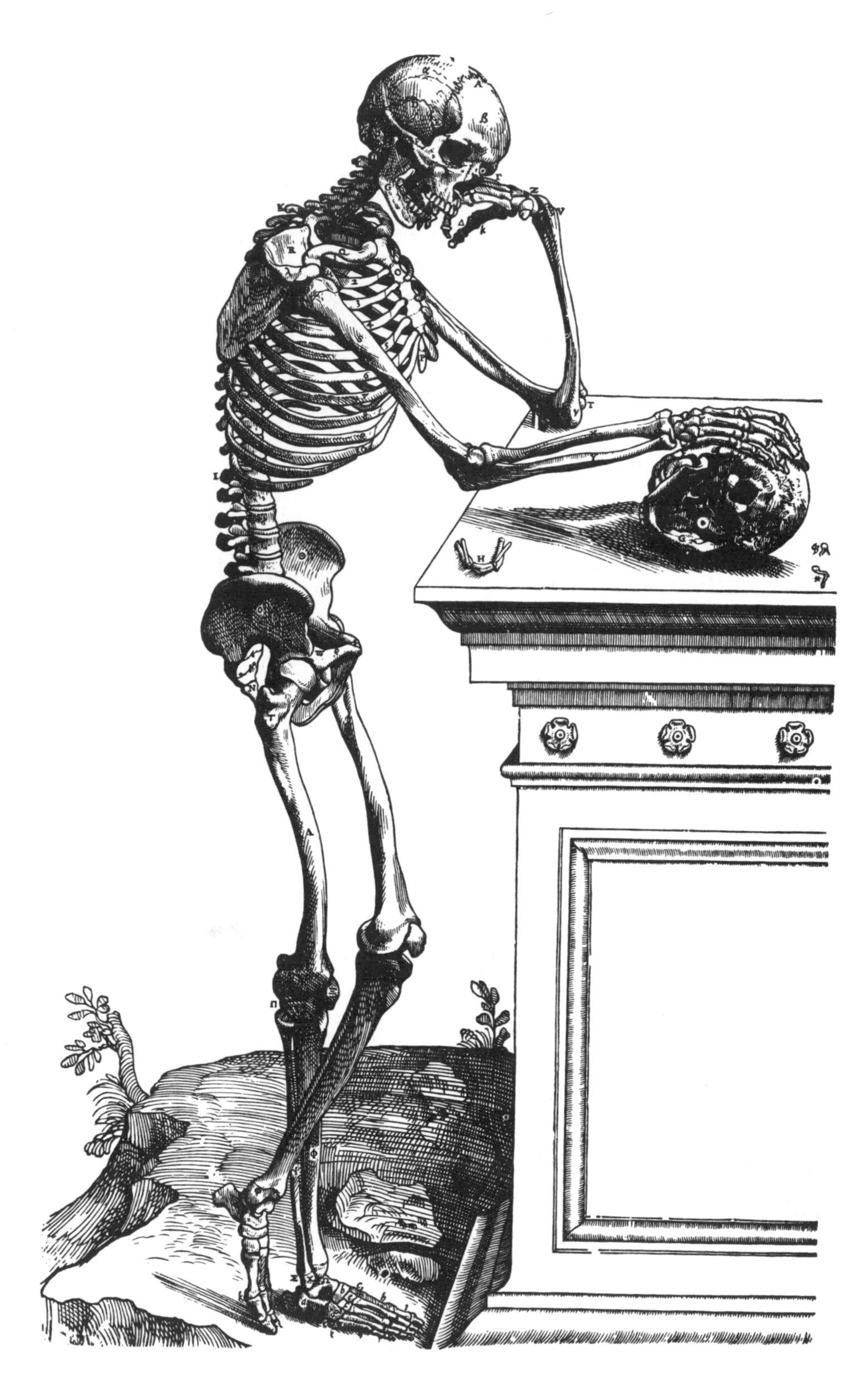

VESALIUS, SKELETON II

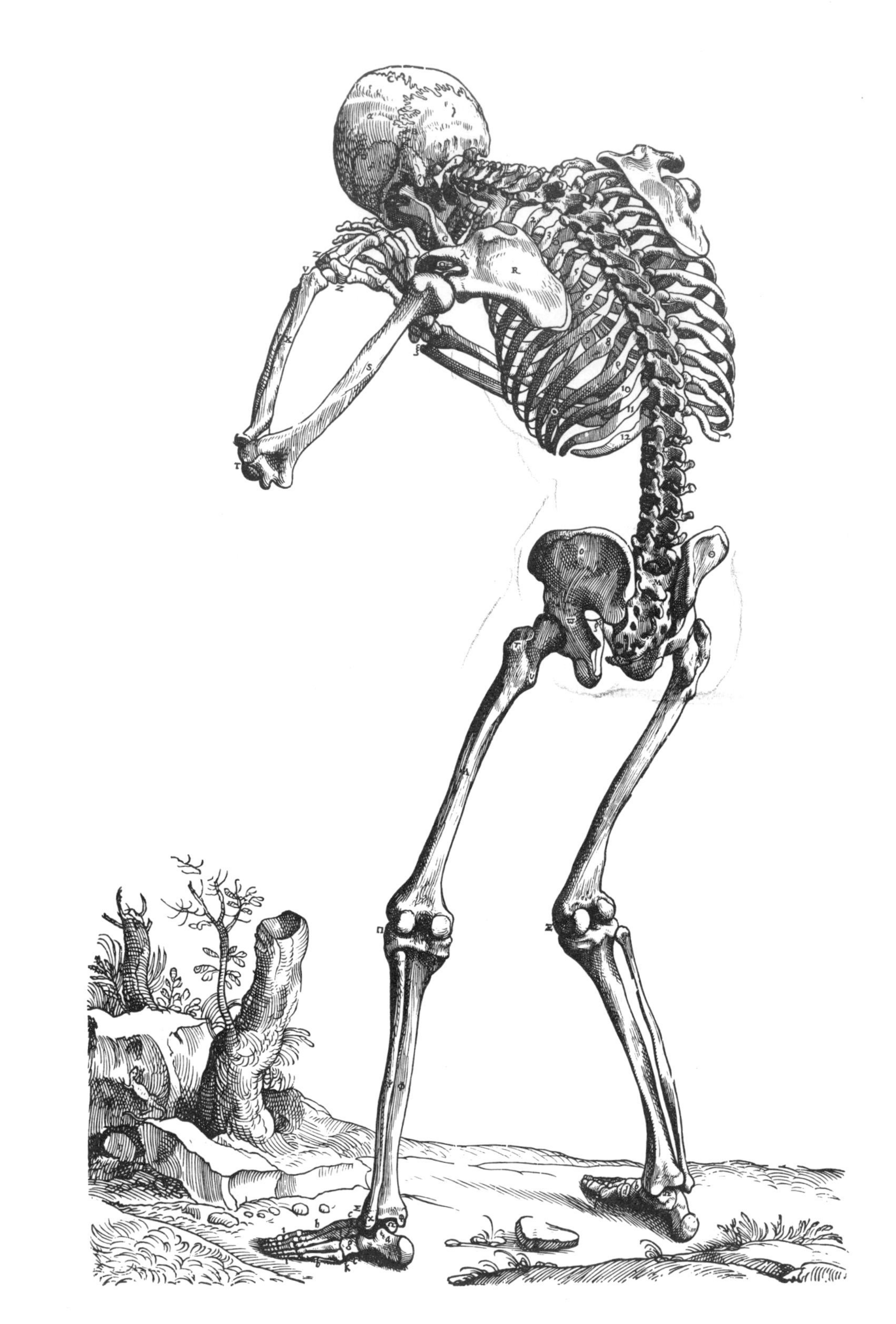

VESALIUS, SKELETON III

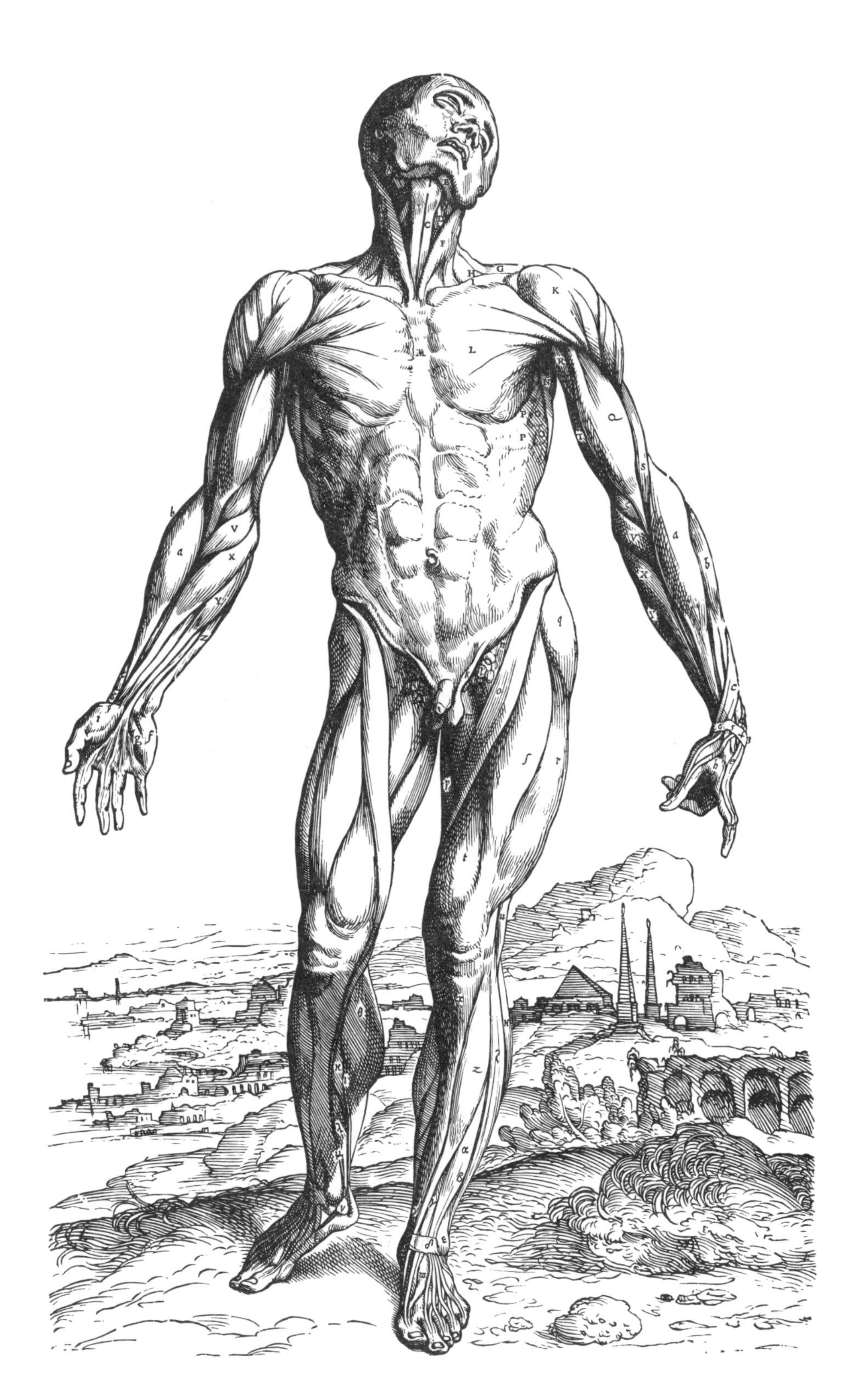

VESALIUS, MUSCLES I

PLATE 145

VESALIUS, MUSCLES II

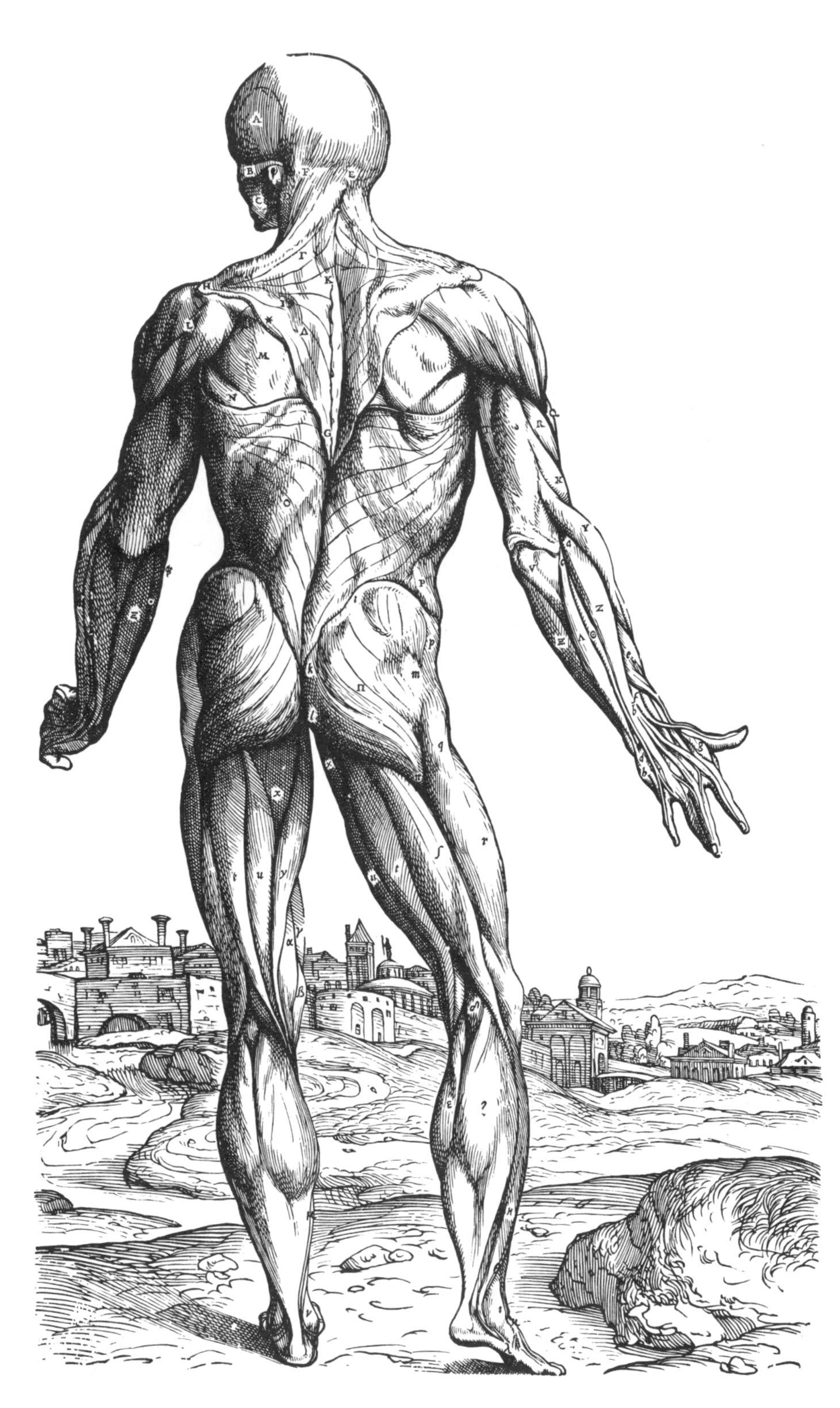

VESALIUS, MUSCLES III

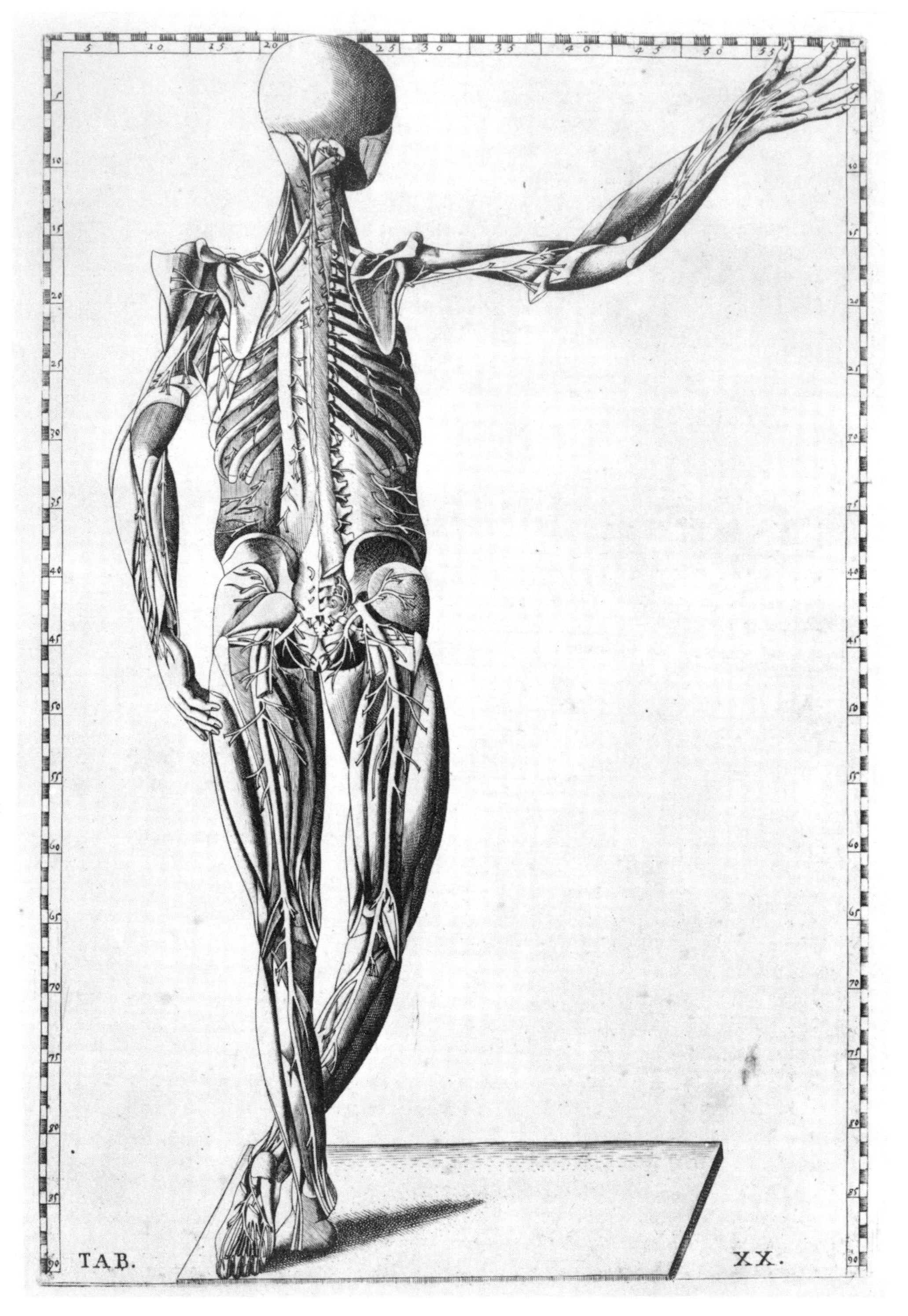

Courtesy of the Metropolitan Museum of Art

EUSTACHIO, PLATE FROM THE "TABULAÉ ANATOMICAE"

PLATE 148

Courtesy of the Metropolitan Museum of Art

MEIER, "APOLLO FLAYING MARSYAS"

Courtesy of the Metropolitan Museum of Art

BORGIONI, "THE DEAD CHRIST"

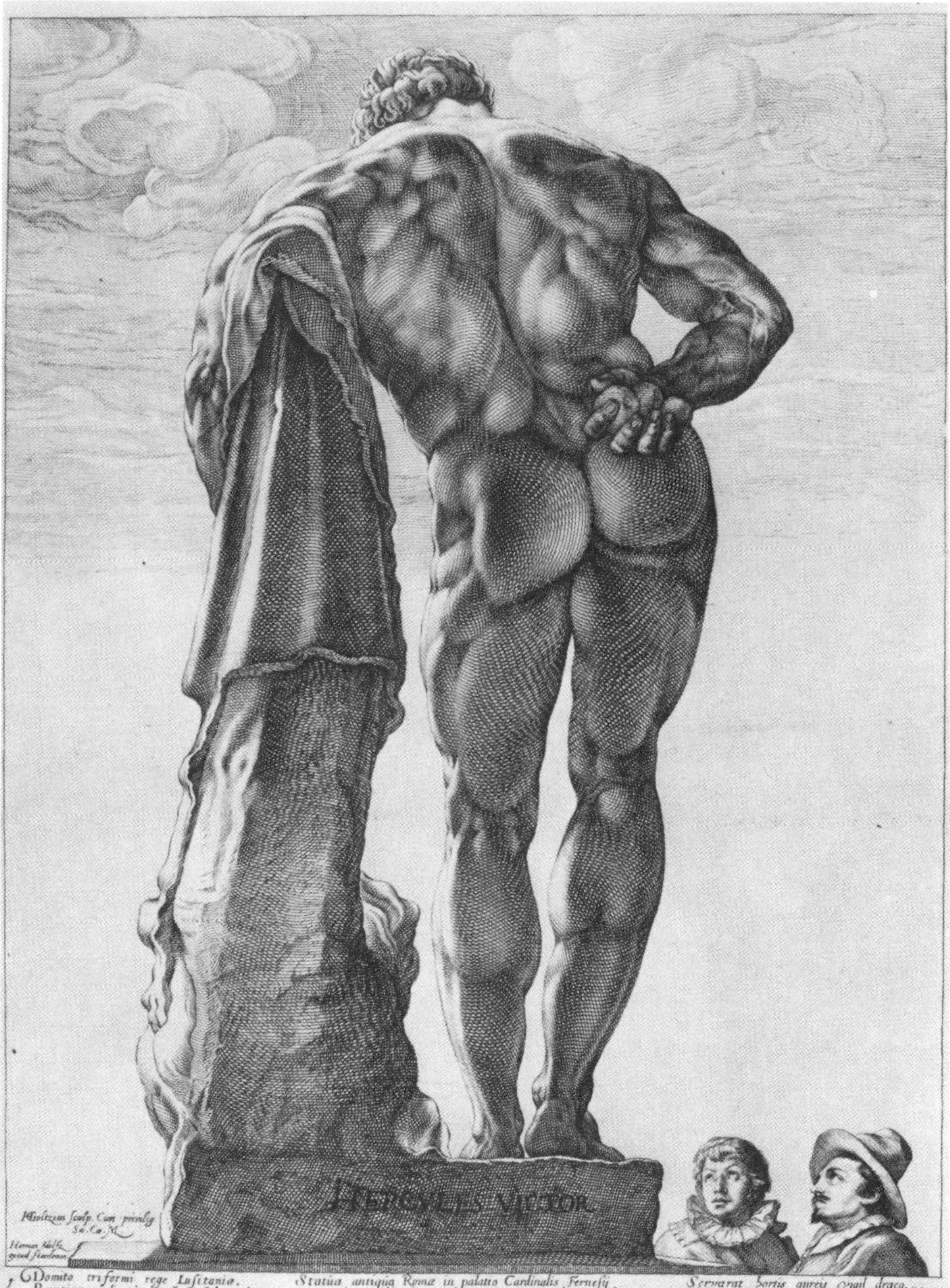

Courtesy of the Metropolitan Museum of Art

GOLTZIUS, THE "FARNESE" HERCULES

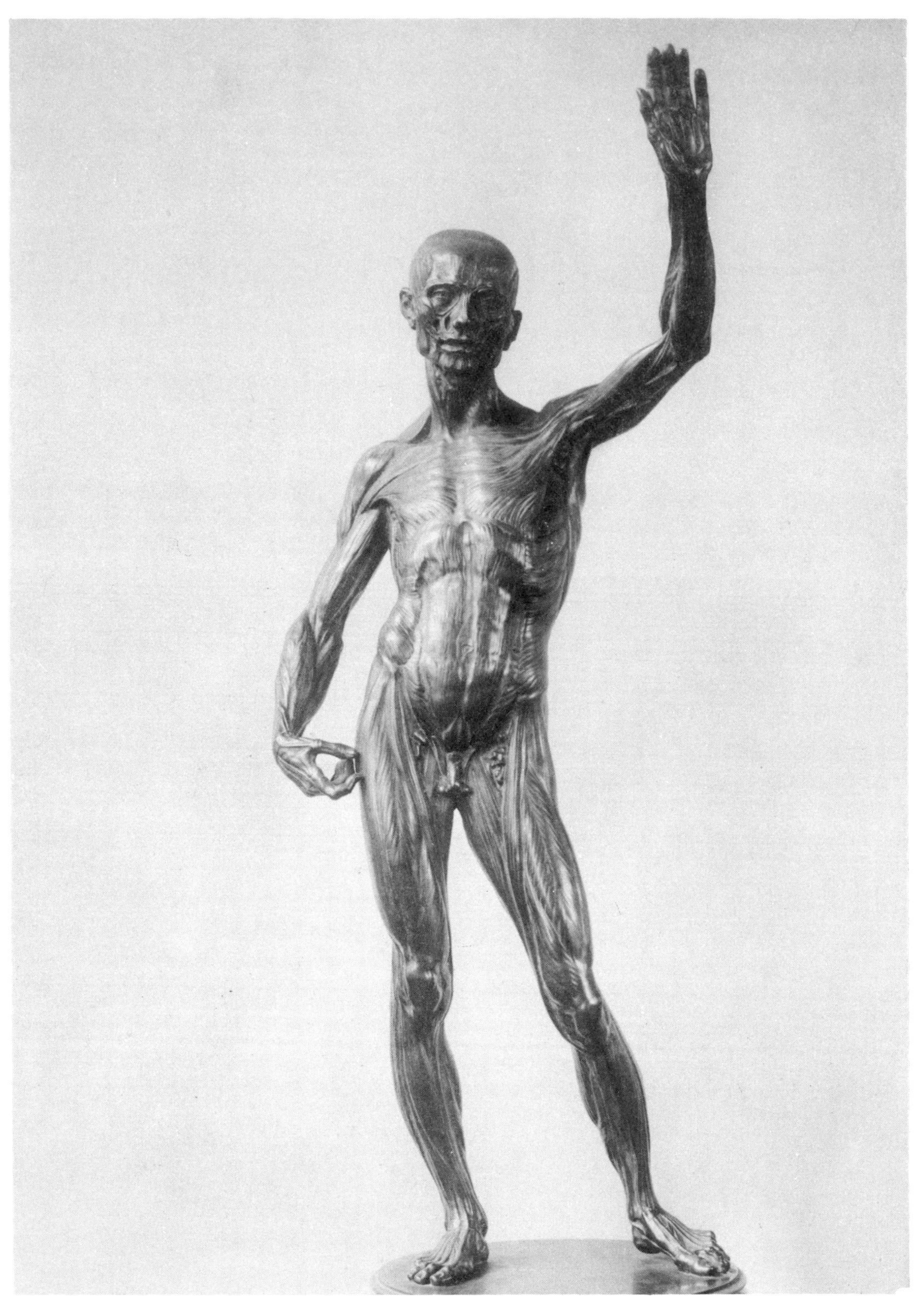

Courtesy of the Museo Nazionale, Florence

CARDI, ANATOMICAL FIGURE

RUBENS, "STUDIES OF VENUS"

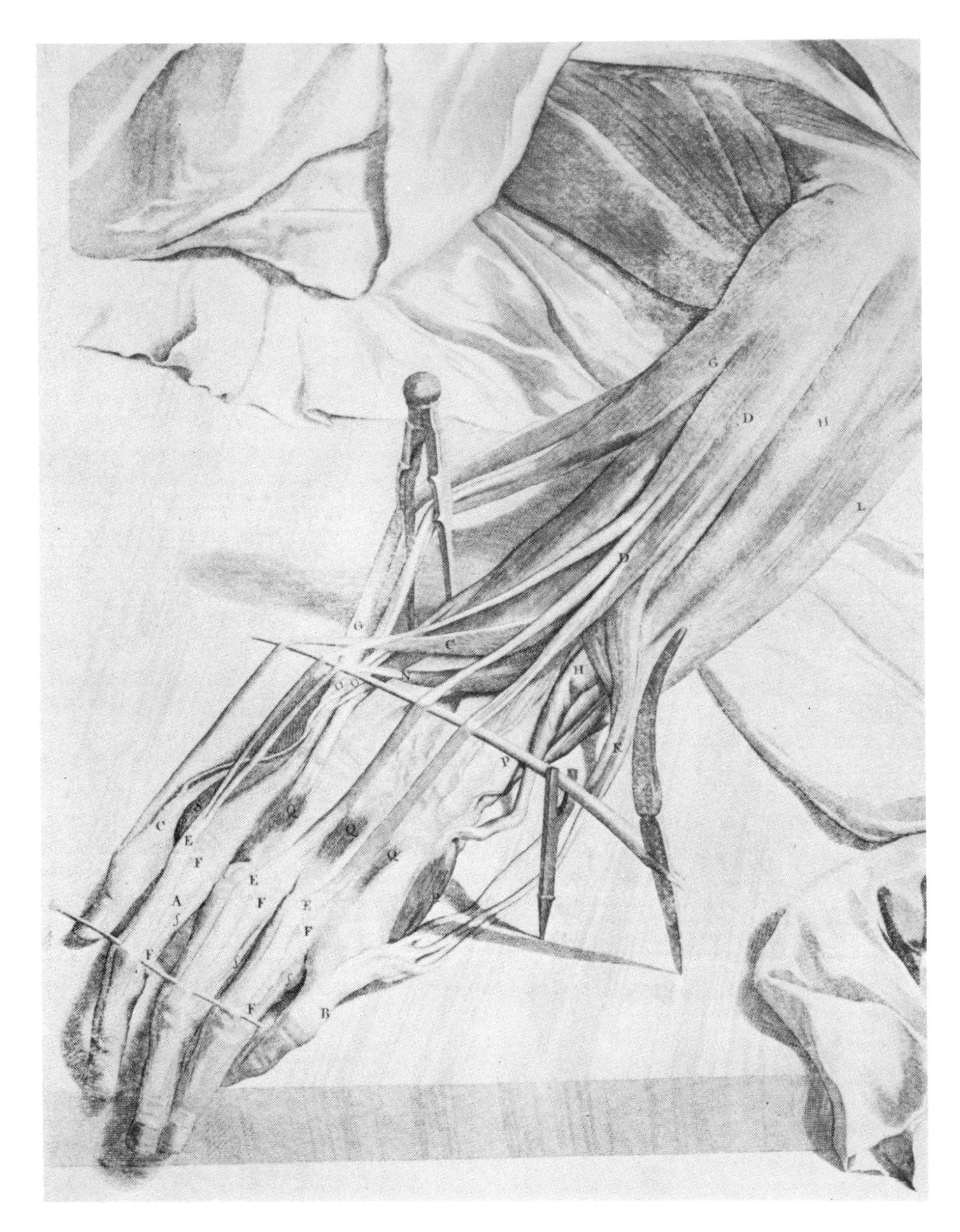

CAMPER, A STUDY OF THE HAND FROM THE ANATOMY OF HUMANE BODIES

PLATE 154

Courtesy of the Metropolitan Museum of Art

GOYA, "THE GIANT"

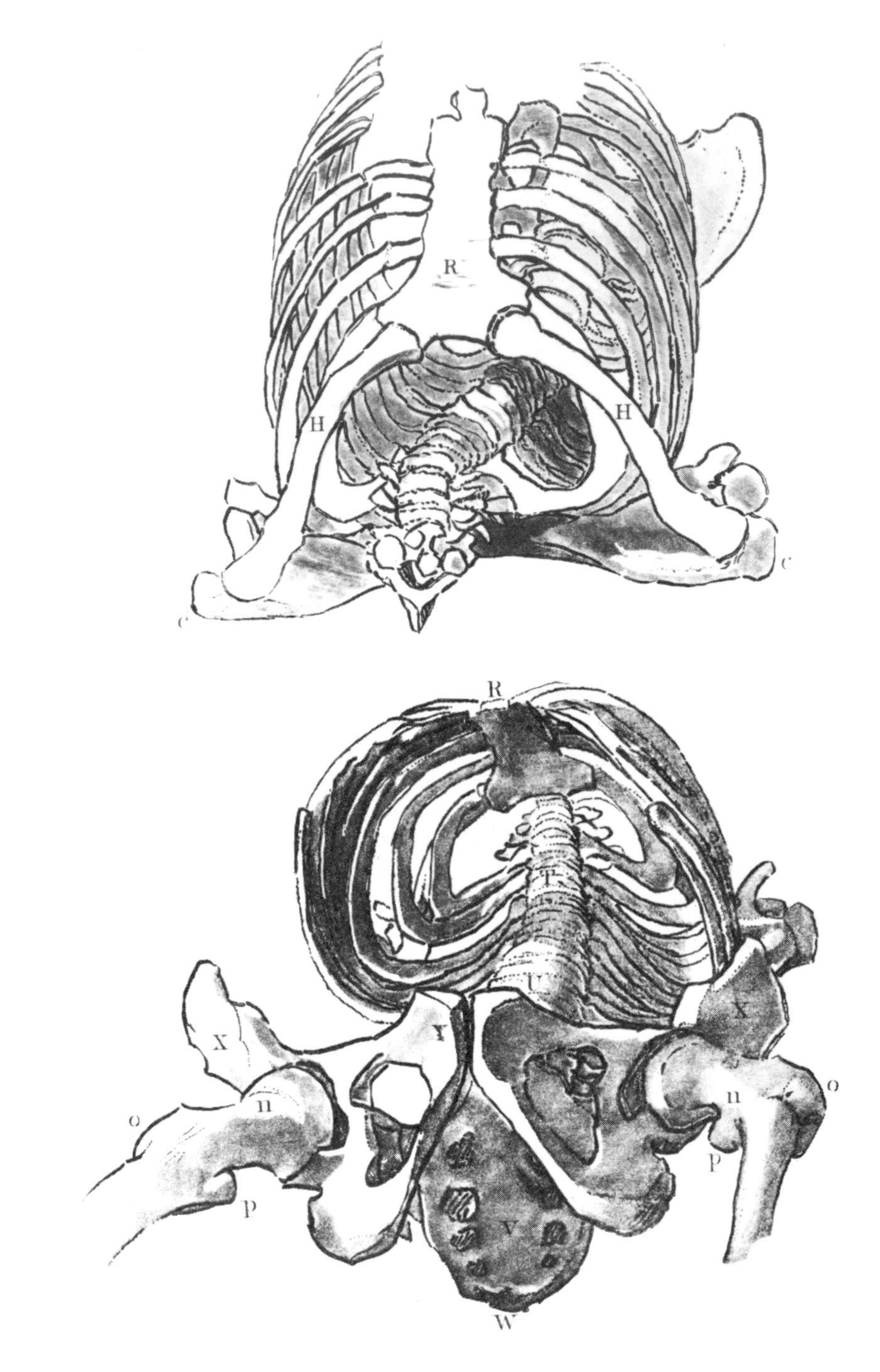

FLAXMAN, VIEWS OF THE THORACIC BASKET

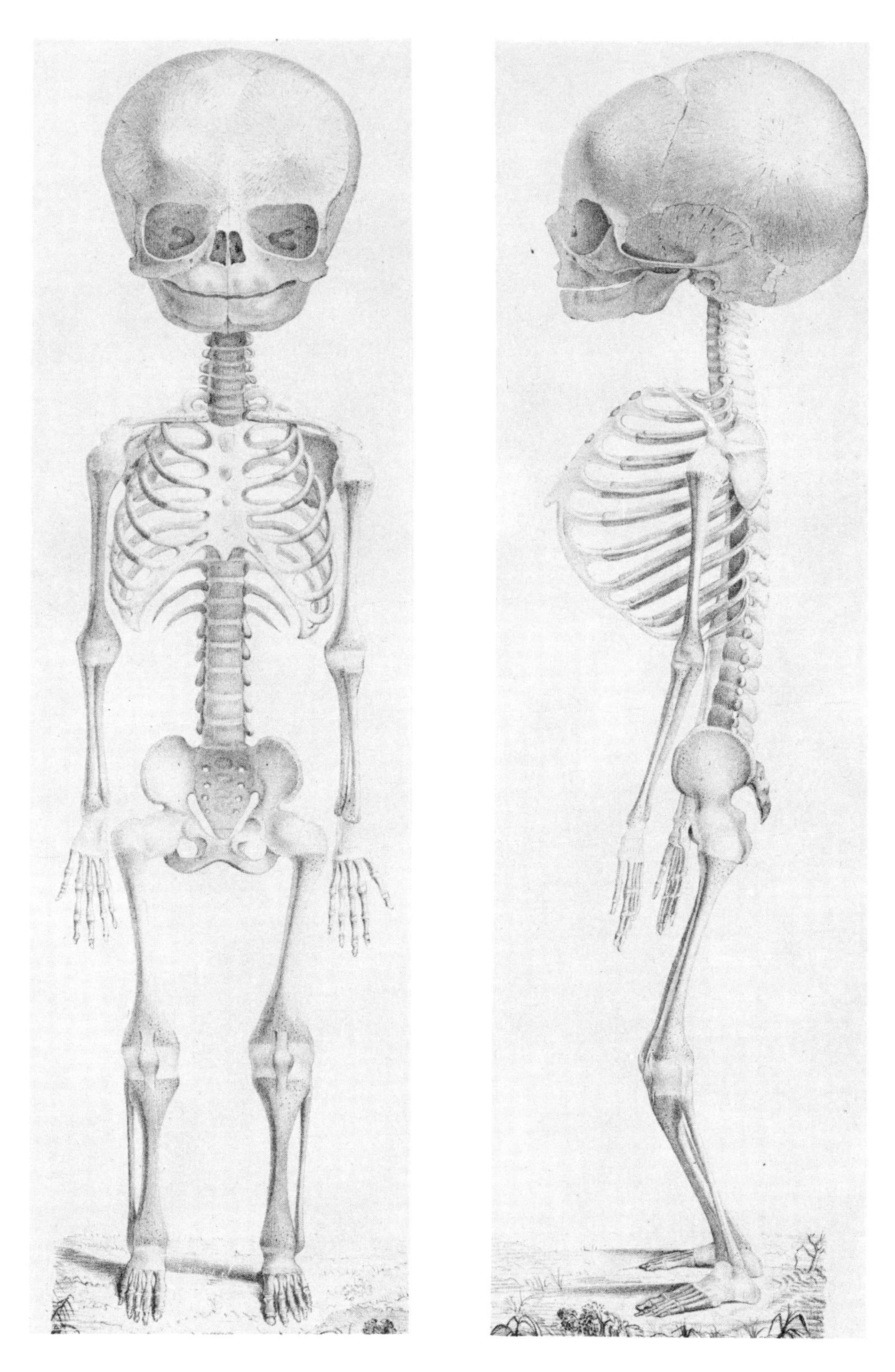

CLOQUET, SKELETON OF INFANT

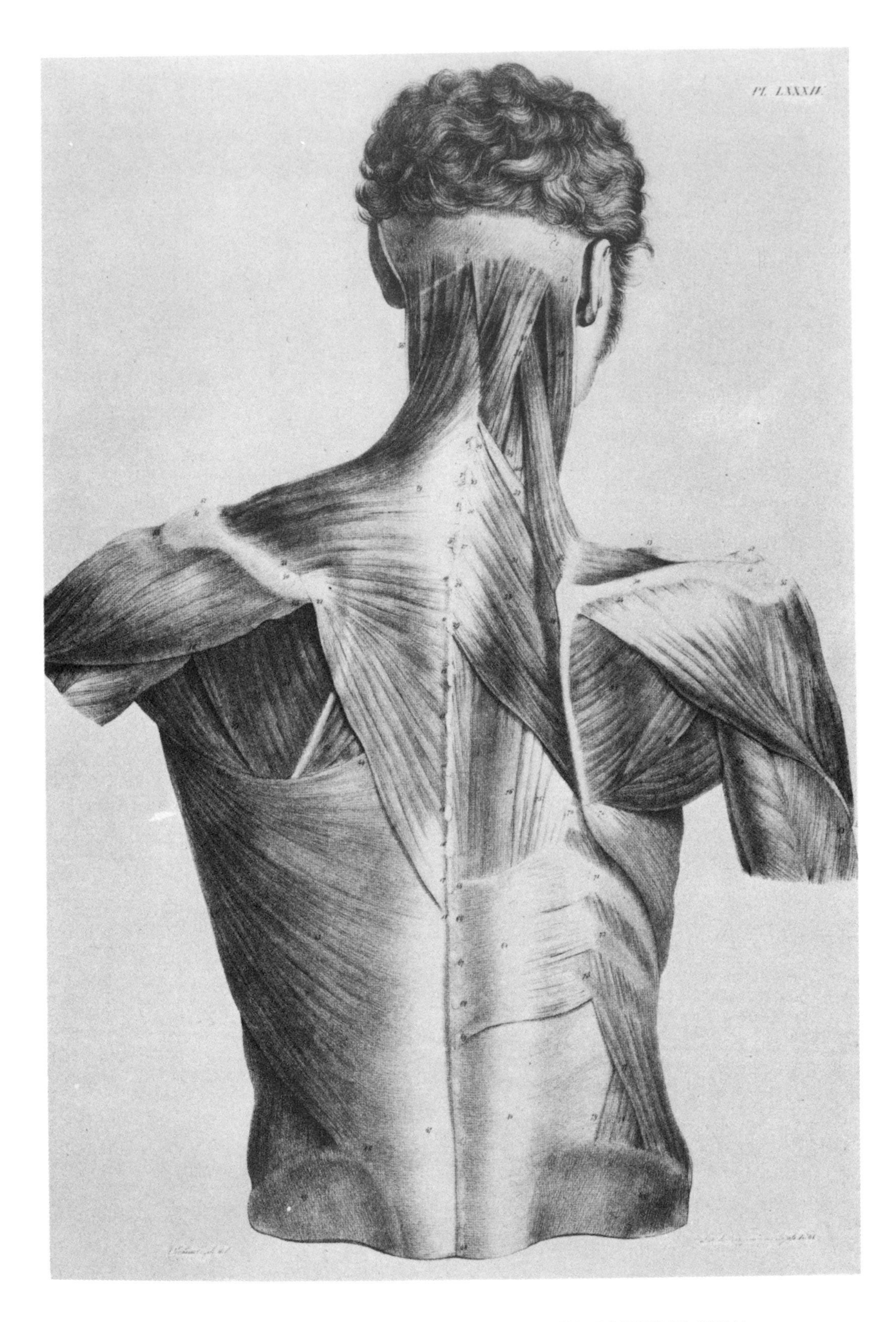

CLOQUET, MUSCULATURE OF TORSO AND NECK, POSTERIOR VIEW

PLATE 158

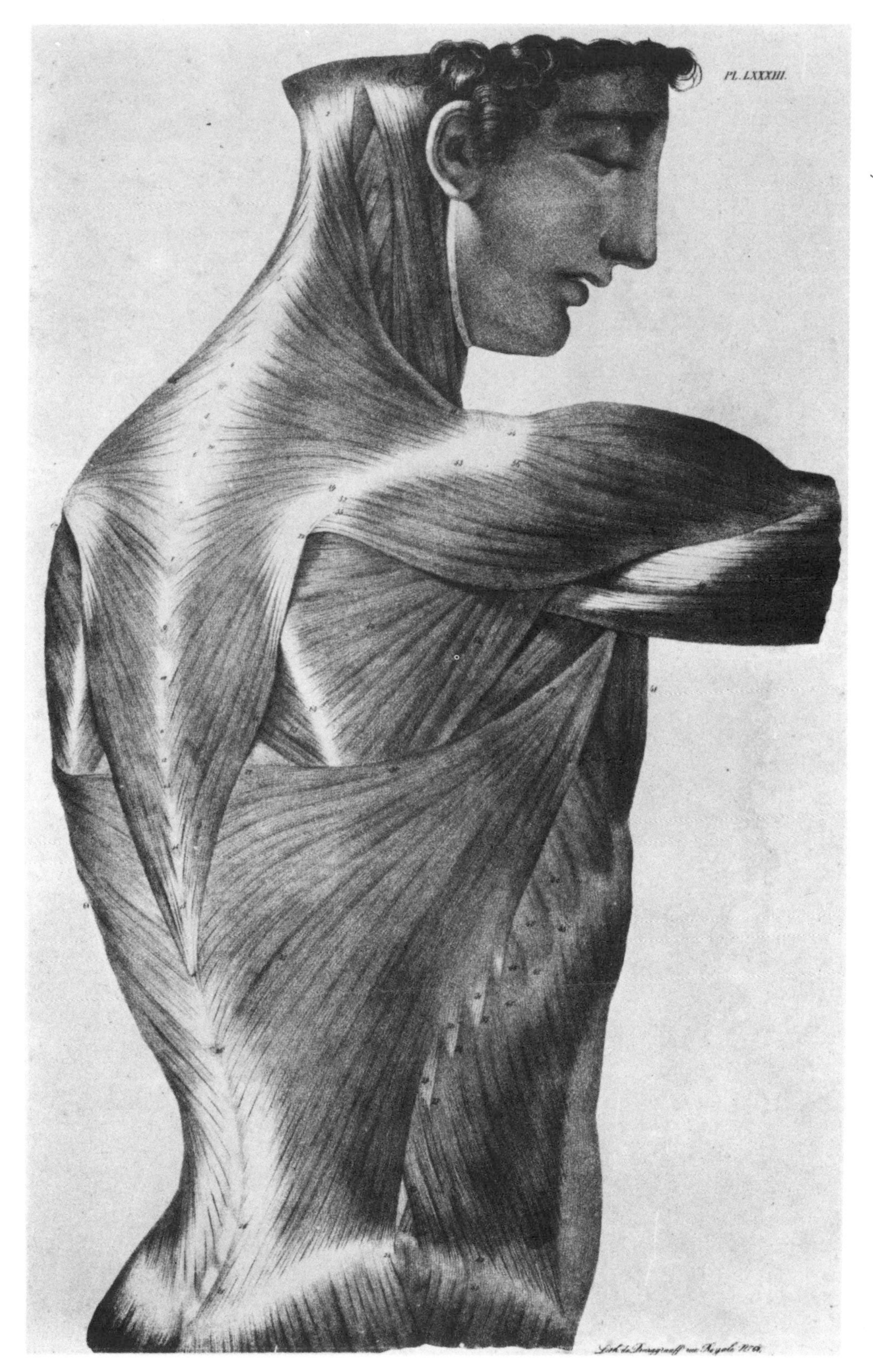

CLOQUET, MUSCULATURE OF TORSO AND NECK, LATERAL VIEW

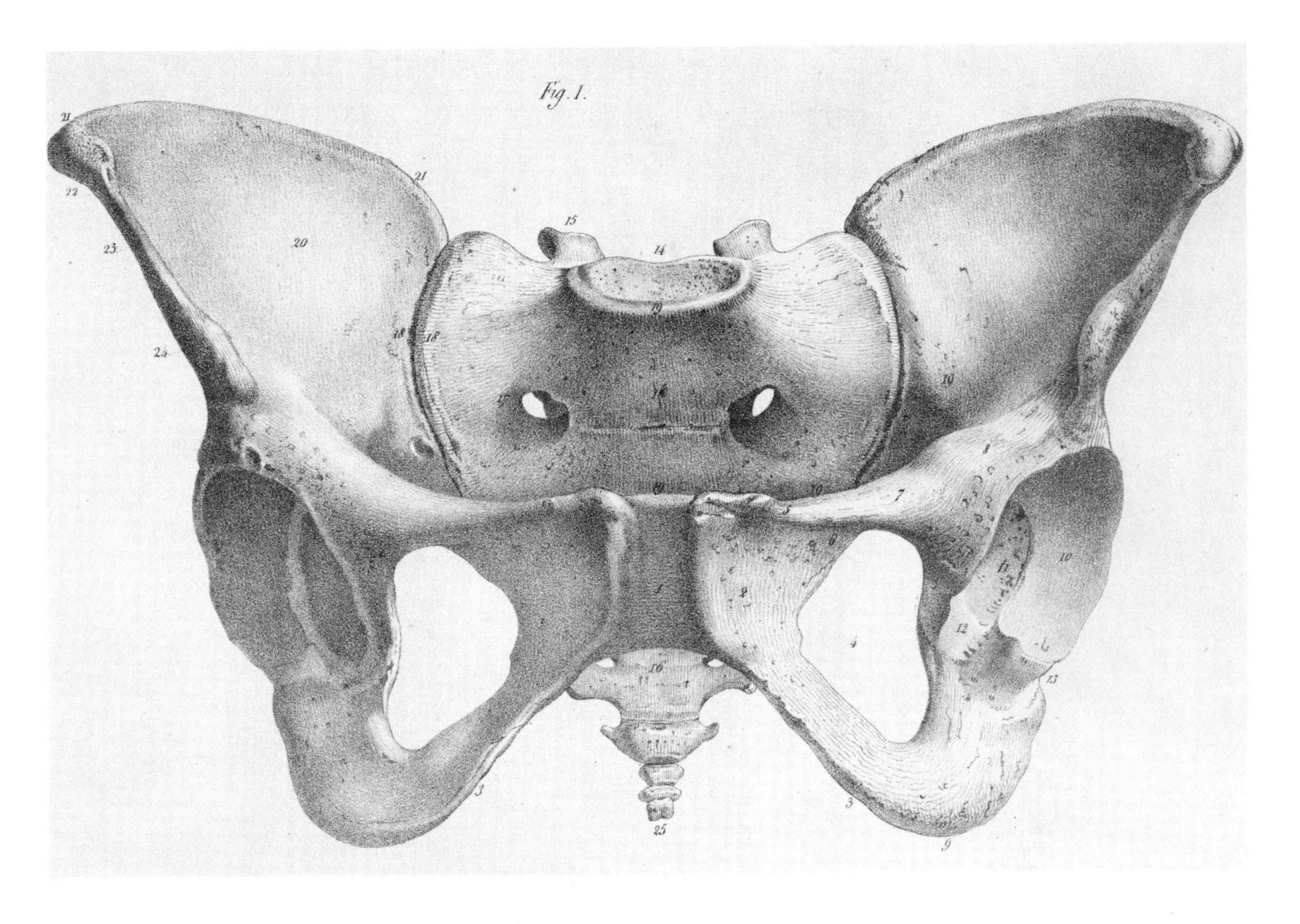

CLOQUET, FEMALE PELVIS, ANTERIOR VIEW

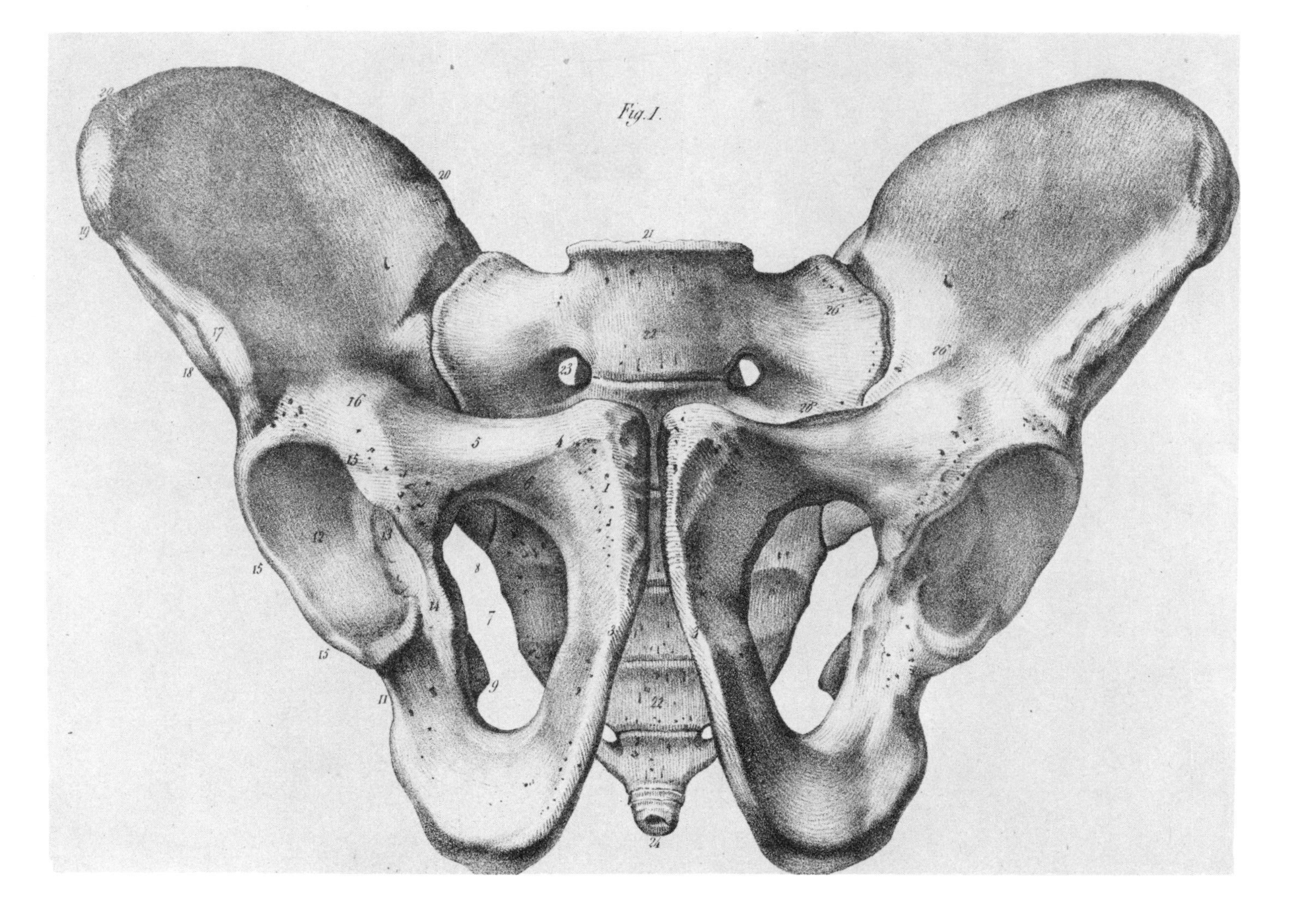

CLOQUET, FEMALE PELVIS, ANTERIOR VIEW FROM BELOW

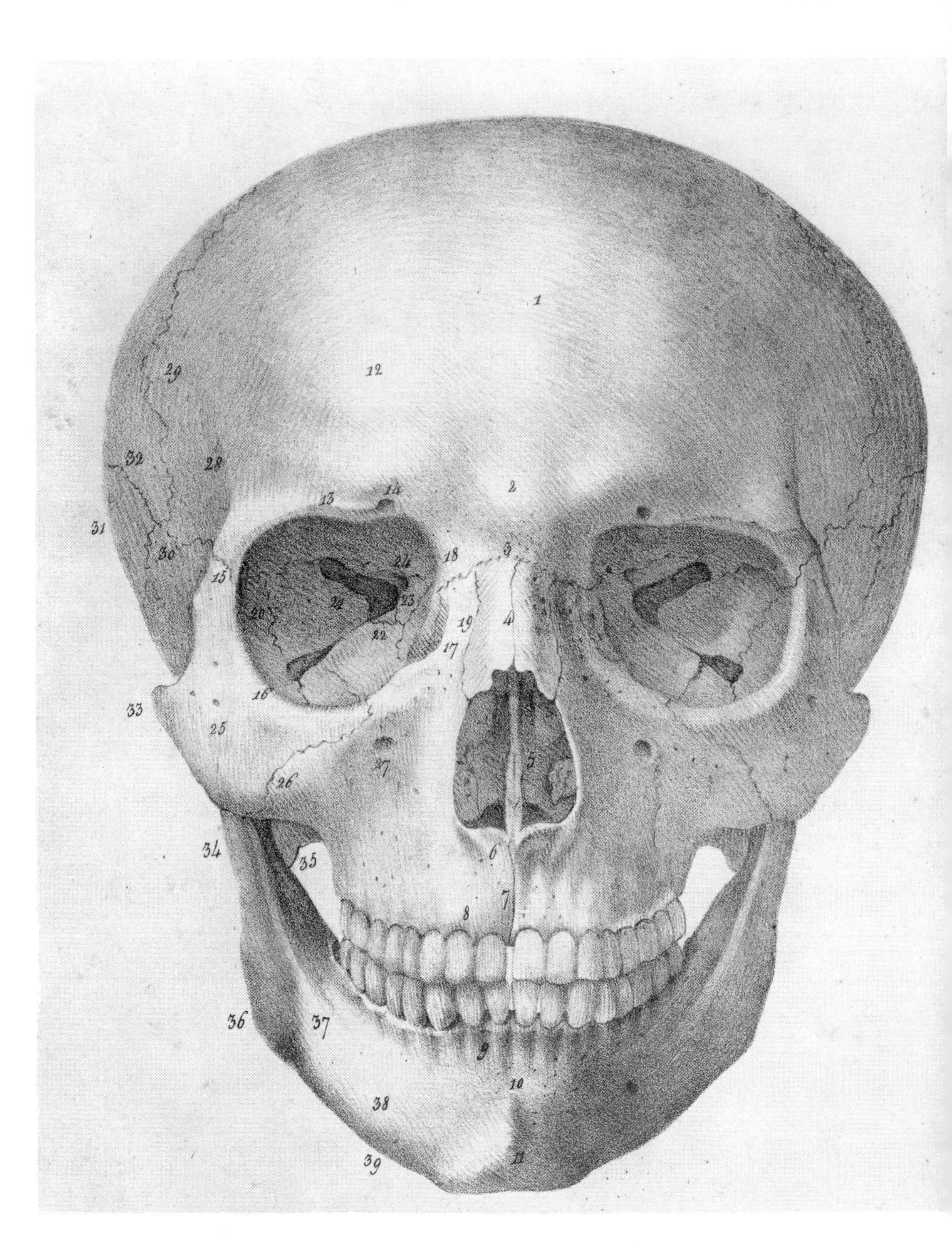

CLOQUET, SKULL, ANTERIOR VIEW

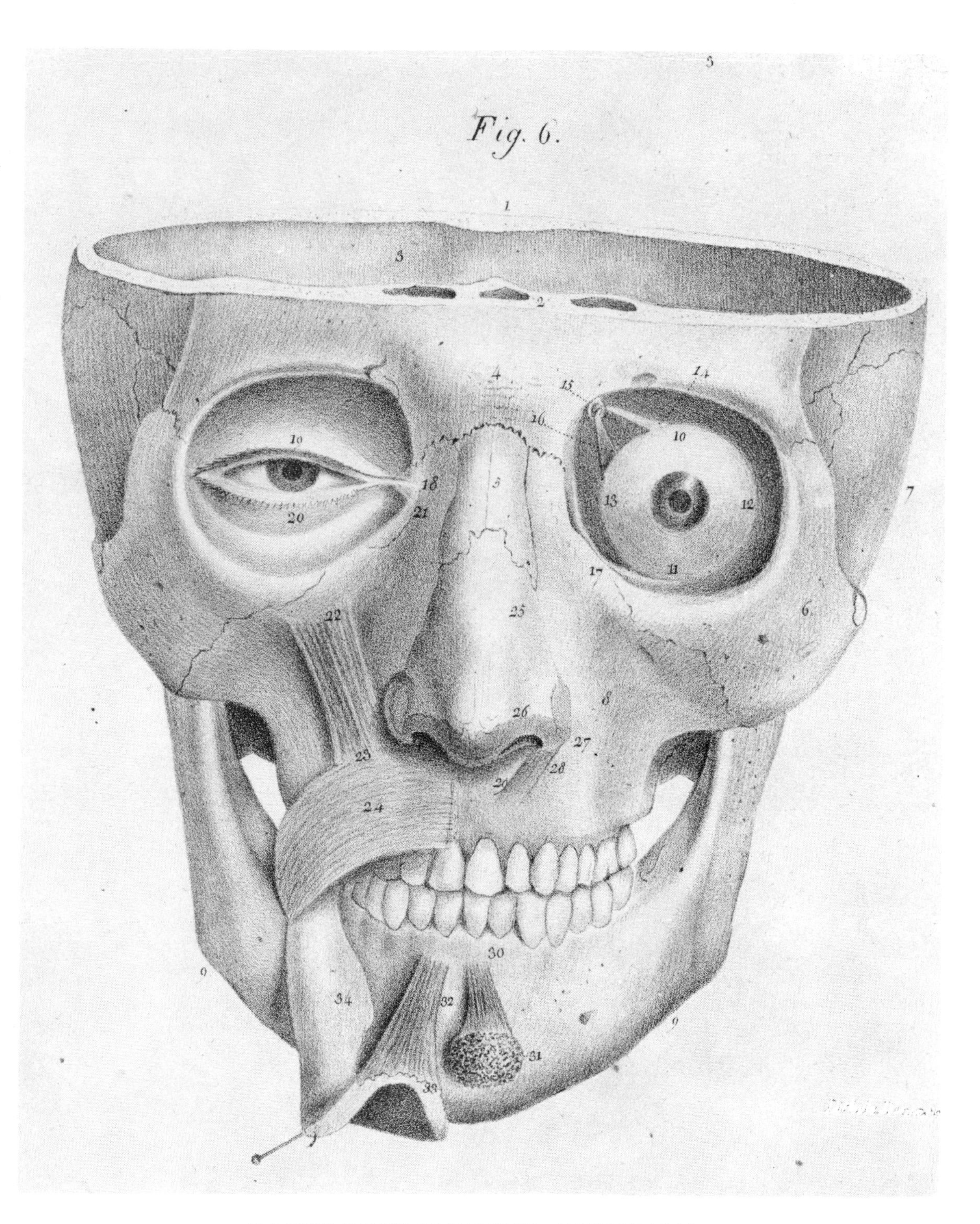

CLOQUET, BONES OF THE SKULL, ANTERIOR VIEW

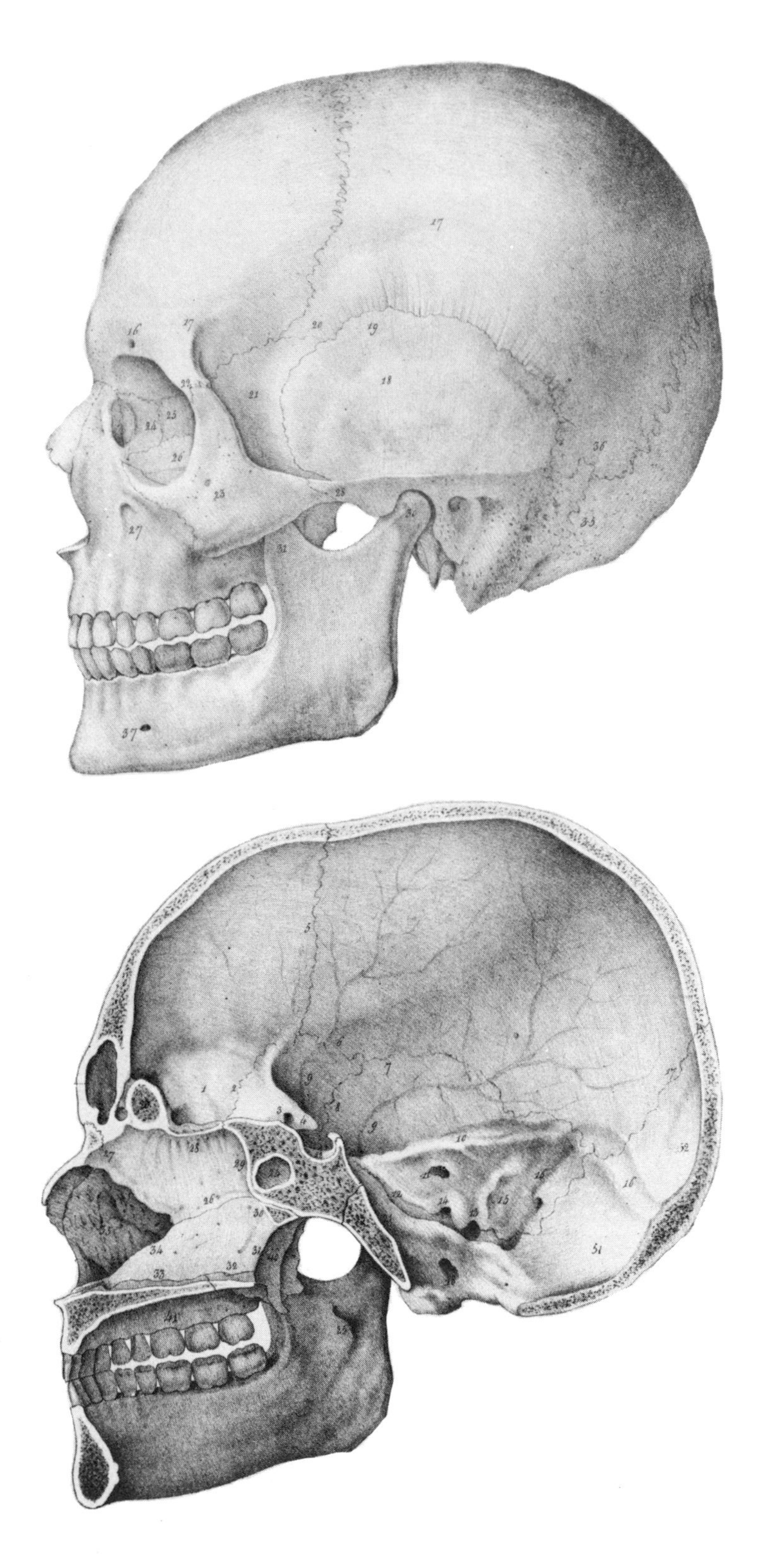

CLOQUET, SKULL, LATERAL VIEW AND SAGITTAL SECTION

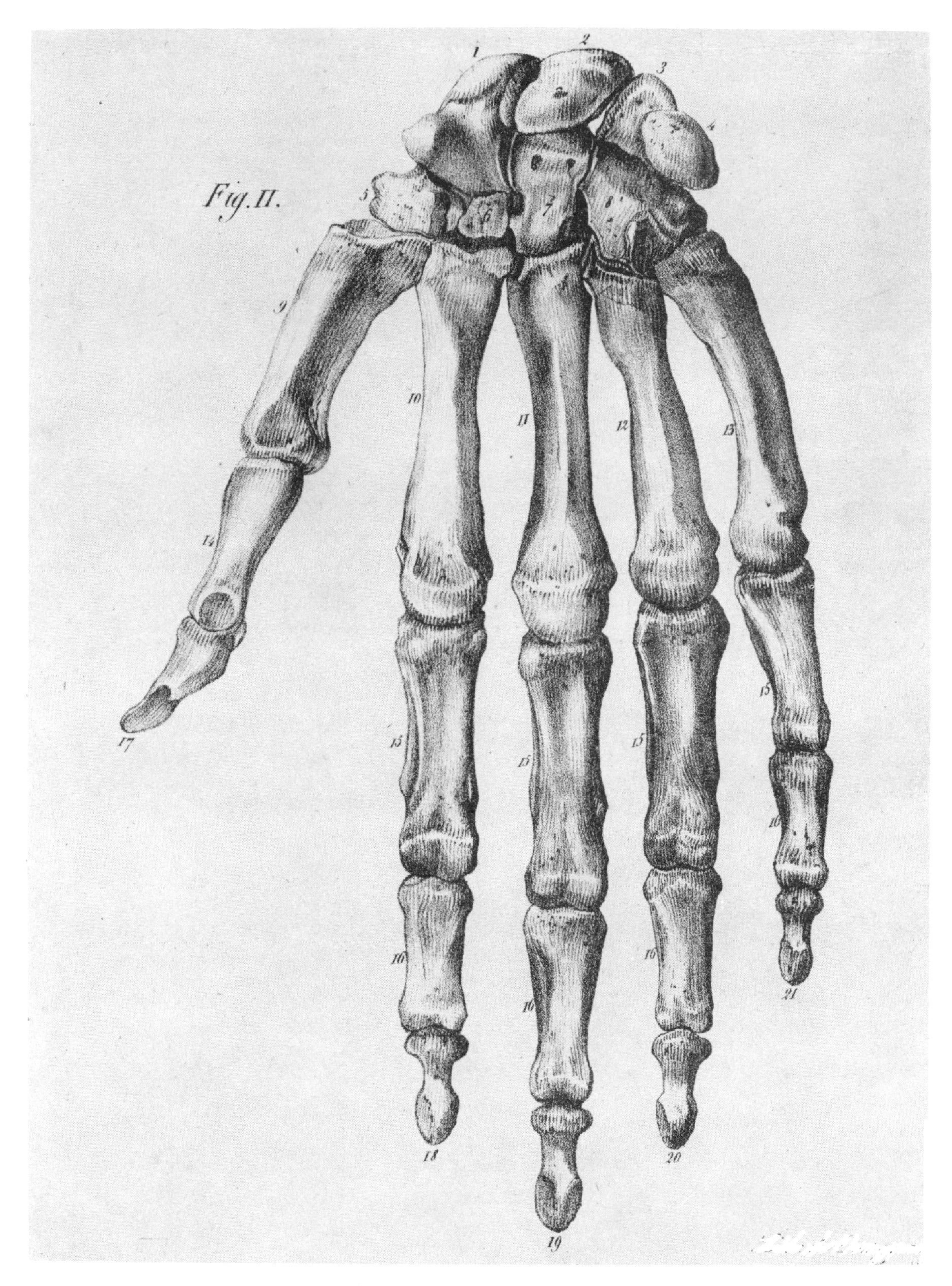

CLOQUET, BONES OF THE HAND

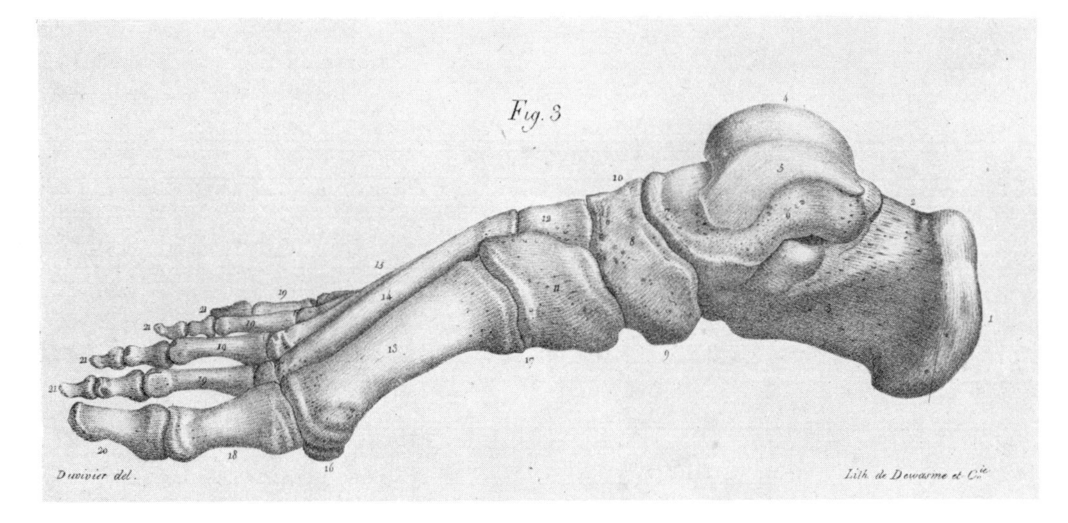

CLOQUET, BONES OF THE FOOT, LATERAL VIEW

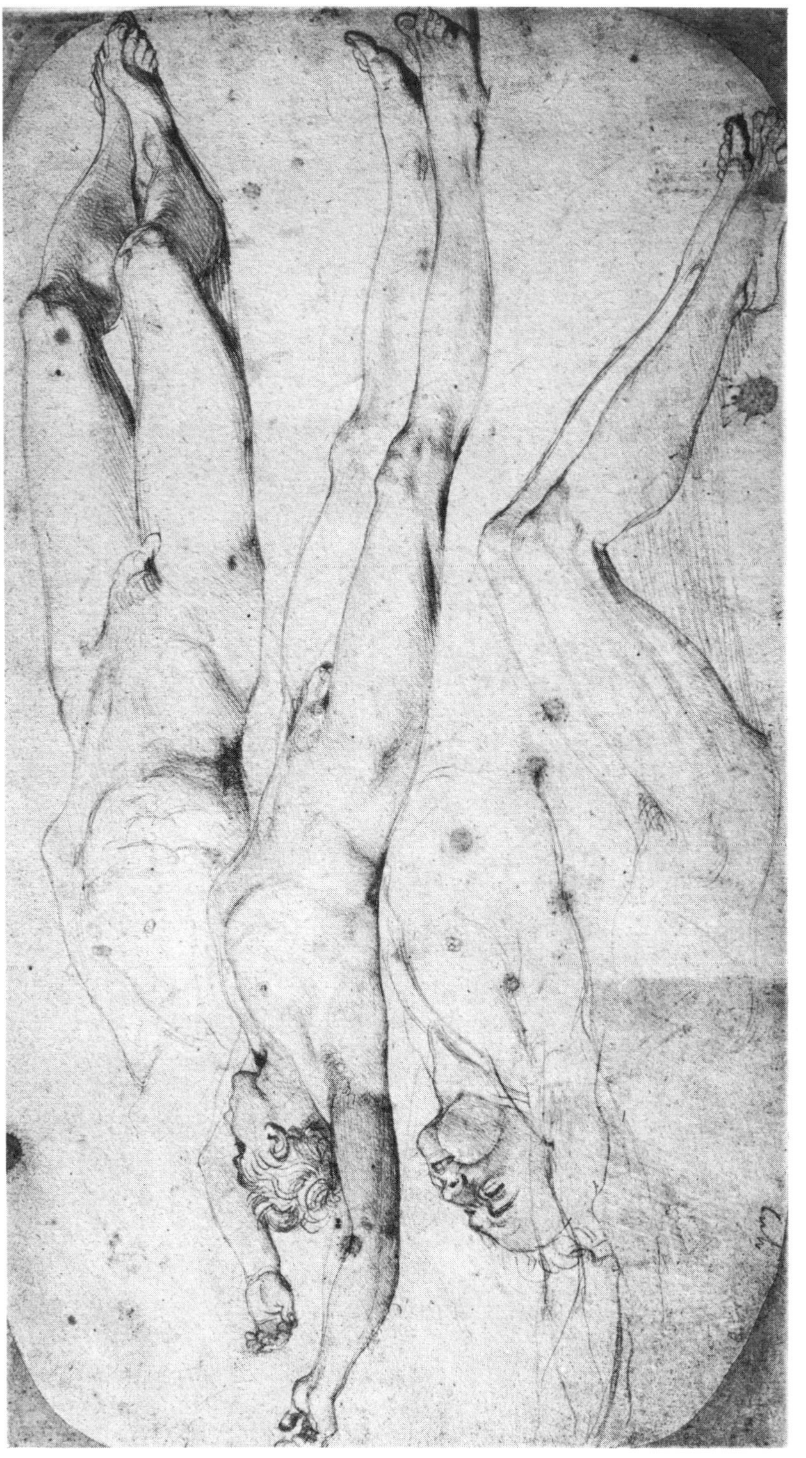

Courtesy of the Metropolitan Museum of Art

INGRES, "STUDIES FOR THE DEAD BODY OF ACRON"

Muscles of The Back.

No. 177.

Integumentary Form.

Muscular Form.

10. Rhomboideus. This muscle arises from the Spinous processes of the seventh cervical and five superior dorsal vertebrae, and is inserted into all that part of the vertebral border of the Scapula situated below its Spine. It is covered, except at its lower third, by the Trapezius. It is the artistic muscle of the Scapular field, as the Coraco-Brachialis is of the Brachial field on the pectoral side of the arm. An imperfect outline of its vertebral border may be seen, in persons in whom the muscles of the back are much developed, through the body section of the Trapezius muscle. The superior angle of the Scapula and its vertebral border being also visible through the Trapezius at points suited to their anatomical relations, and the movements of their parts.

9. Serratus Magnus arises by nine fleshy digitations from the upper border of the eight upper ribs, and is inserted into the whole length of the vertebral border of the Scapula upon the under side of that bone. It is divided into three parts. Its lower third is attached to the inner surface of the inferior angle of the Scapula by an attachment which is partly tendinous and partly muscular. This limb falls below the angle of the Scapula and is covered by the Latissimus Dorsi; through which, notwithstanding the thickness of that muscle, its form is distinctly visible.

Muscles, Tendons and Bones

1. Acromial limb of Trapezius. 2. Wing of Trapezius. 3. Shoulder of Trapezius. 3a Body of Trapezius.
4. Acromial limb of Deltoid. 5. Scapular limb of Deltoid. 6. Infra Spinatus. 7. Teres Major. 8 Teres Minor.
10. Superficial portion of Rhomboideus. 11. 11a Latissimus Dorsi. arises from the spinous processes of the sixth inferior dorsal, from those of the Sacral and Lumbar vertebrae, from the posterior third of the crest of the ilium. See Skeleton sections and from the last four ribs, and is inserted by a strong tendon into the lower part of the inner border of the bicipital groove of the humerus. From its upper border limit it passes outward along the inferior edge of the Rhomboideus, over the inferior angle of the Scapula and the lower third of the Teres major; to disappear, in posterior view, behind the upper third of that muscle. Its contractile fibres arise upon the outer border of the muscles of the loins and sacrum; and, ascending in the manner represented above, cover the flank section of the External oblique muscle, and come upon the outline at 11a. 11a. Its tendon of origin is Aponeurotic; and, in passing outward from the Spinal column, it covers, and forms a sheath for, the muscles of the loins and sacrum. Viz. Sacro-Lumbalis. 12.

13. Flank Section of the External Oblique Muscle.
14. Pelvic Tendon of Latissimus Dorsi.
15. Space Separating the External Oblique from the crest of the Ilium. 16. Crest of the Ilium.
o o The Ribs.

RIMMER, MUSCLES OF THE BACK

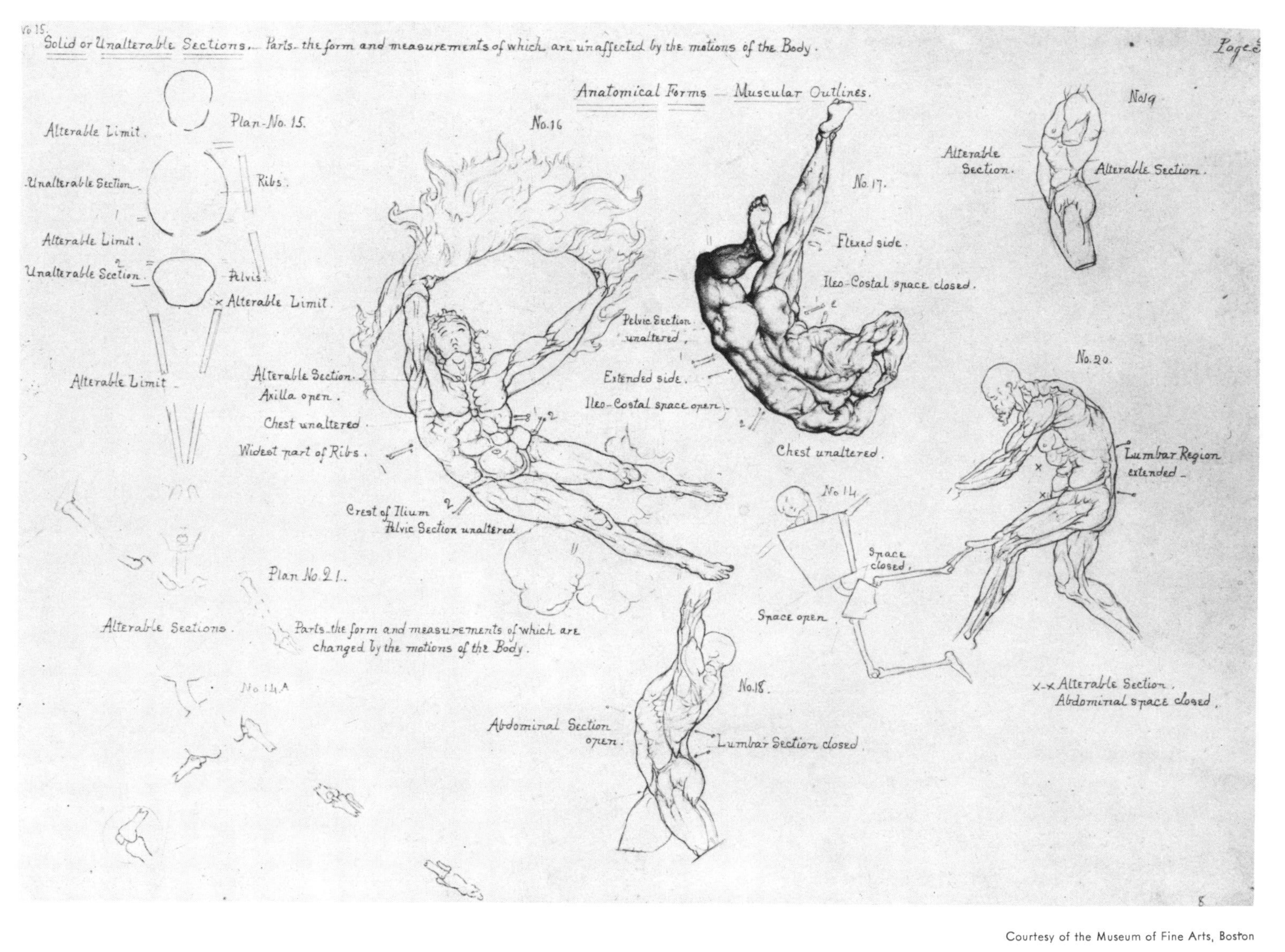

Courtesy of the Museum of Fine Arts, Boston

RIMMER, MUSCLE OF THE TORSO, SOLID AND UNALTERABLE SECTIONS

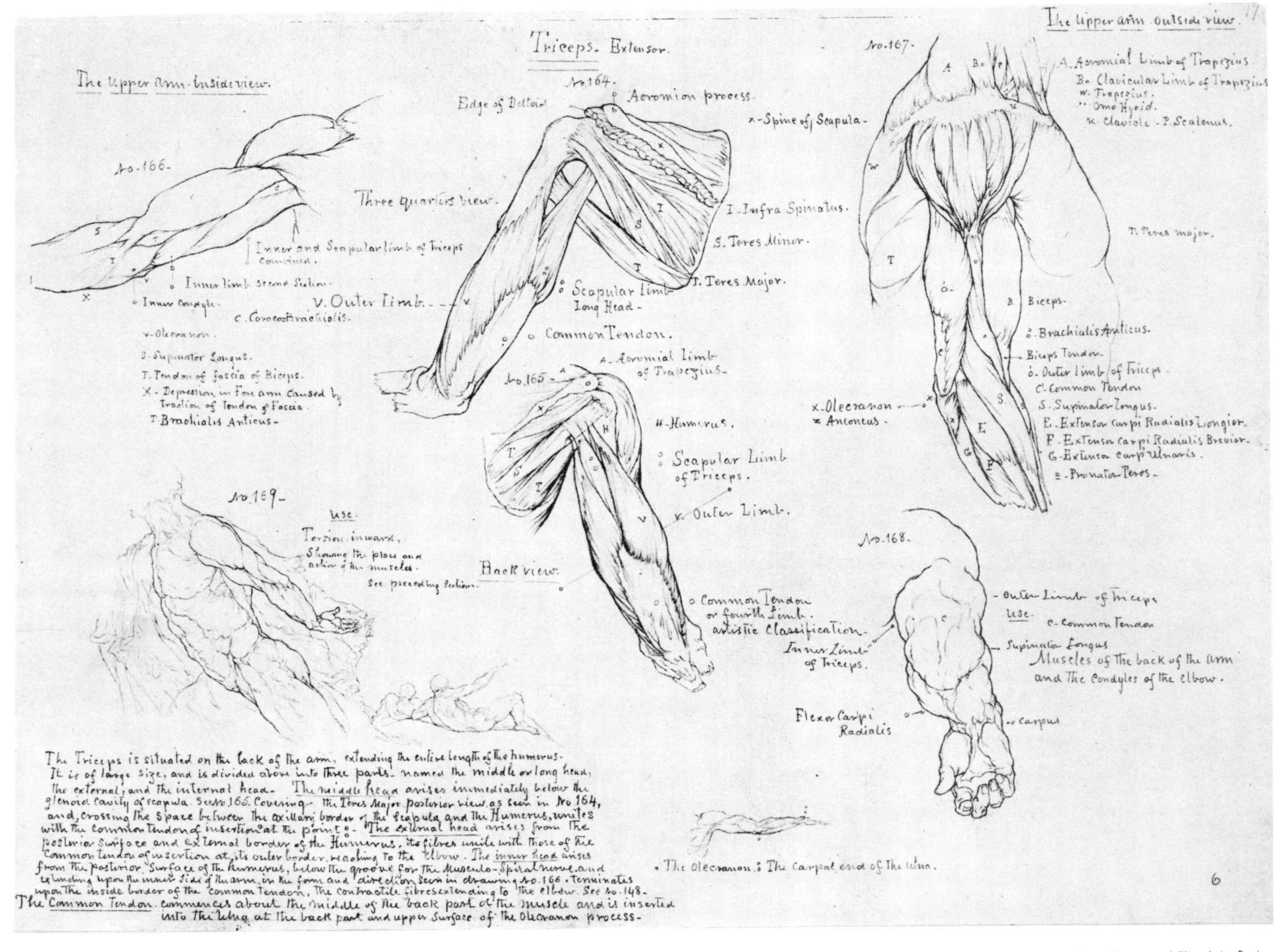

Courtesy of the Museum of Fine Arts, Boston

RIMMER, THE TRICEPS AND MUSCLES OF THE ARM

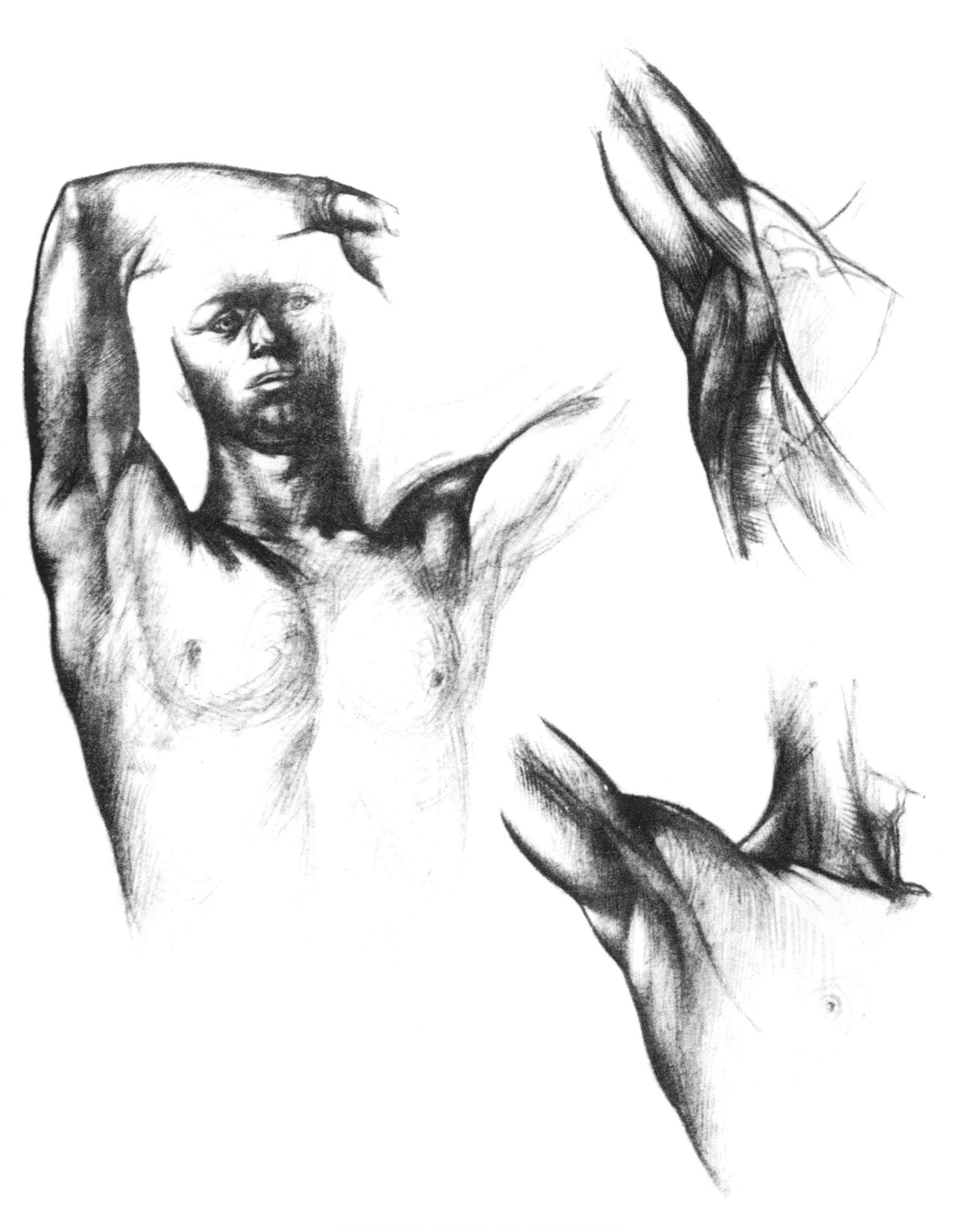

BARCSAY, TRUNK IN MOVEMENT

PLATE 172

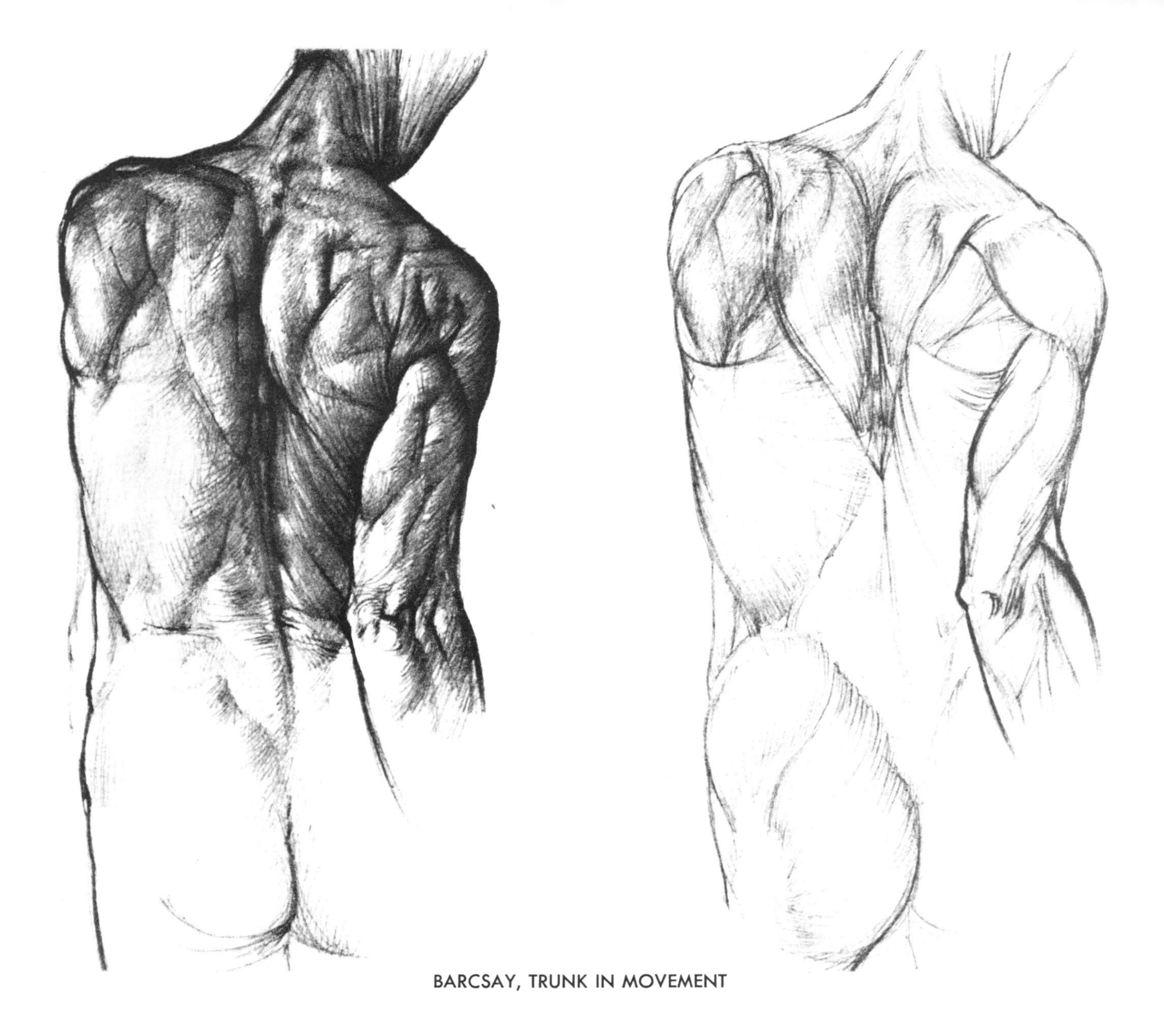

BARCSAY, TRUNK IN MOVEMENT

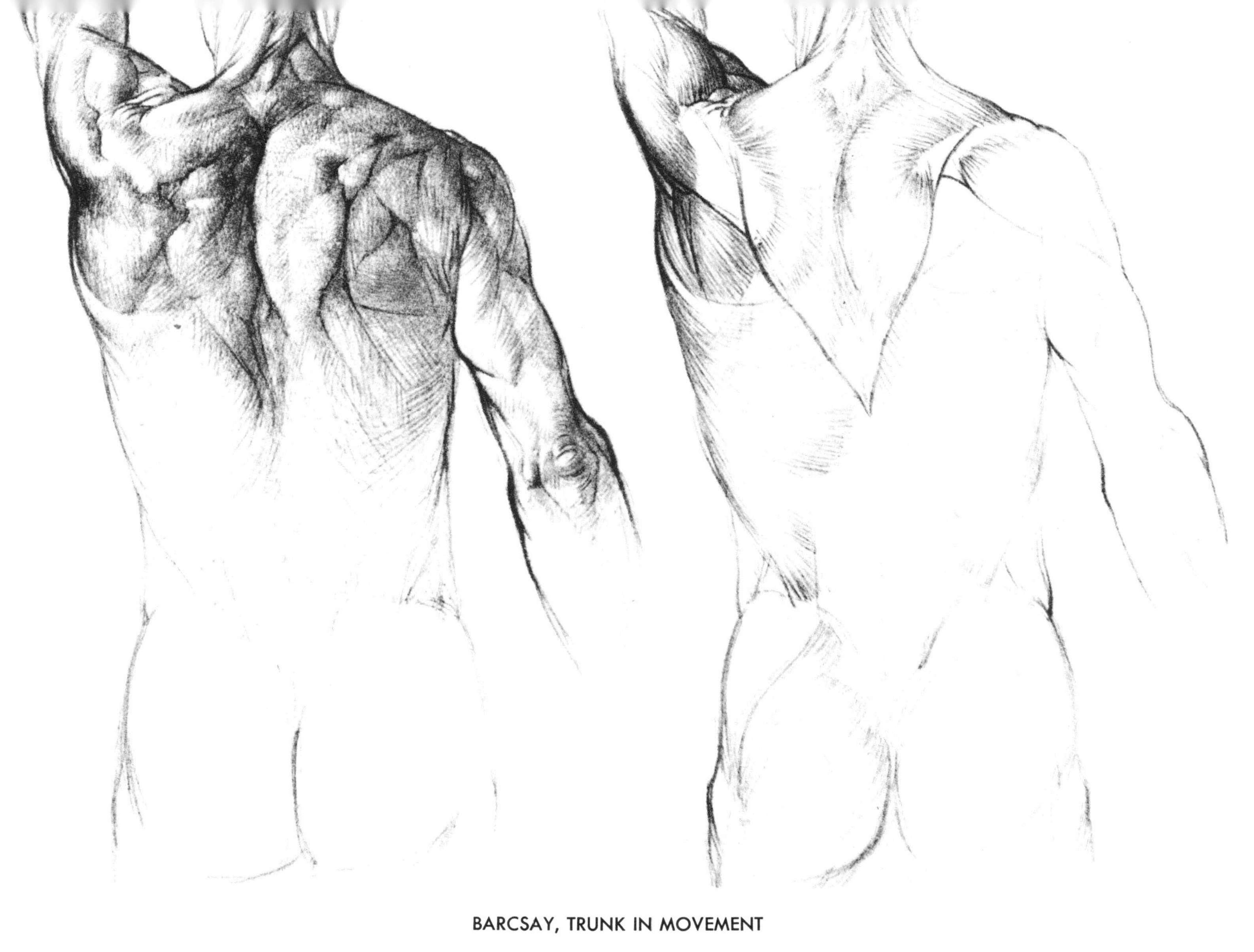

BARCSAY, TRUNK IN MOVEMENT

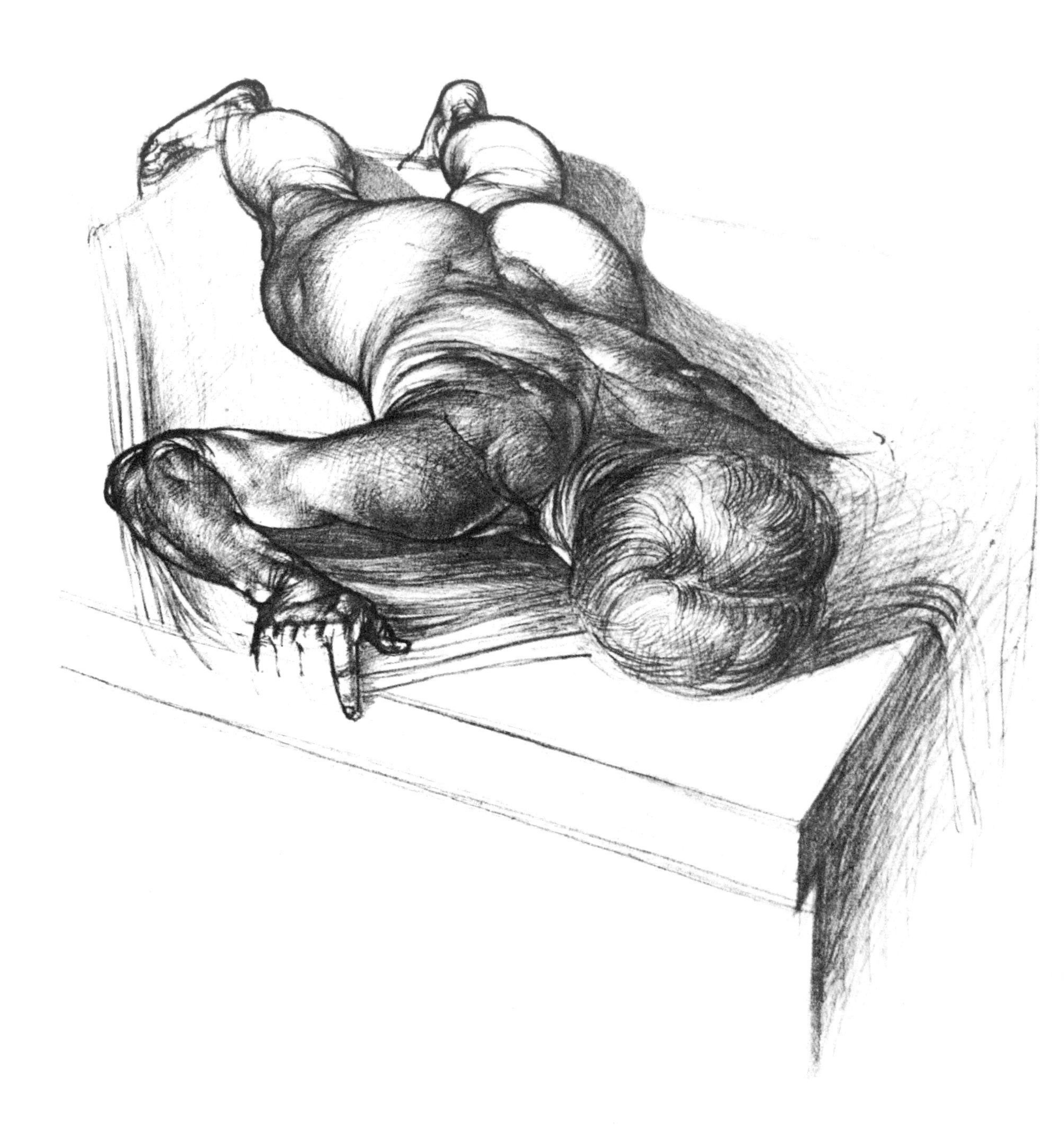

BARCSAY, FORESHORTENED BODY

PLATE 174

BARCSAY, FORESHORTENED BODY

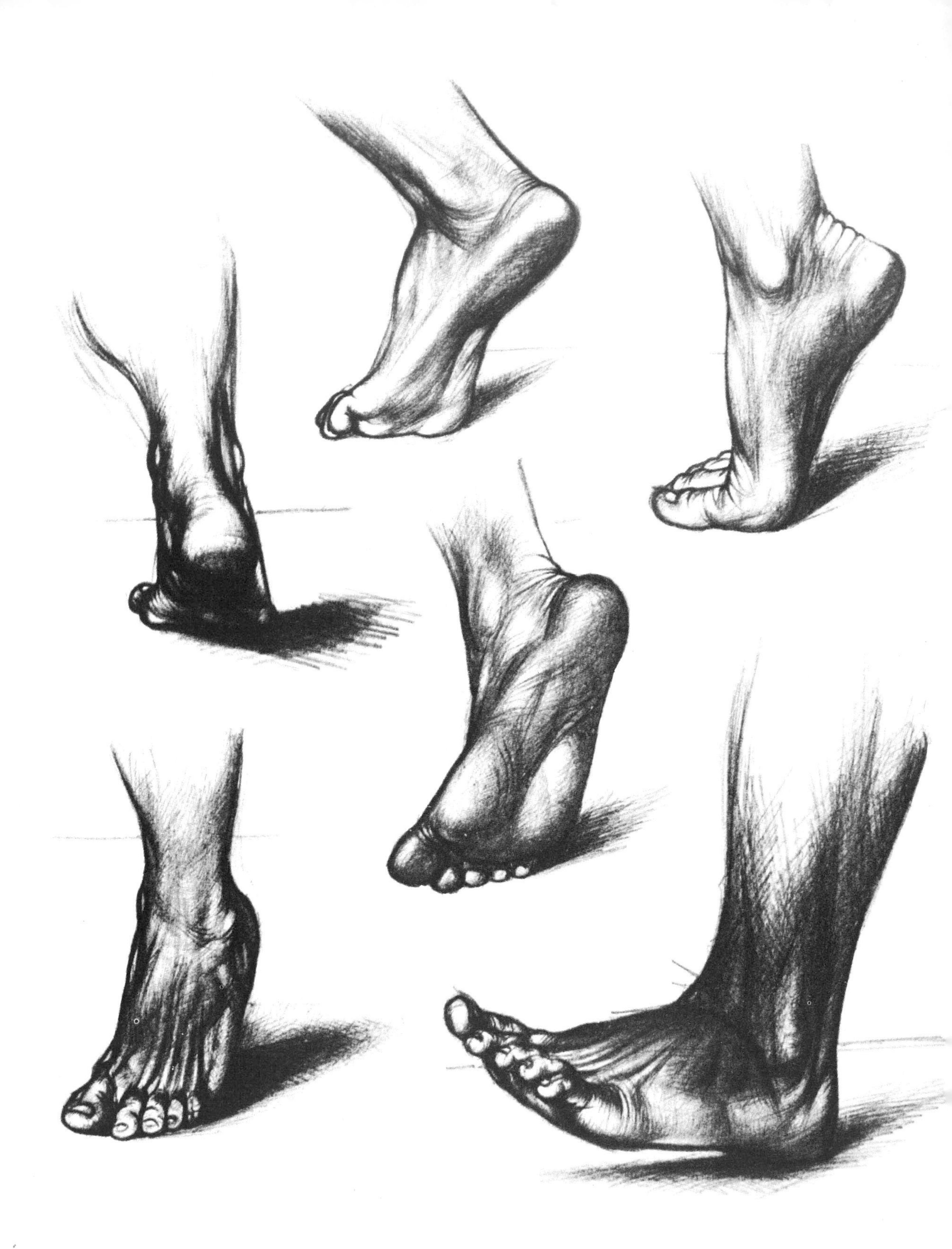

BARCSAY, THE FOOT IN MOVEMENT

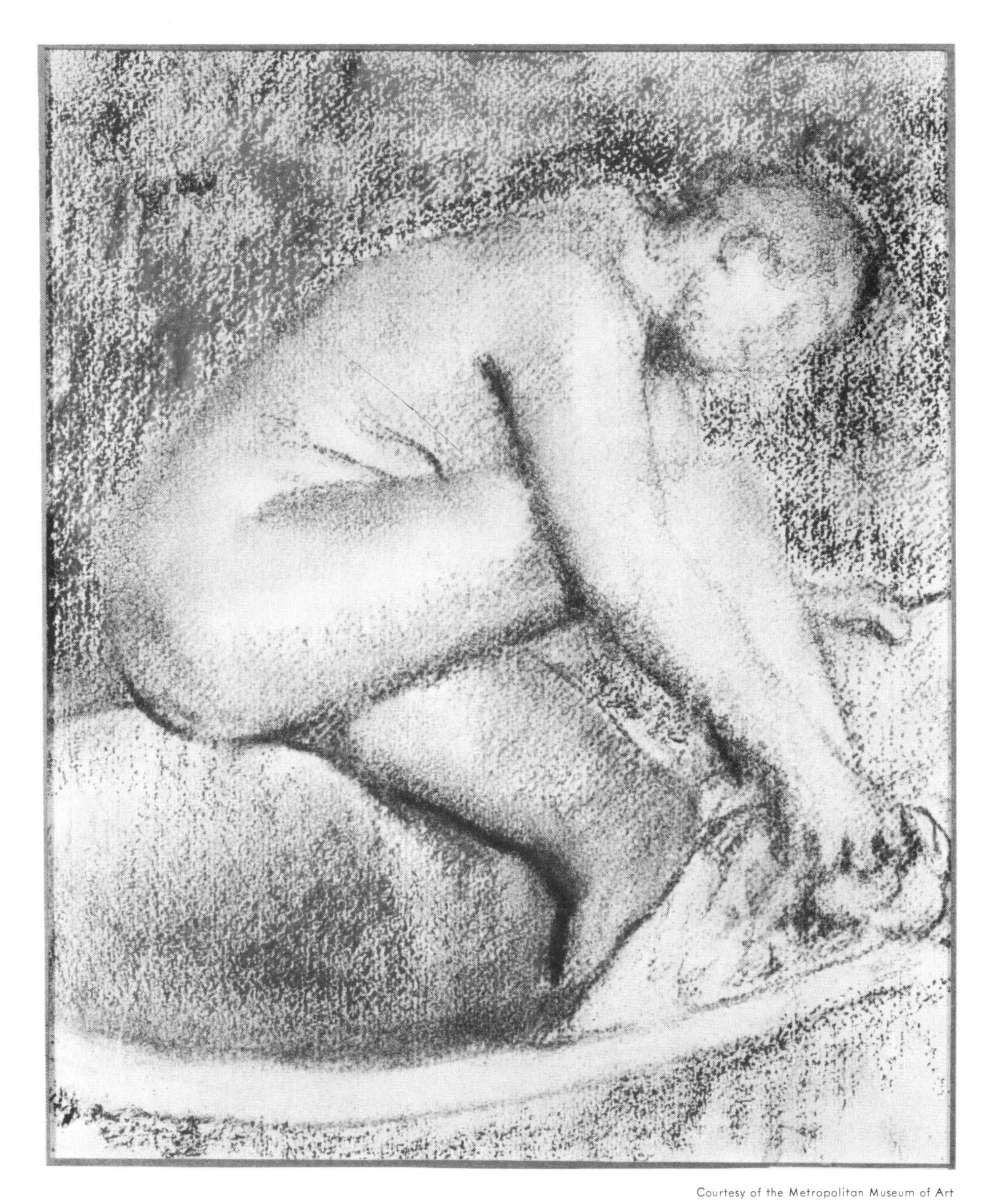

Courtesy of the Metropolitan Museum of Art

DEGAS, "THE BATH"

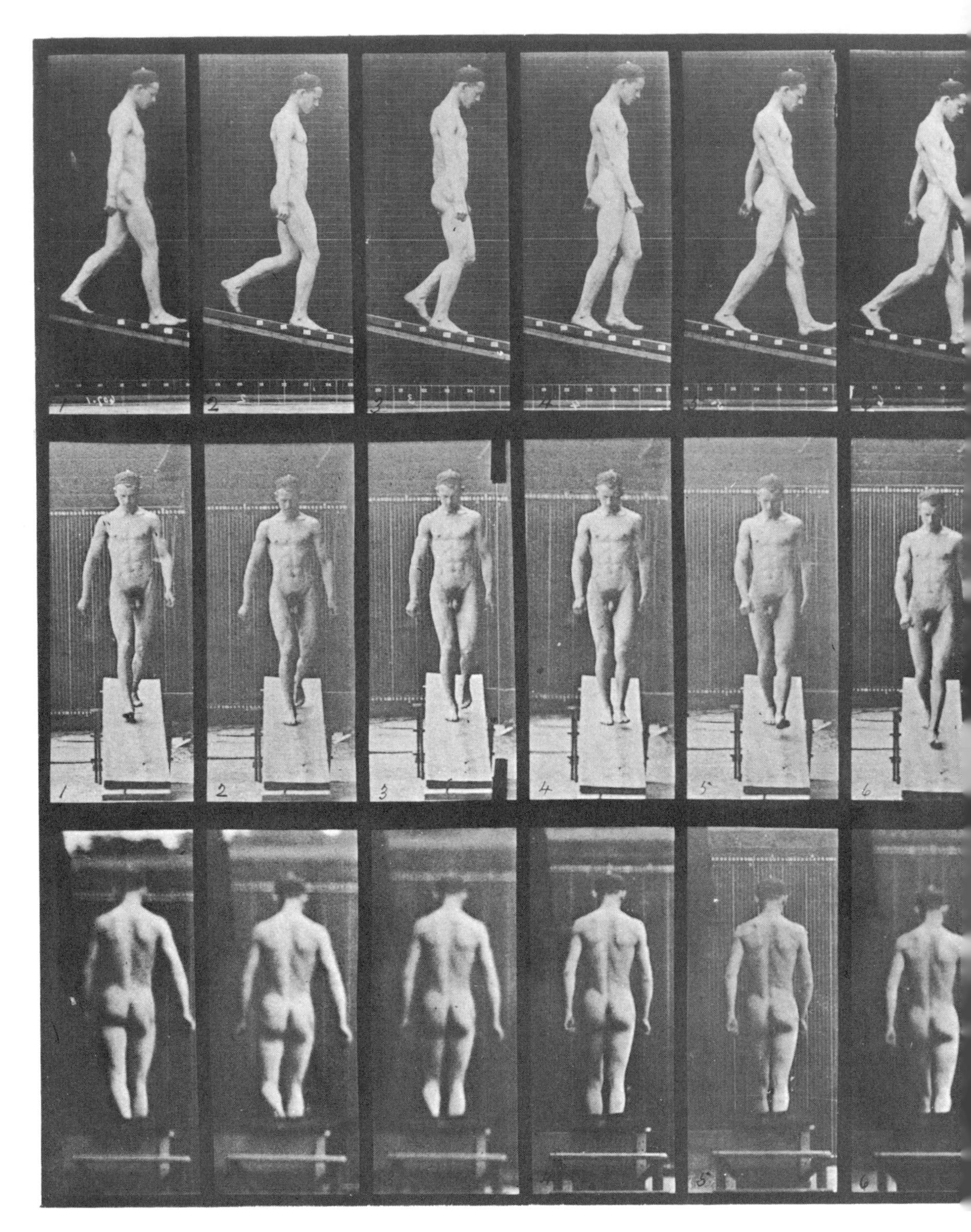

MUYBRIDGE, ATHLETE, WALKING

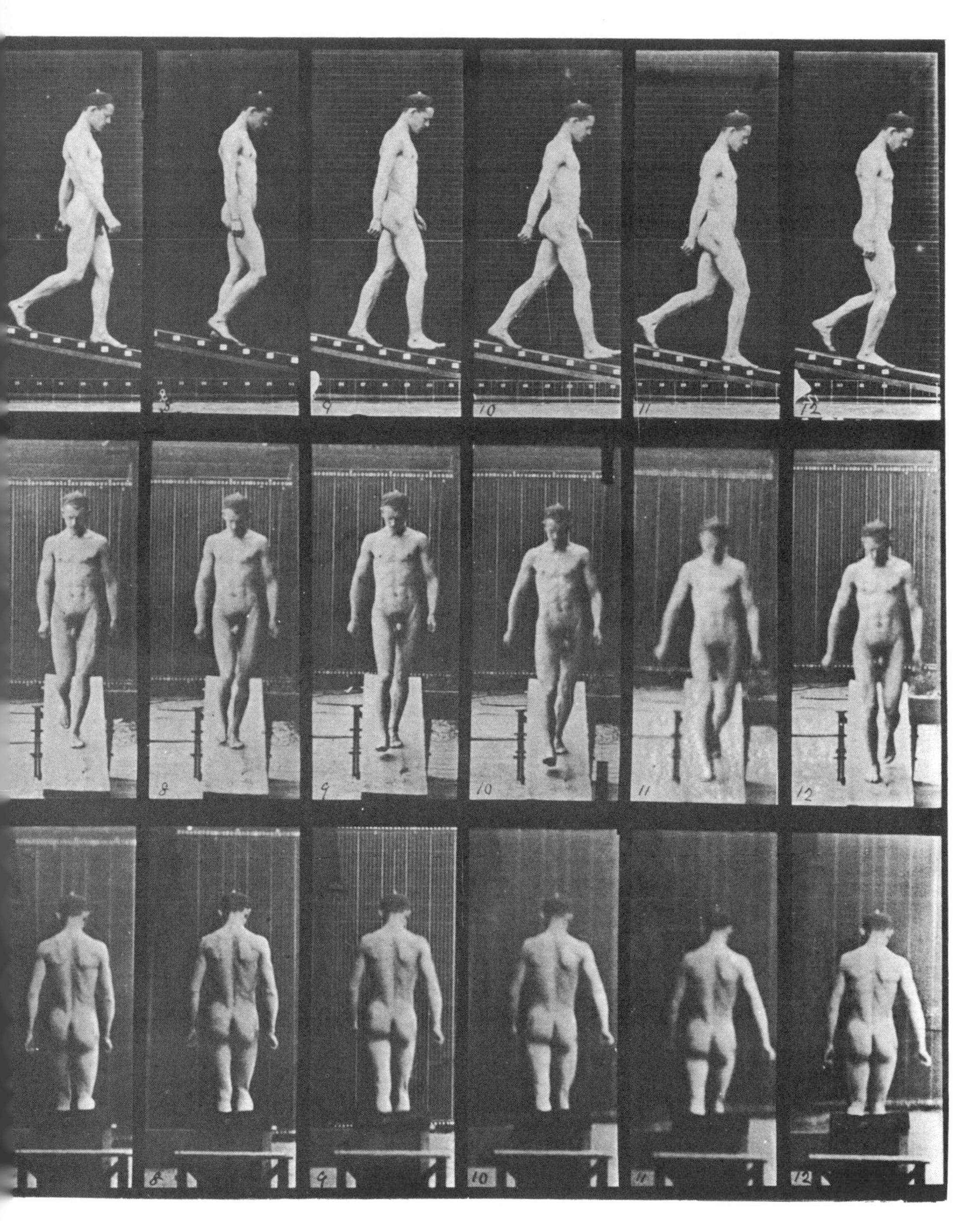

MUYBRIDGE, ATHLETE, WALKING

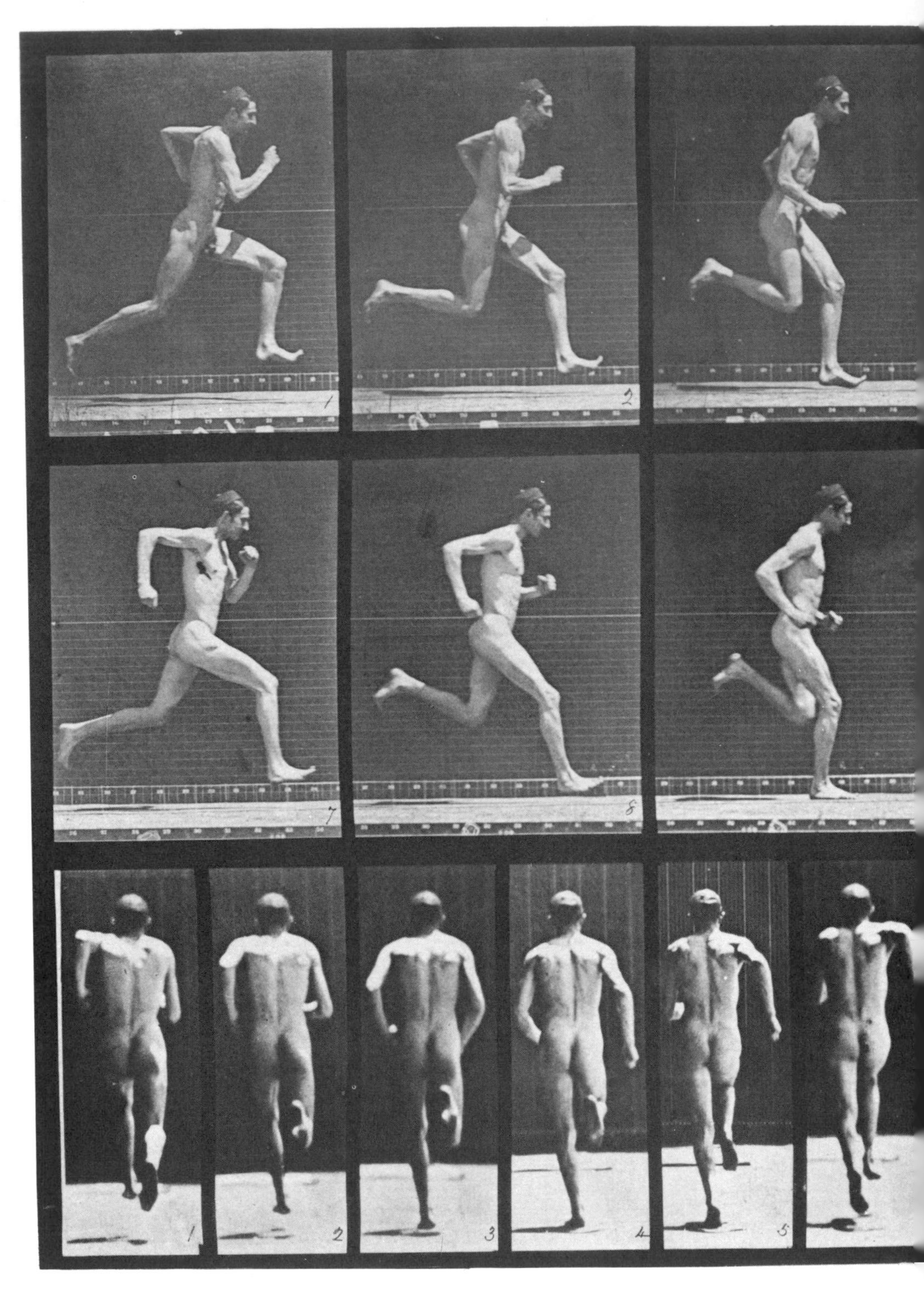

MUYBRIDGE, ATHLETE, RUNNING

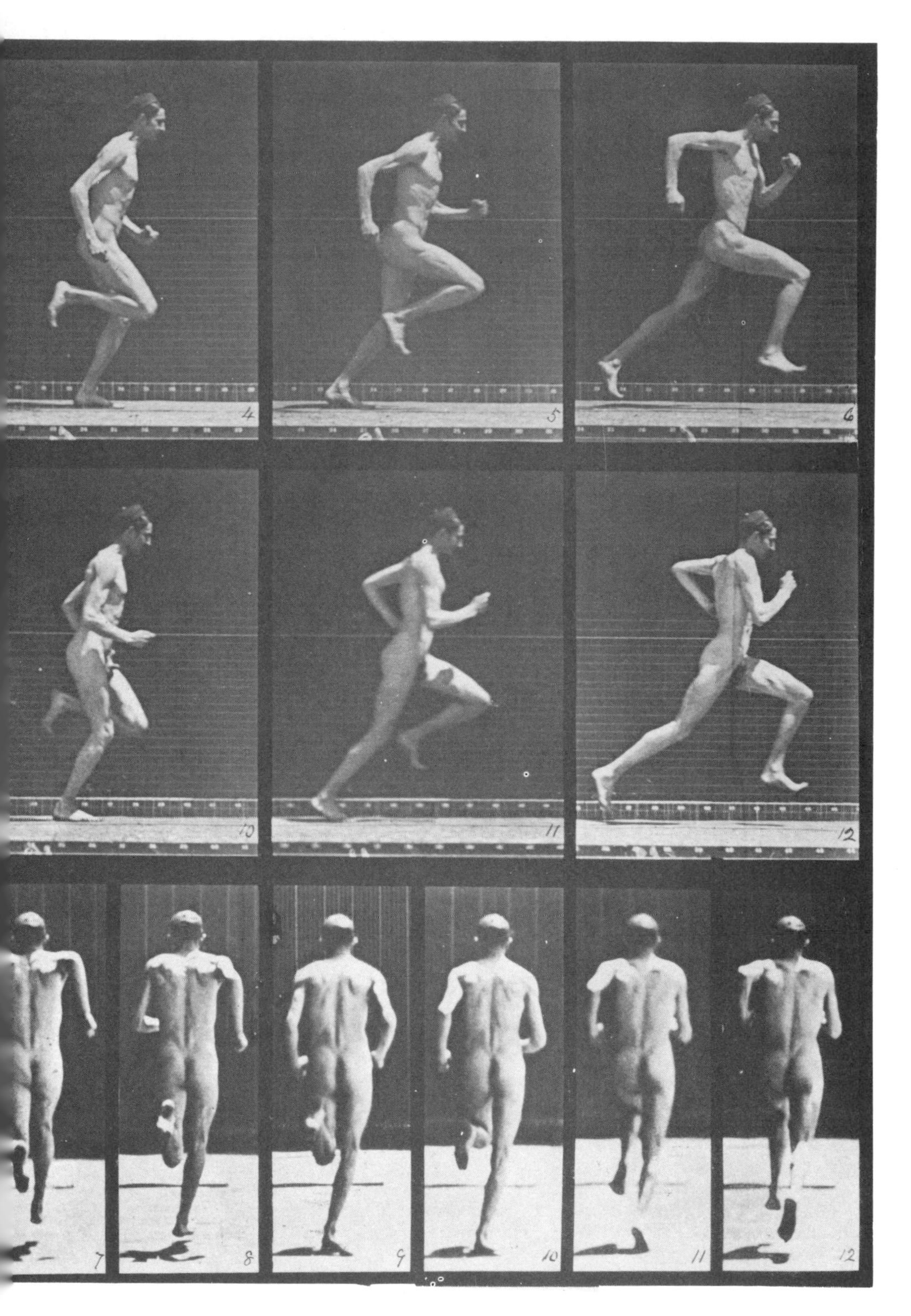

MUYBRIDGE, ATHLETE, RUNNING

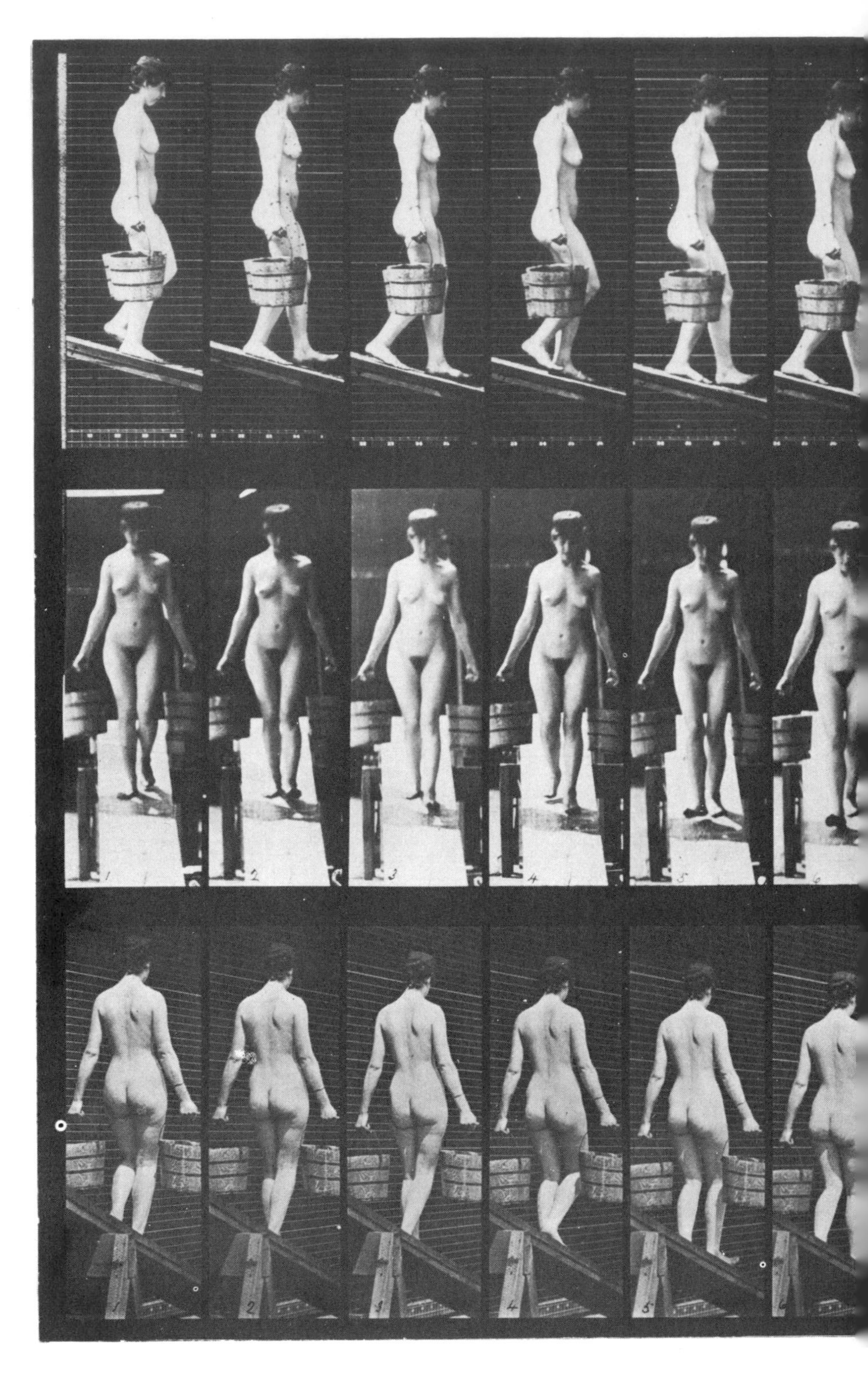

MUYBRIDGE, WOMAN, WALKING

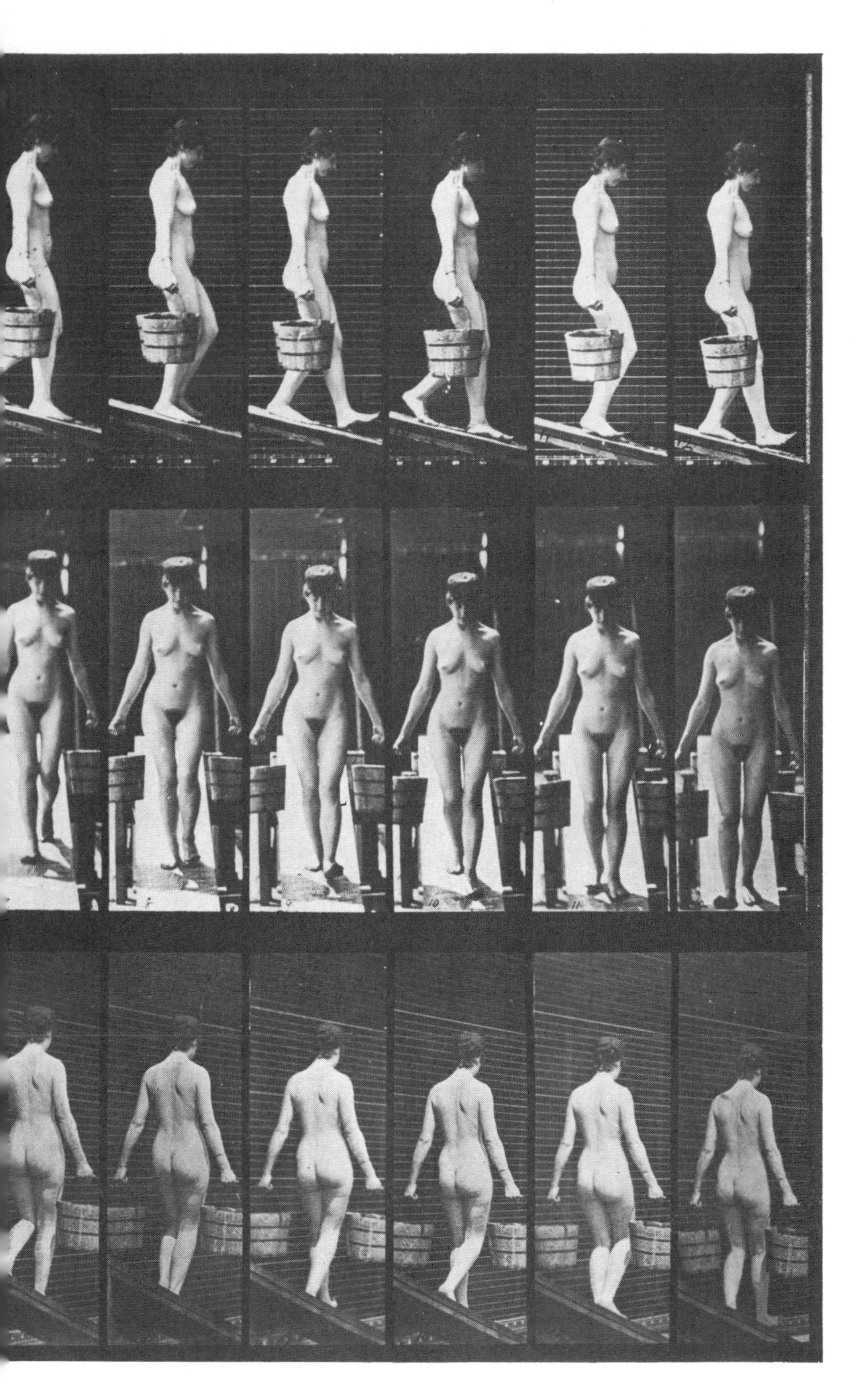

MUYBRIDGE, WOMAN, WALKING

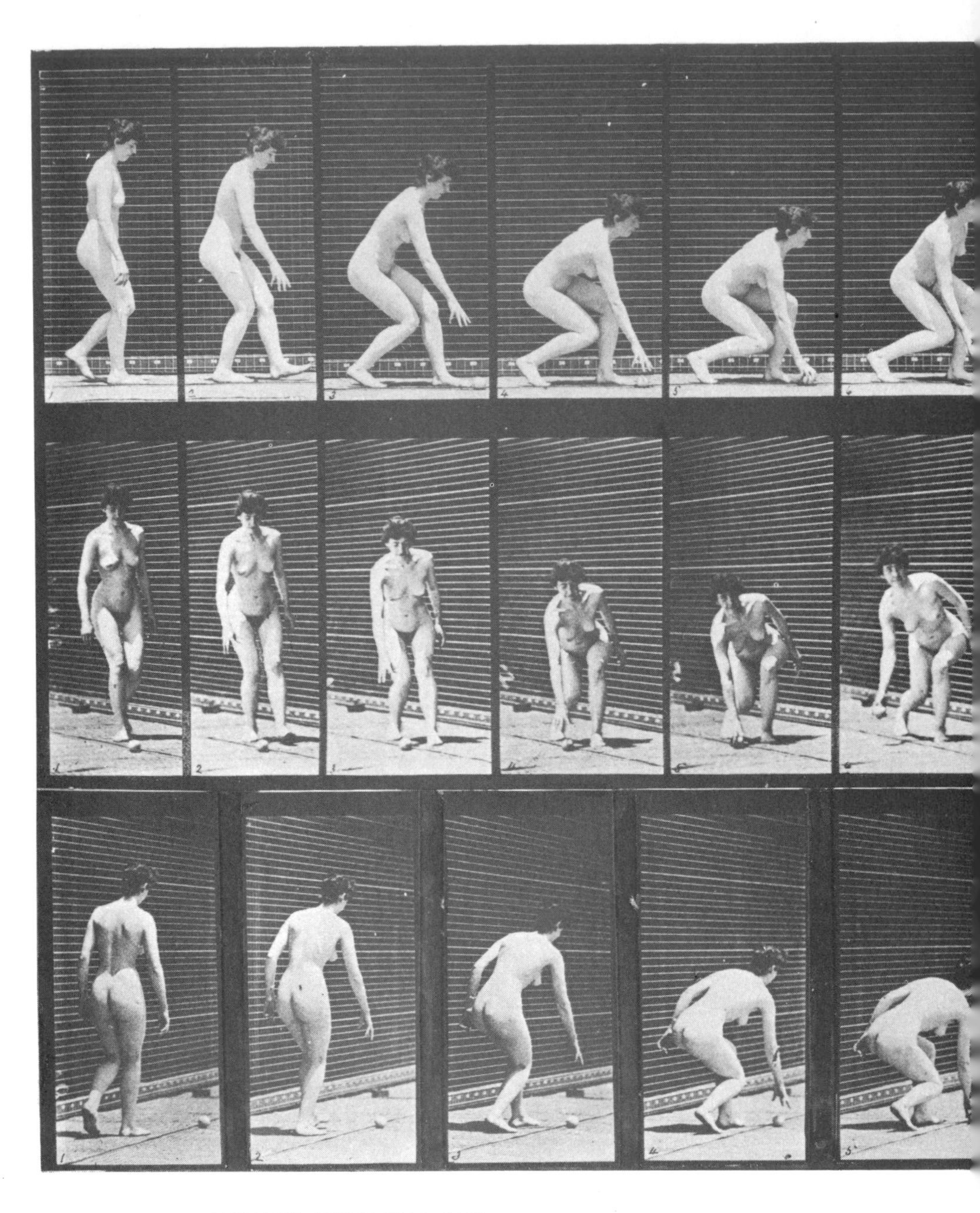

WOMAN, THROWING BALL

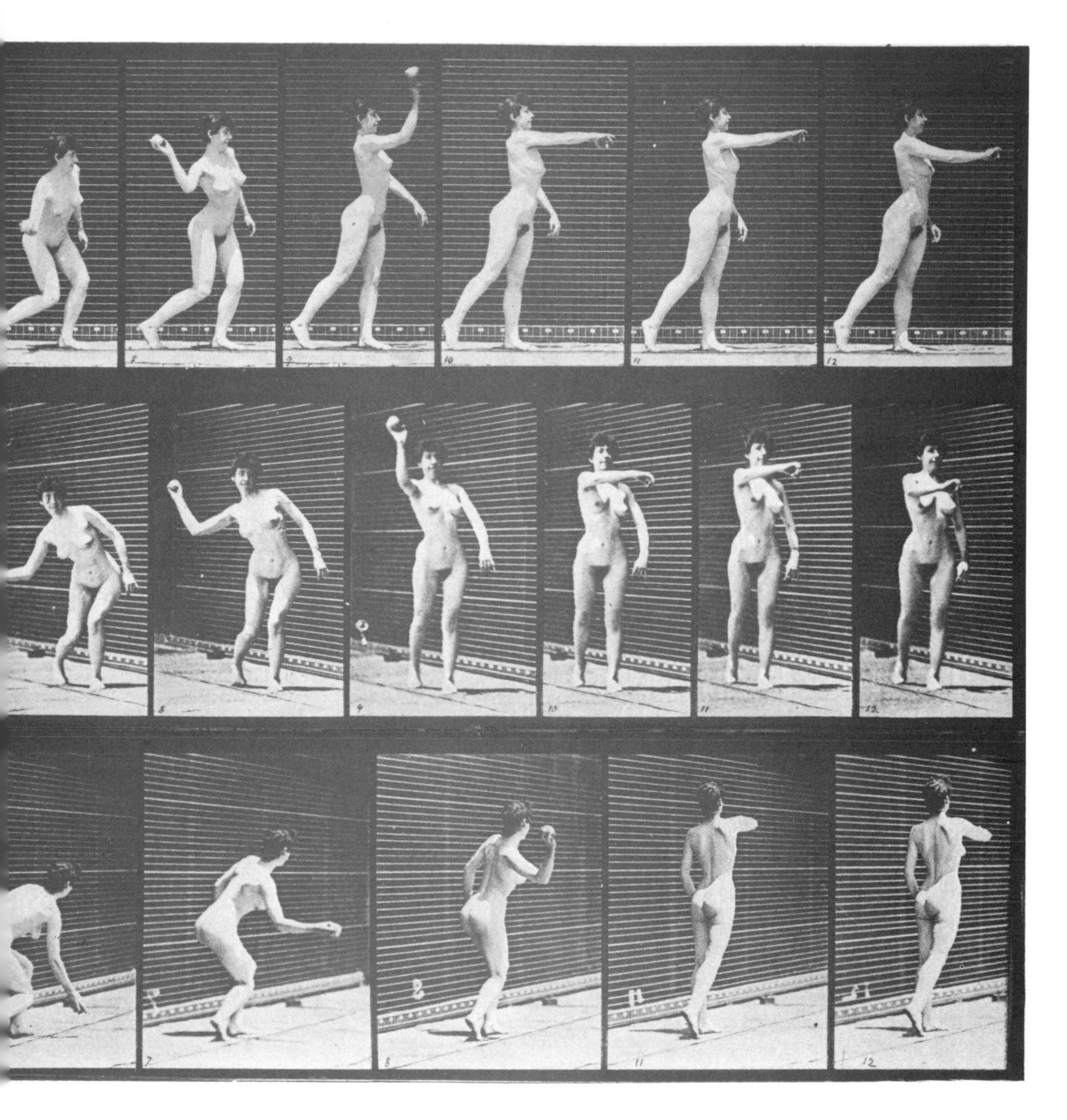

WOMAN, THROWING BALL

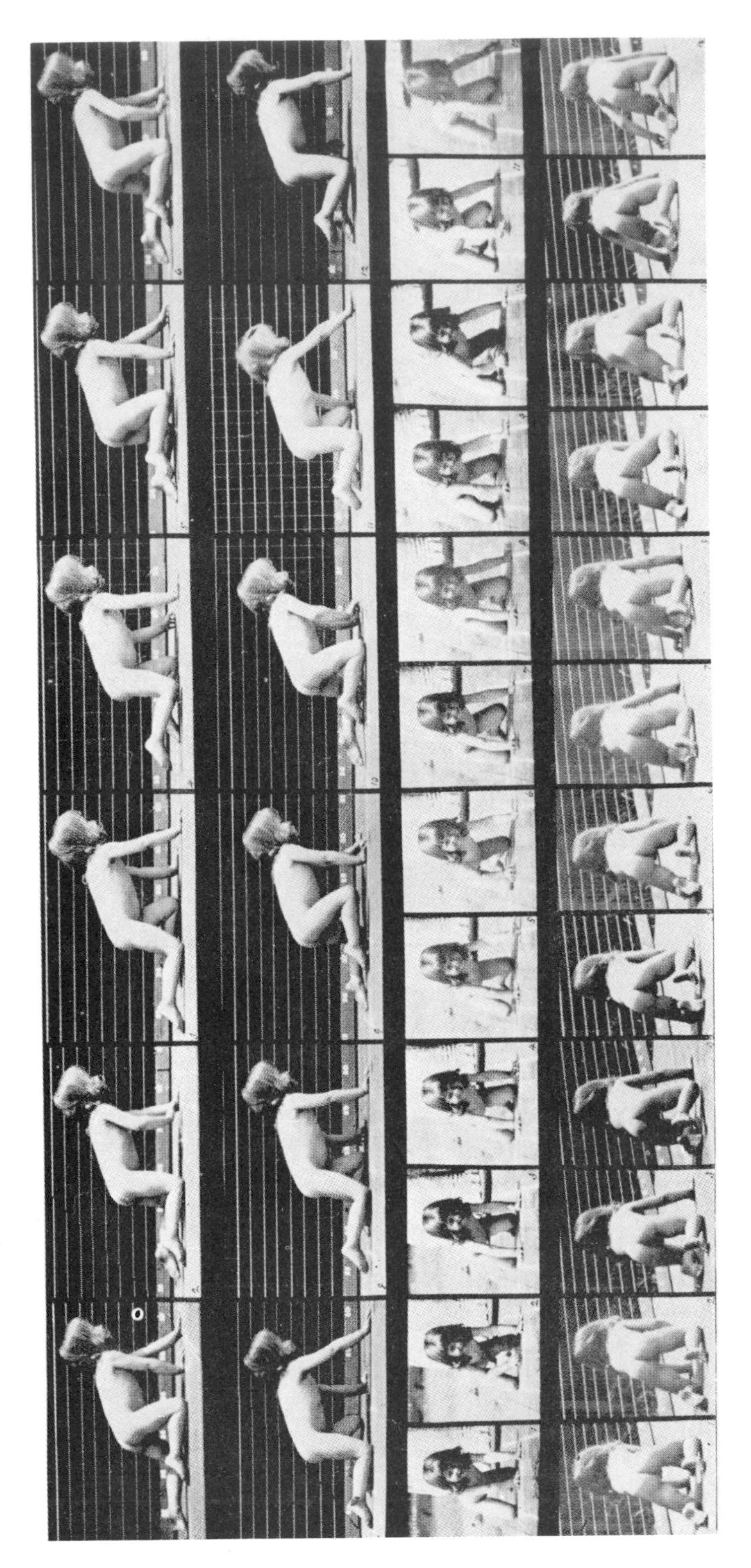

MUYBRIDGE, CHILD, CRAWLING

PLATE 186

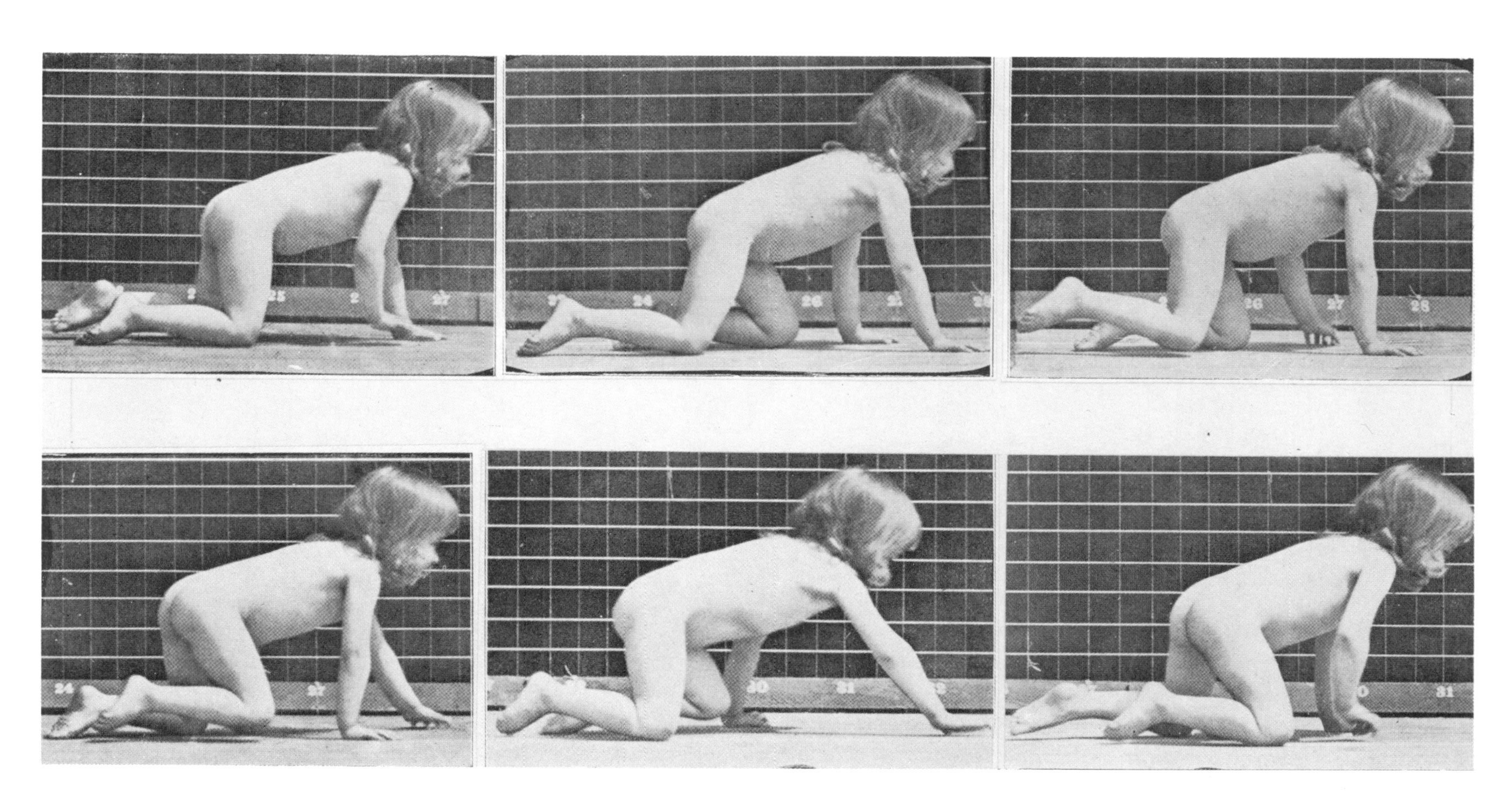

MUYBRIDGE, CHILD, CRAWLING: SELECTED PHASES

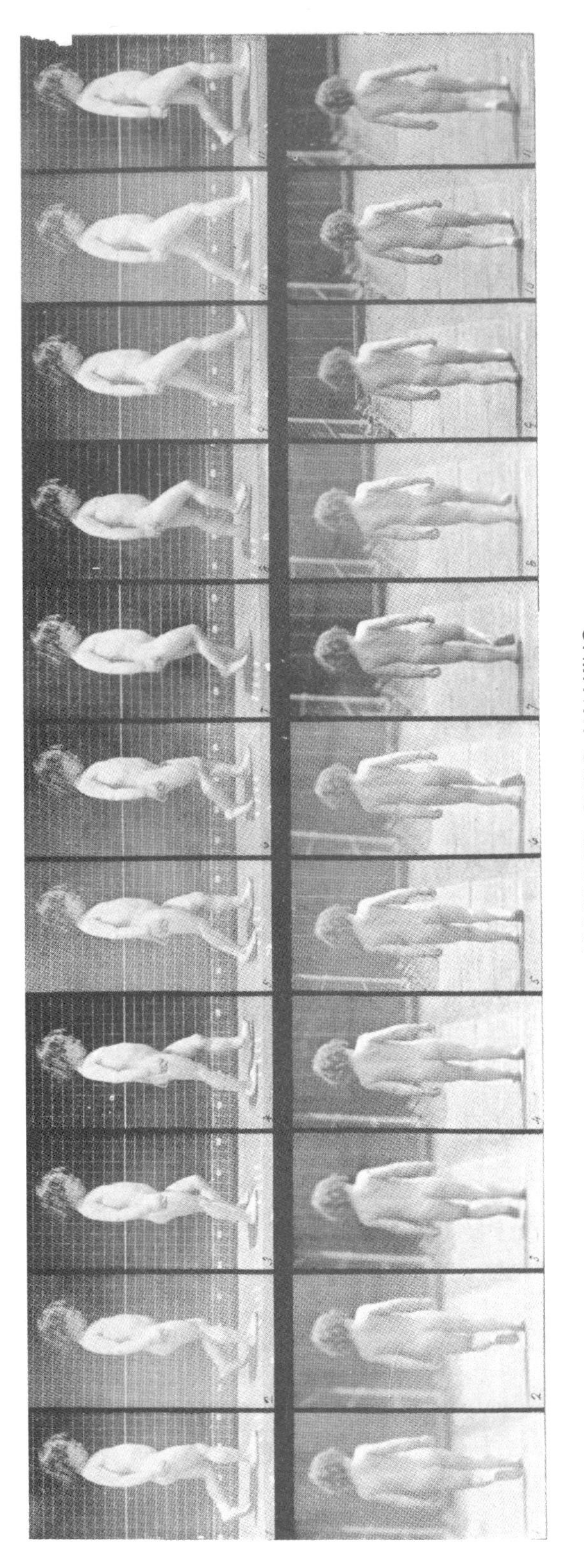

MUYBRIDGE, CHILD, WALKING

PLATE 188

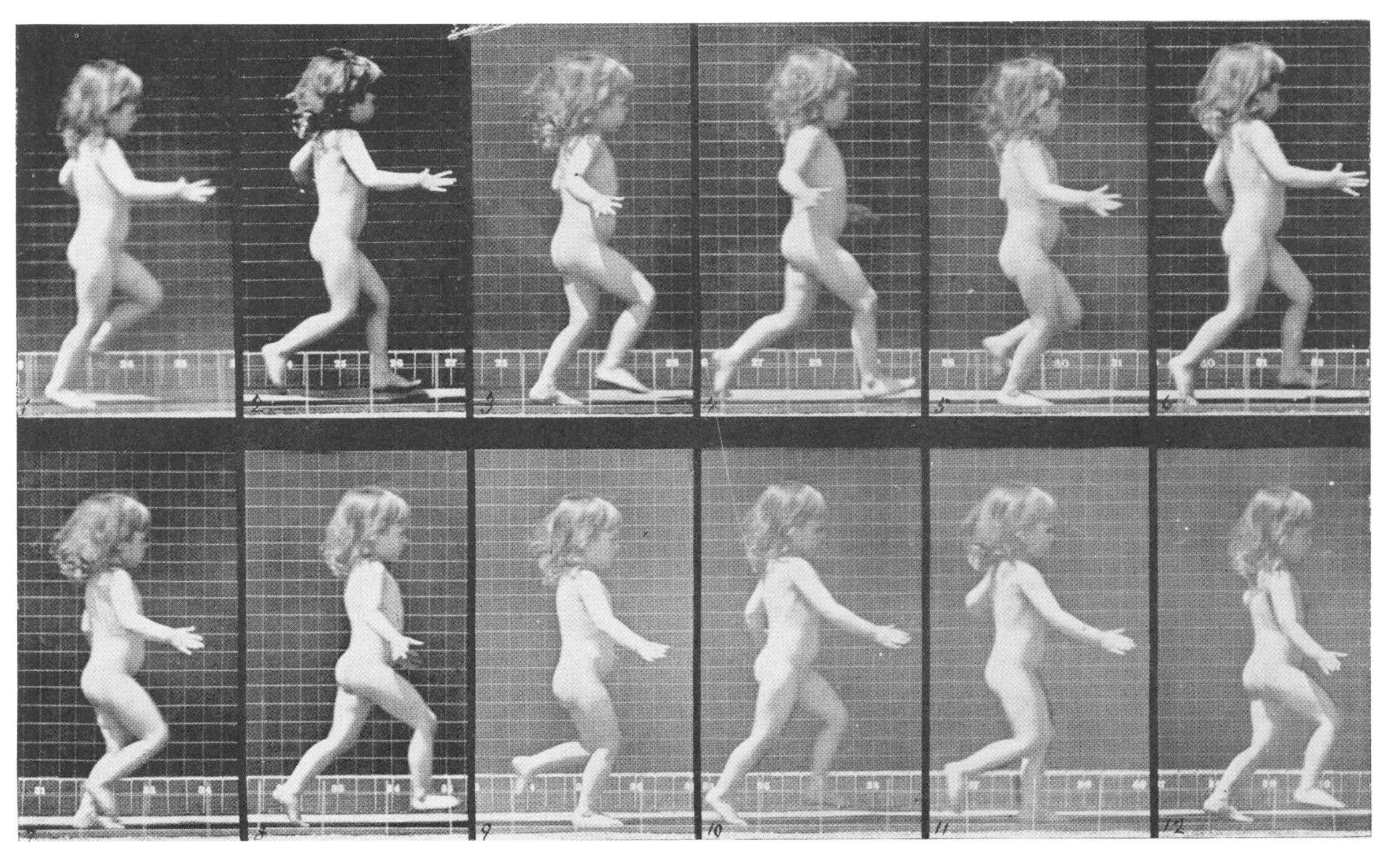

MUYBRIDGE, CHILD, RUNNING

AN ANNOTATED

BIBLIOGRAPHY

OF BOOKS ON HUMAN ANATOMY OF INTEREST TO ARTISTS

By Adolf K. Placzek

This bibliography does not pretend to be complete but aims to suggest, in the various categories it covers, the most important avenues for further study and the wide variety of anatomical material available at different levels of instruction. A visit to one of the more important libraries, such as the New York Public Library, the Metropolitan Museum of Art, or the Columbia University Library, and a review of the books catalogued under Art Anatomy is a rewarding and enriching experience that is strongly recommended to the student. It is hoped that this bibliography will serve as a guide to such individual study.

1. ATLASES AND MEDICAL WORKS

Frohse, Franz. *Atlas of human anatomy*, by Franz Frohse, Max Brödel and Leon Schlossberg. Explanatory text by Jesse Feiring Williams. New York, Barnes and Noble, Inc. 1942.

Convenient small book of descriptive anatomy. Colored plates. No surface description.

Gray, Henry. *Anatomy of the human body* . . . Philadelphia, Lea and Febiger. 1942.

Compact medical textbook for advanced study.

Peck, Stephen Rogers. *Atlas of human anatomy*. New York, Oxford University Press. 1951.

A manual with good photography and clear, concise drawings.

Pfeiffer, L. *Handbuch der angewandten Anatomie* . . . Leipzig, Spamer. 1899.

Interesting for its study of medical proportions and irregularities of the human body.

Sobotta, Johannes. *Atlas of human anatomy* . . . New York, Stechert. 1930-39.

One of the great works. For the art student, only volume 1 (bones, ligaments, joints, muscles, and regions) of particular use.

Spalteholz, Werner. *Hand-atlas of human anatomy*. Philadelphia, J. B. Lippincott. 192-.

Used by generations of medical students. Three volumes of clear and detailed presentations, of which the first two are of use to the art student.

Toldt, Carl. *An atlas of human anatomy*. New York, Macmillan Company. 1926.

"For students and physicians." Of limited use for artists.

Wolff, Eugene. *Anatomy for artists* . . . London, H. K. Lewis and Co. 1933.

Detailed anatomy, good description of surface form, functions and movements.

2. WORKS OF HISTORICAL INTEREST

Albinus, Bernhard Siegfried (1697-1770). *Tabulae anatomicae sceleti et musculorum hominis*. Lugdunae Batavorum. 1747.

Exquisite engravings. Text in Latin and English. Muscles and bones of the male body.

Audran, Gérard (1640-1703). *Des menschlichen leibes Proportionen* . . . Nuremberg, Sandrart. 1686.

Many editions. Proportions and measurements, influenced by Dürer. Based mainly on ancient sculpture.

Buonarroti, Michelangelo (1475-1564). *Michelangelo drawings*, by Ludwig Goldscheider. London, The Phaidon Press. 1951.

Michelangelo, unlike Leonardo, was no scientist; nor did he try, like Dürer, to arrive at set proportions. However, in their perception of form, plasticity and motion of the body, his drawings are unsurpassed.

Camper, Petrus (1722-1789). *Dissertation physique* . . . Autrecht, B. Wild and J. Altherr. 1791.

Proportions, national characteristics and expressions of the human face and head. Good engravings.

Dürer, Albrecht (1471-1528). *Hierinn sind begriffen vier bücher von menschlicher proportion* . . . Nuremberg. 1528.

Many later editions. The great master strove for a "canon" of types—not altogether successful from the classic point of view.

Genga, B. *Anatomia per uso et intelligenza del disegna* . . . Rome, Rossi. 1691.

Large engravings of the male athletic body, based on ancient sculpture.

Jombert, Charles Antoine (1712-1784). *Méthode pour apprendre le dessein . . . représentant différentes parties du corps humain* . . . Paris, L'auteur. 1755.

Engravings, based on the great Renaissance painters. Heads. Surface proportions.

Leonardo da Vinci (1452-1519). *The drawings of Leonardo da Vinci*. With an introduction and notes by A. E. Popham. New York, Reynal and Hitchcock. 1945.

Searching medical analysis of the human body, by the pen of one of the greatest artists—in fact, the almost unique merger of the creative and the scientific mind.

Leonardo da Vinci. *Leonardo da Vinci on the human body; the anatomical, physiological and embryological drawings of Leonardo da Vinci*, by Charles D. O'Malley and J. B. de C. M. Saunders. New York, H. Schuman. 1952.

A reproduction and emendation of Leonardo's research notes and drawings, system by system and region by

region, of the human body.

McMurrich, James Playfair. *Leonardo da Vinci, the anatomist . . .* Baltimore, Williams and Wilkins. 1930.

A full report on Leonardo's achievements as anatomist.

Rubens, Peter Paul (1577-1640). *Die Handzeichnungen von Peter Paul Rubens.* Herausgegeben von Gustav Glück und Franz Martin Haberditzl. Berlin, Bardverlag. 1928.

Details of figures and studies of dramatic motion, with a complete mastery of technique and a perfect understanding of anatomy.

Rubens, Peter Paul. *Théorie de la figure humaine considerée dans ses principes soit en repos ou en mouvement* . . . Paris, C. A. Jombert. 1773.

Engraved by Pierre Aveline, 1702-1760, based on Rubens drawings.

Schadow, Johann Gottfried (1764-1850). *Polyclète; ou théorie des mesures de l'homme* . . . Berlin, The author. 1834.

Based on Petrus Camper. Continued by Physionomie Nationale, 1835. Polyclète contains beautiful plates on proportion of the body, Physionomie Nationale concerns itself with national characteristics, proportions, and expressions of face and head.

Vesalius, Andreas (1514-1564). *The Epitome of Andreas Vesalius* . . . New York, The Macmillan Company. 1949.

A modern translation of the writings of the great Flemish pioneer in anatomy. The original Latin text and the plates are reproduced.

Vesalius, Andreas. *The illustrations from the works* . . . by J. B. de C. M. Saunders and Charles D. O'Malley. Cleveland, The World Publishing Company. 1950.

Excellent reproduction of the plates; translations and annotations.

3. OLDER WORKS ON ANATOMY FOR ARTISTS

Bell, Sir Charles. *The anatomy and philosophy of expression.* London, Henry G. Bohn. 1865 (Fifth edition).

Concerned with the facial expression. Much used and several times reprinted in its time.

Braun, Adolphe Armand. *Hieroglyphic or Greek method of life drawing* . . . London, Postal University. 1918.

Presentation of an individual method, supposedly based on the Greeks. Emphasis on the female body. The photographs of only historical interest.

Duval, Mathias Marie. *Artistic anatomy* . . . translated by Frederick E. Fenton. London, Paris, Cassell and Co. 1848.

Based on lectures delivered at the Ecole des Beaux-arts. A thorough description, little illustration.

Ellwood, George and Yerbury, F. R. *Studies of the human figure, with some notes on drawing and anatomy* . . . London, B. T. Batsford. 1918.

Good descriptive text. The photographs reflect the change of taste.

Hay, David Ramsey. *The geometric beauty of the human figure defined . . . Edinburgh, Blackwood and Sons.* 1851.

An elaborate system of proportion. Little on the human body itself.

Fau, Julien. *The anatomy of the external forms of man, issued for the use of artists, painters and sculptors* . . . London, Hippolyte Ballière. 1849.

Translated from the French edition of 1845. Description of the male body; plates after ancient sculpture (e.g. Laocoön).

Flaxman, John (1775-1826). *Anatomical studies of the bones and muscles, for the use of artists* . . . London, M. A. Nattali. 1833.

Engraved by Henry Landseer after the artist's death. Beautiful large plates of the limbs and parts of the body.

Fripp, Sir Alfred Downing, and Thompson, Ralph R. *Human anatomy for art students . . . and an appendix on comparative anatomy* . . . London, Sealey, Service and Co. 1917.

Good descriptive anatomy, with few illustrations.

Hartley, Jonathan Scott. *Anatomy in art. A practical text book for the art student* . . . New York, Press of Styles and Cash. 1891.

A detailed descriptive analysis of bones and muscles, well illustrated.

Marshall, John. *Anatomy for artists; illustrated by two hundred original drawings by J. S. Cuthbert.* London, Smith, Elder and Co. 1890.

Extensive description, structure as well as surface.

McClellan, George. *Anatomy in its relation to art* . . . Philadelphia, The Author. 1900.

Written by a medical doctor, but with proper emphasis on the visual. Well illustrated. Still very useful, apart from the change in taste with regard to the photographs.

Muybridge, Eadweard. *The human figure in motion. An electro-photographic investigation of the consecutive phases of muscular action.* New York. Dover Publications. Inc. 1955.

Important for the study of motion, consecutive movements, and balance.

Pérard, Victor Semon. *Anatomy and drawing* . . . New York, Pérard. 1928.

Little text. Detailed anatomy, but simplified and clarified for the purposes of sketching.

Richer, Paul Marie Louis Pierre. *Anatomie artistique; description des formes extérieures du corps humain* . . . Paris. 1890.

Line drawings of rare clarity and simplicity, in the classic taste. "The body at rest and in its principal movements." Good text.

Rimmer, William. *Art anatomy.* Dover Pub.Inc. 1962.

Interesting mainly as a record of teaching methods of the 19th century. Plates after sketches of William Rimmer, 1816-1879, sculptor, painter, physician and art teacher, with his remarks.

Roth, Ch. *The student's atlas of artistic anatomy* . . . New York, Westermann. 1891.

The male athlete's body analyzed by muscles and bones. Limited scope.

Smith, John Rubens. *A key to the art of drawing the human figure* . . . Philadelphia, S. M. Stewart, 1831.

An early American art teacher, concerned with proportions and perspective. Good plates.

Thomson, Arthur. *A handbook of anatomy for art students.* New York. Dover Publications, Inc. 1964.

By an art teacher. Few illustrations, but extensive and

valuable textual explanations.

Vanderpoel, John H. *The human figure.* New York. Dover Publications, Inc. 1958.

The parts of the body, visually rather than analytically.

4. RECENT BOOKS

Andrews, Sloan. *Anatomy and figure construction for the fashion and figure artist.* New York, Manhattan Art Studios. 1935.

Barcsay, Jeno. *Anatomy for the Artist.* Budapest, Corvina. 1956.

Based on Greek proportions and Renaissance examples, with modern instruction for quick sketching.

Best Maugard, Adolfo. *The simplified human figure. Intuitional expression.* New York, Alfred A. Knopf. 1936.

Geometrical-schematic approach. Stress on simplification.

Bradbury, Charles Earl. *Anatomy and construction of the human figure.* New York, McGraw-Hill Co. 1949.

Equally concerned with structure and representation. Competent text. Well illustrated.

Bridgman, George B. *Constructive anatomy.* New York, Bridgman Publishers. 1936.

Sketches of the various parts of the body, from the structural-mechanic point of view. "Man as machine."

Bridgman, George B. . . . *Complete guide to drawing from life.* New York, Sterling Publishing Co. 1952.

The new, broader edition. Added information on drawing techniques.

Cox, George James. *Art and "the life"; a book on the human figure, its drawing and design.* Garden City, N. Y., Doubleday, Doran and Co. 1933.

Main value: the stimulating text. Techniques for the artist rather than analysis of the body. A good chapter on books about anatomy.

Dobkin, Alexander. *Principles of figure drawing.* Cleveland, World Publishing Co. 1948.

Different techniques with illustrations from old and modern painters and master draftsmen.

Doust, Len A. *A manual on drawing the human figure.* London and New York, F. Warne and Co. 1936.

Simple, with emphasis on the visual, primarily for pencil sketching.

Dunlop, James M. *Anatomical diagrams for the use of art students* . . . New York, Macmillan Co. 1946.

Clear and precise colored diagrams, with short descriptions; useful for structural perception.

Eisele, Louis A. *Figure drawing for fashion and costume designers* . . . Pelham, N. Y. Bridgman Publishers. 1939.

Simplified schemes of the female body.

Giusti, George. *Drawing figures* . . . New York and London, The Studio Publishing Co. 1944.

Touching on many aspects—historical, technical, anatomical.

Farris, Edmond J. *Art students' anatomy.* New York. Dover Publications, Inc. 1961.

A thorough introduction. Photographs, anatomical drawings, roentgenograms.

Hurley, Faye. *Simplified anatomy of the human figure* . . . New York, Library Associates. 1945.

Elementary handbook for beginners.

Lenssen, Heidi. *Art and anatomy.* New York, J. J. Augustin. 1944.

"Studies of the human figure by masters of the Middle Ages and the Renaissance, together with anatomical drawings by contemporary American artists."

Loomis, Andrew. *Figure drawing for all it's worth.* New York, The Viking Press. 1946.

A textbook for art students, instructive on technique. Visual approach.

Marsh, Reginald. *Anatomy for artists.* New York, American Artists Group. 1945.

Based on older sources. Useful illustrations. Visual approach.

Meyner, Friedrich. *Künstleranatomie.* Leipzig, E. A. Seemann. 1951.

Sketches and diagrams of muscles, bones and details. Good photographs, also historical reproductions. Emphasis on the male body.

Moses, Walter Farrington. *Artistic anatomy* . . . Los Angeles, Borden Publishing Co. 1939.

A simplified but still thorough presentation. Little text.

Refregier, Anton. *Natural figure drawing, with photos and drawings* . . . New York, Tudor Publishing Co. 1948.

Ranging over a wide field in a light style.

Tomasch, Elmer J. *The ABC's of anatomy* . . . New York, The William-Frederick Press. 1947.

Elementary handbook with stress on the solidity of volumes rather than the linear quality.

5. SOURCES OF BIBLIOGRAPHICAL REFERENCE ON ANATOMY

Choulant, Ludwig. *History and Bibliography of Anatomic Illustration.* (Translated by Mortimer Frank.) Chicago, Illinois, The University of Chicago Press. 1920.

An extensive historical survey of the field.

Cox, George James. *Art and "the life"; a book on the human figure, its drawing and design.* Garden City, N. Y., Doubleday, Doran and Co. 1933.

Contains an excellent chapter on books about anatomy, with extended discussion of several.

Duval, Mathias Marie, and Bical, Albert. *L'anatomie des maîtres.* Paris, A. Quantin. 1890.

Thirty plates of anatomical drawings by Leonardo, Michelangelo, Géricault, and others. Historical introduction and notes.

Duval, Mathias Marie, and Cuyer, Edouard. *Histoire de l'anatomie plastique. Les maîtres, les livres et les écorchés.* Paris, Société française d'éditions d'art. 1898.

Well illustrated and more recent than Choulant, the German edition of which came out in 1852.

Singer, Charles. *A Short History of Anatomy and Physiology; from the Greeks to Harvey.* New York. Dover Publications. Inc. 1957.